Synthesis, Characterization, and Theory of Polymeric Networks and Gels

Synthesis, Characterization, and Theory of Polymeric Networks and Gels

Edited by

Shaul M. Aharoni

Allied-Signal, Inc.
Morristown, New Jersey

SPRINGER SCIENCE+BUSINESS MEDIA, LLC

Library of Congress Cataloging-in-Publication Data

Synthesis, characterization, and theory of polymeric networks and gels
  / edited by Shaul M. Aharoni.
       p.   cm.
    "Proceedings of an American Chemical Society Division of Polymeric
  Materials Science and Engineering symposium on synthesis,
  characterization, and theory of polymeric networks and gels, held
  April 5-10, 1992, in San Francisco, California"--T.p. verso.
    Includes bibliographical references and index.
    ISBN 978-1-4613-6314-9   ISBN 978-1-4615-3016-9 (eBook)
    DOI 10.1007/978-1-4615-3016-9
    1. Polymer networks--Congresses. 2. Colloids--Congresses.
  I. Aharoni, Shaul M. II. American Chemical Society. Division of
  Polymeric Materials: Science and Engineering.
  QD382.P67S96 1992
  668.9--dc20   92-26765
                                                                     CIP

Proceedings of an American Chemical Society Division of Polymeric Materials Science and
Engineering symposium on Synthesis, Characterization, and Theory of Polymeric Networks and
Gels, held April 5–10, 1992, in San Francisco, California

ISBN 978-1-4613-6314-9

© 1992 Springer Science+Business Media New York
Originally published by Plenum Press in 1992
Softcover reprint of the hardcover 1st edition 1992

PREFACE

Polymer science is a technology-driven science. More often than not, technological breakthroughs opened the gates to rapid fundamental and theoretical advances, dramatically broadening the understanding of experimental observations, and expanding the science itself. Some of the breakthroughs involved the creation of new materials. Among these one may enumerate the vulcanization of natural rubber, the derivatization of cellulose, the giant advances right before and during World War II in the preparation and characterization of synthetic elastomers and semi-crystalline polymers such as polyesters and polyamides, the subsequent creation of aromatic high-temperature resistant amorphous and semi-crystal-line polymers, and the more recent development of liquid-crystalline polymers mostly with main-chain mesogenicity. Other breakthroughs involve the development of powerful characterization techniques. Among the recent ones, the photon correlation spectroscopy owes its success to the advent of laser technology, small angle neutron scattering evolved from nuclear reactors technology, and modern solid-state nuclear magnetic resonance spectroscopy exists because of advances in superconductivity.

The growing need for high modulus, high-temperature resistant polymers is opening at present a new technology, that of more or less rigid networks. The use of such networks is rapidly growing in applications where they are used as such or where they serve as matrices for fibers or other load-bearing elements. The rigid networks are largely aromatic. Many of them are prepared from multifunctional wholly or almost-wholly aromatic kernels, while others contain large amount of stiff difunctional residus leading to the presence of many main-chain "liquid-crystalline" segments in the "infinite" network.

Many behavioral aspects of flexible branched polymers, in the pre-gel state as well as in the final network, lend themselves to description by the fractal model wherein the exhibit features consistent with non-integer dimensionality. Such a behavior is strongly manifested by pre-gel and post-gel rigid and semi-rigid networks and gels, organic as well as inorganic.

This volume was developed from a symposium under the same name sponsored by the Division of Polymeric Materials: Science & Engineering Inc. at the 203rd meeting of the American Chemical Society in San Francisco, CA., April 5 - 10, 1992. The book is divided into four parts, each contain-ing theoretical and experimental contributions. The part entitled Fractal Aspects of Polymer Networks and Gels starts with two theoretical chapters by Daoud and Vilgis, followed by two experimental chapters. This part concludes with a chapter by Sokolov and Blumen, theoretically treating the effects of mixing on reaction kinetics. A theoretical contribution by Edwards and another bky Erman, Mark and associates appear at the head and middle of the second part, Rigid and Semi-Flexible Networks and Gels. The part entitled Networks and Gels in Force Fields contains theoreticl contributions by Termonia, Weiner and Gao, and Mark, Erman and co-workers, interspersed with experimental chapters. The last part, Flexible Networks and Gels, starts with a theoretical chapter on industrially useful networks by Dusek and Somvarsky, and continues with several experimental ones.

In selecting the above chapters, I was guided by a desire to increase the visibility, especially among theoreticians, of the emerging technologies of rigid and semi-rigid networks, and to present more than only the conventional views in the older field of flexible networks. If this volume does not succeed at least in part of this mission, the failure is entirely mine.

Finally, I thank my wife Alice for her patience, support and encouragement., without which this book would not have come into being.

Shaul M. Aharoni

Polymer Science Laboratories
Allied-Signal Inc. Research & Technology
Morristown, New Jersey 07962-1021, USA

September 1992

CONTENTS

FRACTAL ASPECTS OF POLYMER NETWORKS AND GELS

RIGID AND SEMIFLEXIBLE NETWORKS AND GELS

FRACTAL PROPERTIES OF BRANCHED POLYMERS

M. Daoud

Laboratoire Leon Brillouin (C.E.A.-C.N.R.S.)
C.E.N. Saclay 91191 Gif/Yvette; France

1. Introduction

It has been known for a long time that amorphous polymers[1] behave as what is called nowadays fractals[2]. This was found for linear polymers and also for branched structures[3] in the vicinity of the sol-gel transition. The latter was studied some years ago, and it was realized very early that the distribution of masses is very broad, and becomes infinitely polydisperse near the gelation threshold. More recently, excluded volume effects were taken into account[4-6],

and it was shown that each of the branched polymers in a sol is a fractal. In the following, we wish to show some consequences of such polydispersity on the (average) fractal behavior. As we shall see, although the study of branched macromolecules seems to be analogous to that of linear chains, it differs from it in a basic way. For instance, the measured dimension D_{eff} of a dilute solution is not the actual fractal dimension D of every polymer, but is a function of both D and of the polydispersity index τ, to be introduced below. This implies that a single measurement does not necessarily give the actual dimension of fractal objects, and one has to check for possible effects of polydispersity. The origin of such effects is that the mass distribution function cannot be reduced to a single average mass, but to at least two masses, namely N_w and N_z that diverge in different ways near the gelation threshold. In what follows, we will discuss some consequences of this for the static properties. Another important consequence, for the rheological characteristics of these systems is that in the same way as they cannot be reduced to a single mass, they also have a distribution of relaxation times that is very broad[7][8] and may be described by at least two times. As we shall see, there is still discussion about the divergences of these times near the sol-gel transition, and there might be more than one universality class for the dynamics in the reaction bath. In what follows, we will remind briefly the main properties of the polymers and of the distribution of molecular weights in section 2. Then we will discuss the static properties of dilute solutions. Section 4 is devoted to the semi-dilute solutions. Finally, the distribution of relaxation times and its consequences for the rheological properties both in the reaction bath and in dilute solutions will be discussed in section 5.

2. The distribution of masses

Randomly branched polymers are made by letting multifunctional units react. We will assume that such reaction is carried in the absence of solvent. The case when solvent is present is assumed to be the same as the present one and solvent is added, because equilibrium is assumed to be reached. This implies among others that diffusion is negligible, and that only chemical reactivity is important. A further assumption is that there is equal reactivity of all the

Synthesis, Characterization, and Theory of Polymeric Networks and Gels
Edited by S.M. Aharoni, Plenum Press, New York, 1992

1

functional units, whatever their state of reaction. As a consequence of these assumptions, percolation should be an adequate model to describe the formation of branched polymers. We know from the early work of Flory and Stockmayer that this describes the sol-gel transition and that the macromolecules that constitute the sol are very polydisperse, and that the distribution P(N)of masses has at least two characteristic masses, namely N_z, and N_w that diverge at the threshold in such a way that their ratio also diverges. More recently[9,10], it was shown that at a distance $\varepsilon \equiv p-p_c$ from the gel point, the probability $P(N,\varepsilon)$ of finding a macromolecule made of N units is

$$P(N,\varepsilon) \sim N^{-\tau} f(\varepsilon N^{\sigma})$$

(1)

where τ and σ are percolation[11] exponents that depend on space dimension and will be discussed below. Note that only the polymers of the sol are considered in $P(N,\varepsilon)$, and that the gel fraction is excluded from it. The characteristic masses that we mentioned above are moments of this distribution

$$N_w \sim \int N^2 P(N,\varepsilon) \, dN$$

(2)

and

$$N_z \sim \frac{1}{N_w} \int N^3 P(N,\varepsilon) \, dN$$

(3)

It was also assumed that each of the polymers in the distribution is a fractal, with fractal dimension D_p that was calculated by several techniques[12-14] and is very close to its Flory estimate $D_p = 5/2$. Thus the mass N of every polymer is related to its radius R(N) by

$$N \sim R(N)^{D_p}$$

(4)

Finally, the exponent τ of the distribution, equation (1), is related to the fractal dimension D_p of the macromolecules in the reaction bath

$$\tau = 1 + \frac{d}{D_p}$$

(5)

where d is the dimension of space. The latter relation may be interpreted by calculating the probability P(R) of finding a polymer with radius R, R being measured with a logarithmic scale. One finds

$$P(R) \sim R^{-d}$$

(6)

This implies that whatever the size of the macromolecules in a given class, the distribution of masses is such[15], in the reaction bath, that corresponding polymers are in a C* situation. Because they are fractals, there is still plenty of space that is left, so that smaller polymers in a class of smaller size are also in a C* situation. This description is valid from the largest size N_z to the smallest polymers, so that space is completely filled with monomers. In this respect, one may say that the distribution of sizes is fractal: whatever the length scale one is considering, one always observes polymers in a C* situation.

Finally let us mention that the exponent σ that appears in relation (1) may be eliminated if one evaluates N_z as a function of N_w instead of ε. One finds

$$N_z \sim N_w^{Dp/(2Dp-d)} \approx N_w^{5/4}$$

(7)

In all the following, we will always assume that the distribution of masses is given and quenched. Once the synthesis is made, the distribution is not changed by the dilution that is made. Only the fractal dimension changes , as we will see.

3. Dilute solutions

Let us assume that we synthesized branched polymers below the gel point. The reaction is quenched. Then an excess solvent is added so that the resulting solution is dilute, and the various polymers are far apart from each other . We assume the solvent to be good. Because the solution is dilute, the screening effects of the excluded volume interaction that were present in the reaction bath disappear, and each of the polymers swells. Therefore, the fractal dimension changes from the value D_p it had in the reaction bath to a value D_o .

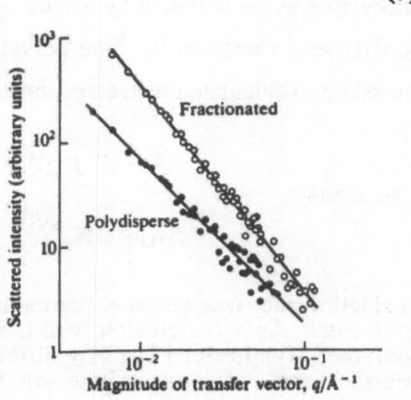

The latter was calculated in a Flory approximation[4][6], and is D_o =2 for d=3. Thus, if one had a monodispersed solution, the scattered intensity $S_1(q)$ in the intermediate regime in a small angle neutron scattering experiment would be

$$S_1(q) \sim q^{-D_o}$$

(8)

The solution however is polydisperse, as discussed above, and what is measured is an average over the entire distribution of masses[16][17]. When this averaging procedure is made, we find, in the same regime that the total scattered intensity is

$$S_t(q) \sim q^{-D_o(3-\tau)} \approx q^{-8/5}$$

(9)

The corresponding experiment was attempted recently on Polyurethane dissolved in deuterated toluene. Figure 1 shows the results for a fractionated-monodisperse- and a polydisperse samples. The first one leads to D_o. Combining this result with the second , one determines the exponent τ for the distribution function. It was found by Bouchaud et al.[18] that $D_o = 1.98\pm0.03$ and $\tau = 2.2 \pm 0.05$, in reasonable agreement with the values quoted above. More generally, equation (9) may be written in the scaled form

$$S_t(q) \sim N_w\, f(qR_z)$$

(10)

where R_z is the z averaged radius of gyration of the polymers

$$R_z^2 \sim \frac{1}{N_w} \int N^2 R^2(N)\, P(N,\varepsilon)\, dN$$

(11)

and f(q) a function that is regular for small argument and singular for large argument, in such a way that one recovers equation (9) above. It is possible to calculate the above integral and to eliminate ε in favor of N_w. One finds, in agreement with (9)

$$R_z^2 \sim N_w^{5/4} \tag{12}$$

The latter relation was checked first experimentally by Leibler and Schosseler[19] on Polystyrene cross-linked by irradiation. Similarly, Candau et al.[20] checked that the diffusion coefficient varies as R_z^{-1} on polystyrene cross-linked with divinyl benzene. Finally, it is possible to calculate the average intrinsic viscosity[16]

$$<[\eta]> = \int [\eta(N) \; N \; P(N,\varepsilon) \; dN \tag{13a}$$

one finds

$$<[\eta]> \sim N_w^{3/8} \tag{13b}$$

a relation that was tested experimentally, with reasonable agreement, by Patton et al.[20] on polyester. As a conclusion to this section it seems that for a large number of branched polymers synthesized in many different ways, the percolation distribution of sizes applies. Furthermore, when in a dilute solution, swelling occurs, and there results a change in the fractal dimension. As a consequence of the polydispersity, however, any scattering experiment does not lead directly to the fractal dimension of the polymers, but to an effective dimension that is a function of both the fractal dimension D_0 and of the polydispersity exponent τ. Therefore, a single experiment is not sufficient to determine the fractal dimension, and one needs two independent measurements, such as for instance the radius of gyration (or the scattered intensity in the intermediate scattering range) and the intrinsic viscosity in order to check that polydispersity is not affecting the results and changing the exponents.

4. Semi-dilute solutions

The entire approach in last section is valid as long as the various polymers are far from each other. Above an overlap[22] concentration C*, there is interpenetration of the large polymers by the small ones, and gets progressively back to the melt situation described in section 1 as concentration is increased. Because of the interpenetration, screening effects are present, and there appears a screening length ξ. The various characteristic properties may be calculated by scaling arguments. One needs to know the molecular weight dependence of the overlap concentration C*. This is determined by a space filling condition:

$$C* \int R^3(N) \; P(N,\varepsilon) \; dN \sim 1 \tag{14a}$$

leading to

$$C* \sim N_w^{-3/8} \tag{14b}$$

where the exponent in the latter relation is another combination of the exponents D_0 and τ. Equation (14b) was checked experimentally by Adam et al. by measuring the second virial coefficient for dilute solution as a function of N_w by elastic light scattering.

For concentrations larger than C*, the smaller polymers in the distribution interpenetrate the larger ones and screen the excluded volume interactions. Therefore, one has to split the distribution $P(N,\varepsilon)$ into two parts. The small masses are far from each other and behave as in dilute solutions. The larger ones are screened and behave as in the reaction bath, if one defines the adequate statistical unit. We define a blob as the larger mass g_z in the smaller part of the distribution. Therefore all polymers with mass smaller than g_z and parts of large macromolecules made of a smaller number of units behave as in the previous section. The radius ξ of a blob corresponds to the screening length for the excluded volume interaction, and is a local property that has to be independent of the average molecular weight in the distribution. We expect it to be a function of the monomer concentration C only. Assuming that

$$\xi \sim C^{-a} \tag{15}$$

and insisting that it is of the order of the average radius of gyration at C^*, we get

$$\xi \sim C^{-5/3} \tag{16a}$$

Note that the same averaging as in section 2 is present for the blob.
The radius of gyration of a large polymer is the same as in the reaction bath if one takes the blob as the statistical unit. This is made of g_z monomers and is such that

$$\xi \sim g_z^{1/2} \tag{16b}$$

therefore the radius of a polymer is

$$R(N) \sim \{ \frac{N}{g_z} \}^{2/5} \xi \tag{17a}$$

calculating the average radius following the lines of the previous section, we get

$$<R_z \sim N_w \ C^{-1/3} \tag{17b}$$

The latter two relations may also be found by writing down scaling relations for the characteristic distances:

$$L_z \sim N_w^{5/8} \ F_L \ (C/C^*) \tag{18}$$

where the unknown function $F_L(x)$ behaves as a power law in the semi dilute range, with an exponent that depends on the length we are considering. Relations (16) and (17) were not observed experimentally so far.

5. Dynamics

5.1. Reaction bath

The first characteristics of the sol-gel transition is the divergence of the viscosity η of the sol below the gelation threshold, and the vanishing of the modulus G of the gel above it. Two extra exponents are defined for these quantities

$$\eta \sim \varepsilon^{-s} \tag{19}$$

and

$$G \sim \varepsilon^{\mu} \tag{20}$$

Whereas percolation describes the static properties of the polymers and gel, there seems to be two universality classes for s and μ[34-40]. These were discussed very recently by Arbabi and Sahimi[23] who managed to summarize in a single model the various results that were conjectured. Roughly speaking, one finds both the Zimm and Rouse limits[24], respectively with and without hydrodynamic interactions. In the former limit, it was found[25-29]

$$s_1 = \nu - \beta/2 \tag{21a}$$

$$\mu_1 = (d-2)\nu + \zeta \tag{22a}$$

5

whereas in the latter, the exponents are[30][31]

$$s_2 = 2v-\beta \qquad (21b)$$

$$\mu_2 = vd \qquad (22b)$$

with v, β and d being respectively the exponents for the radius of the largest polymers, the gel fraction and space dimension. The exponent z corresponds to the elastic path and was calculated in the Flory approximation by Roux[32] and by Family and Coniglio[33]

$$\zeta = D_p/2 \qquad (23)$$

where D_p is the fractal dimension of the polymers in the reaction bath, relations (4) and (5), and is related to the previous exponents

$$D_p = d - \beta/v \qquad (24)$$

In what follows, we will keep the notation for s and μ in order to describe both cases. Viscosity and modulus are the complex and real parts of a single response function, the complex viscosity $\bar{\eta}(\omega)$, or of the complex modulus $\bar{G}(\omega)$, such that

$$\bar{G}(\omega) = i\omega\,\bar{\eta}(\omega) \qquad (25)$$

Following Efrös and Schklovskii[41], it was possible to write a scaling relation[42] for the complex modulus:

$$\bar{G}(\omega) \sim \varepsilon^\mu\, f_\pm(i\omega\,T) \qquad (26)$$

with

$$T \sim \varepsilon^{-s-\mu} \qquad (27)$$

For high frequencies, one is probing restricted regions of space. Therefore the modulus should be independent of ε. One finds

$$\bar{G}(\omega) \sim (i\omega)^{\mu/(s+\mu)} \qquad (28)$$

Therefore both the real and complex parts of the modulus should be proportional to each other, and the loss angle d should be a constant

$$\delta = \frac{\pi}{2}\frac{\mu}{s+\mu} \qquad (29)$$

Relations (27) to (29) were checked experimentally by several teams. It remarkable that the ratio $r = \mu/(s+\mu)$ is experimentally the same for all cases. Using relations (21) and (22), one finds respectively $r=2/3$ and $r = 3/4$ for the Rouse and Zimm approximations respectively. Experimentally[35][42-44], this ratio is approximately $r \approx .7$, in agreement with both approximations.

Therefore, the scaling relation for the complex modulus, and viscosity, seems to be satisfied, and we may proceed to the distribution of relaxation times $H(\tau)$. The latter is directly related to the complex viscosity[45]

$$\bar{\eta}\,(\omega) \sim \int \frac{H(\tau)}{1-i\omega\tau}\ d\tau$$

(30)

Knowing the scaling form for the viscosity, one gets the scaled[7] form of the distribution of relation times:

$$H(\tau) \sim \tau^{-\mu/(s+\mu)}\ h(\tau/T)$$

(31)

where T is the longest relaxation time, relation (27). Calculating the moments of the distribution, we find that it is characterized by two diverging times

$$T_1 \sim \frac{\int \tau\ H(\tau)\ dLn\tau}{\int H(\tau)\ dLn\tau} \sim \varepsilon^{-s}$$

(32)

and

$$T \sim \frac{\int \tau^2\ H(\tau)\ dLn\tau}{\int \tau\ H(\tau)\ dLn\tau} \sim \varepsilon^{-s-\mu}$$

(33)

The longest time T was measured by Axelos and Kolb[43] in their frequency dependence of the complex modulus. The shorter one was measured by Sorensen et al.[46] and by Martin et al.[47] with quasi elastic light scattering experiments. The main result here is that the distribution of times is very wide, and decreases as a power law cut-off at large times by an exponentially decreasing function. Whereas the exponent in the power law decrease is essentially the same for all kinds of hydrodynamic assumptions, the behavior of the characteristic times T_1 and T depends explicitly on s and μ.

5.2. Dilute solutions

In a dilute solution, hydrodynamic interactions are present, and the characteristic time for any polymer is related to its radius by the classical expression

$$T(N) \sim R^3(N)$$

(34)

Therefore the largest relaxation time T_d is related to the largest radius. Using relations (34) and (12) we get

$$T_d \sim N_w^{15/8}$$

(35)

The distribution of relaxation times $H_d(\tau)$ may be directly related to the distribution of masses. Because the various masses are far apart from each other, we may assume that their relaxations are independent. Therefore we have

$$P(N) \, dN = H_d(\tau) \, dLn\tau$$

(36)

thus we find

$$H_d(\tau) \sim t^{-4/5} \, h(\tau/T_d)$$

(37)

Again, the distribution decreases as a power law with an exponential cut off at large times. It may also be characterized by two diverging moments

$$T_{1d} \sim N_w^{3/8}$$

(37a)

and

$$T_d \sim N_w^{15/8}$$

(37b)

Note that because in the reaction bath the various polymers are strongly coupled to each other, a relation similar to (36) does not hold.

5.3. Non exponential relaxations.

Because of the wide distribution of relaxation times, non classical relaxations are expected[48][49] except for extremely large times, when the distribution may be neglected and only the largest time (T or T_d respectively for the reaction bath or the dilute solutions) comes into play. For shorter times, one has to take into account the distribution of times. As an example, we consider the stress relaxation s(t) after steady shear in the reaction bath, without any solvent. We have

$$\sigma(t) = \overset{\circ}{\gamma} \int^{\infty} H(\tau) \, \tau \, e^{-t/\tau} \, d \, Ln\tau$$

(38)

For times larger than the longest time T, we know that the distribution is decreasing exponentially. Neglecting the power laws, we get

$$\sigma(t) = \overset{\circ}{\gamma} \int^{\infty} e^{-\tau/T} \, e^{-t/\tau} \, d\tau \qquad (t \gg T)$$

(39)

Performing the integral by a saddle point method, we get

$$\sigma(t) = \overset{\circ}{\gamma} e^{-\{t/T\}^{1/2}}$$

(40)

For shorter times, t≪T, relation (39) may be split into two parts, for relaxation times shorter and larger than t. We assume that only the short relaxation times act, and that the large ones have not the time to act. We get

$$\sigma(t) \approx \overset{\circ}{\gamma} \int^{t} H(\tau) \, e^{-t/\tau} \, d\tau \; + \; \overset{\circ}{\gamma} \int_{t}^{\infty} H(\tau) \, d\tau$$

(41)

$$\sigma(t) \approx \overset{o}{\gamma}\{ \ \eta \ + \int\limits_{0}^{t} H(\tau) \ [e^{-t/\tau} -1] \ d\tau \tag{42}$$

$$\sigma(t) \approx \overset{o}{\gamma}\{ \ \eta \ - t^{s/(s+\mu)}\} \tag{43a}$$

$$\sigma(t) \approx \sigma_0\{ \ 1 - (\frac{t}{T_1})^{s/(s+\mu)}\} \qquad (\ t \ll T \ ) \tag{43b}$$

with $s_o = \overset{o}{\gamma} \ \eta$, and where we used relation (30) for T_1. Thus we find that for times shorter than the longest relaxation time, there is a power law relaxation, whereas for larger times, the relaxation behaves as a stretched exponential. A similar calculation was carried out in dilute solutions for the birefringence relaxation in a dielectric relaxation experiment[49]. It was found that for large times the relaxation is

$$B_N(t) \sim \exp[-(\ t/N^{3/2})^{2/5}] \tag{44}$$

for a monodisperse solution, assuming a Zimm hydrodynamics, and

$$B(t) \sim \exp[-(\ t/N_z^{3/2})^{1/4}] \tag{45}$$

for a polydisperse solution.

6. Conclusion

We considered randomly branched polymers. Their main characteristics are that first, every polymer is a fractal, and second that the distribution of molecular weights is extremely large, and decreases as a power law. Because of this, the averages that are measured are related to but different from the fractal dimension. The measured exponents depend both on the fractal dimension of every macromolecule, and on the exponent t of the distribution. This implies that measuring $S(q) \sim q^{-\alpha}$ in a scattering experiment does not necessarily imply that the fractal dimension is α. One has first to check that the polydispersity effects are not too important, and do not lead to an effective fractal dimension. The main results we found were that the fractal dimension of a polymer is $D_p =5/2$ in the reaction bath, in agreement with percolation. We also found that in a dilute regime the swelling of the polymers leads to a fractal dimension $D_0 =2$. Foe semi dilute solutions, one has to split the distribution into two parts. The small masses are swollen and behave as in a dilute regime. The large poymers are penetrated by the small ones and screening of the excluded volume interactions takes place. Therefor the large scale behavior of these large masses is similar tho that in the reaction bath. The cross over mass corresponds to the blob.

Similarly, there is a very wide distribution of relaxation times, that is characterized by two diverging times, T and T_1. In the reaction bath, both depend on the exponents s and μ for the viscosity and the modulus close to the threshold, and are still under discussion. It seems that there are at least two universality classes, as discussed recently by Arbabi and Sahimi. In a dilute solution, it is assumed that hydrodynamic interactions are present, and only one universality class is observed. Because of this large distribution of times, non exponential relaxations are found for times that are not very large compared to the largest time T. For times smaller than T, it is found that there is a power law relaxation, whereas for times larger than T, a stretched exponential behavior should be observed.

Finally, we stress that if the basic object that is cross-linked is a linear chain instead of a small chemical unit, the observed exponents are not always those that we just discussed: If one cross-links chains from a melt, it was shown that the classical Flory-Stockmayer theory should be valid[51]. When cross-linking is made from a semi-dilute solution, the width of the critical region, where percolation exponents are observed increases[52]. More precisely, it was shown that the width ε^* of the region where critical exponents are found is

$$\varepsilon^* \sim \{ZC^{5/4}\}$$

(46)

where Z is the number of monomers in the initial chains, and C the monomer concentration. A second discussion concerns the possible differences that may be observed for the viscoelastic behavior. We know from the studies on dielectric mixtures, that when one is not at the threshold, it is relatively easy to observe for the complex viscosity an exponent that is not the critical exponent but rather an effective medium exponent[53] with $s = \mu = 1$, so that the distribution of relaxation times instead of being given by relation (31), is rather

$$H_{em}(\tau) \sim \tau^{-1/2} f(\tau/T')$$

(47)

where the longest time T' and the average time T_1' are rather than relations (32) and (33)

$$T' \sim \varepsilon^{-2} \sim N_w^2$$

(48)

and

$$T_1' \sim N_w$$

(49)

Note that this implies that the complex modulus writes then

$$\bar{G}(\omega) \sim \varepsilon^\mu g(i\omega T')$$

(50)

and, for frequencies larger than 1/T', has a frequency dependence that is different from what we discussed above. Assuming that it depends only on w, we find

$$\bar{G}(\omega) \sim (i\omega)^{1/2} \qquad (\omega T' \gg 1)$$

(51)

The latter relation was observed by Chambon and Winter[54] among others. Therefore, it is possible to find effective exponents between the limiting values 1/2 and 0.7, describing cross-over situations. Exponents smaller than the lower limit were observed recently by Antonietti[55] and are less easy to understand by the above considerations.

References

1. P.J. Flory, 1953, Introduction to Polymer Chemistry, Cornell University Press, Ithaca.
2. B. Mandelbrot, 1977, The Fractal Geometry of Nature, Freeman, San Francisco .
3. B. H. Zimm, W. H. Stockmayer, J. Chem. Phys. 17:1301, (1949) .
4. J. Isaacson, T. C. Lubensky, J. de Phys., 42: 175, (1981).
5. P. G. de Gennes, C.R. Acad. Sci.(Paris), 291: 17, (1980).
6. M. Daoud, J. F. Joanny, J. Physique , 42: 1359, (1981).
7. M. Daoud , J. Phys A21: L973, (1988).
8. J.E. Martin, J.P. Wilcoxon, Phys. Rev. Lett., 61: 373, (1988).
9. D. Stauffer, J. Chem. Soc. Faraday Trans. II , 72 : 1354, (1976).
10 . P. G. de Gennes J. de Phys. Lett. , 37 , 1, (1976).
11. D. Stauffer, Introduction to Percolation Theory, Taylor and Francis, London, (1985).
12. H. Nakanishi, H.E. Stanley, Phys. Rev. B 22: 2466, (1980).

13. B. Derrida, J. Vannimenus, J. Physique Lett. 41: 473, (1980). See also B. Derrida, D. Stauffer, H.J. Herrmann, J. Vannimenus, J. Physique Lett. 44 :701,(1983).
14. H. Nakanishi, H.E. Stanley, Phys. Rev. B 22 : 2466 (1980).
15. M. E. Cates, J. de Phys. Lett. 38 : 2957 , (1985) .
16. M. Daoud , F. Family , G. Jannink , J. de Phys. Lett. 45 : 119 , (1984) .
17. J. E. Martin, B. J. Ackerson, Phys. Rev. A31 :1180 (1985).
18. E. Bouchaud, M. Delsanti, M. Adam, M. Daoud, D. Durand, J. Physique Lett. 47 : 1273, (1986).
19. L. Leibler , F. Schosseler , in Physics of Finely Divided Matter , Springer Proc. Phys. 5 , Springer verlag, Berlin, 135 , (1985).
20. S. J. Candau, M. Ankrim, J. P. Munch, P. Rempp, G. Hild, R. Osaka, in Physical Optics of Dynamical Phenomena in Macromolecular Systems, W. De Gruyter,Berlin,p.145(1985).
21. E. V. Patton , J. A. Wesson , M. Rubinstein , J. C. Wilson , L. E. Oppenheimer, Macromolecules, 22: 1946, (1989).
22. M. Daoud, L. Leibler, Macromolecules 41 : 1497, (1988).
23. S. Arbabi, M. sahimi, Phys. Rev. Lett., 65 : 725, (1990).
24. M. Doi , S. F. Edwards , the theory of polymer dynamics , Oxford Science publications , (1986) .
25. J. Kertesz , J. Phys. A16 : L471 , (1983) .
26. A. Coniglio , H. E. Stanley , Phys. Rev. Lett. 52 : 1068 , (1984) .
27. P. G. de Gennes , J. de Phys. Lett. 37 : 1 , (1976) .
28. H.J. Herrmann, D.C. Hong, H.E. Stanley, J. Phys. A 17 : L261, (1984).
29. M. J. Stephen , Phys. Rev. B17:4444 , (1978) .
30. P. G. de Gennes , J. de Phys. Lett. 40 : 197 , (1979) .
31. M. Daoud, A. Coniglio, J. Phys. A 14 : L301, (1981)
32. 86. S. Roux, C.R. Ac. Sci. (Paris) 301, 367 (1985).
33. F. Family, A Coniglio, J. Physique Lett.46 : 9 (1985).
34. C. Allain, L. Salome, Macromolecules, 20 : 2597, (1987).
35. M. Rubinstein, R. H. Colby, J. R. Gillmor, in "Space-time organization in Macromolecular fluids, Springer series in Chemical Physics 51 : 66, (1989).
36. M. Adam, M. Delsanti, J. P. Munch, D. Durand, J. Physique 48, 1809(1987).
37. M. Adam, M. Delsanti, D. Durand, Macromolecules18, 2285 (1985).
38. B. Gauthier-Manuel, E. Guyon, J. Physique Lett. 41, 503 (1980).
39. M. Adam, M. Delsanti, D. Durand, G. Hild, J. P. Munch, Pure Appl. Chem. 53 : 1489, (1981).
40. J. E. Martin , D. Adolf, J. Wilcoxon , Phys. Rev. Lett, 61 : 2620, (1988), Phys. Rev. A39 : 1325, (1989).
41. A. L. Efros , B. I. Shklovskii , Physica Status Solidi , B76 , 475 , (1976).
42. D. Durand, M. Delsanti, M. Adam, J. M. Luck, Europhys. Lett. 3 : 297, (1987).
43. M. Axelos, M. Kolb, Phys. Rev. Lett., 64 :1457, (1990).
44. D. Adolf, J. E. Martin , J. Wilcoxon, Macromolecules, 23 :527, (1990).
45. J.D. Ferry, Viscoelastic Properties of Polymers, (1980) J. Wiley, N.Y.
46. W. F. Shi, W. B. Zang, C. M. Sorensen, unpublished.
47. D. Adolf, J. E. Martin , Macromolecules, 23 : 3700, (1990).
48. M. Adam, M. Delsanti, J. P. Munch, Phys Rev. Lett. 61 : 706, (1988).
49. M. Daoud, J. Klafter, J. Phys. A23, L981, (1990).
50. J. E. Martin, J. Wilcoxon, J. Odinek, Phys Rev. A43 : 858, (1991).
51 P. G. de Gennes, J. Physique Lett.,38 : 355 , (1977) .
52 . M. Daoud , J. de Phys. Lett. 40 : 201 , (1979) .
53. I. Webman, J. Jortner, M. H. Cohen, Phys. Rev. B16 : 2593, (1977).
54. F. Chambon, H. H. Winter, J. Rheology, 31 : 683, (1987).
55. M. Antonietti, private communication.

NOVEL NETWORK STRUCTURES: FRACTAL-RIGID-FLEXIBLE NETWORKS

T. A. Vilgis

Max-Planck-Institut fur Polymerforschung
Ackermannweg 10, 6500 Mainz, FRG

I. Introduction

It is realised recently that the theory of rubber elasticity and gels offers new directions, where novel incrediences play an important role. After the great success of phantom type theories[1-3] i.e., non—interacting chains, the discussions in the past decade has been dominated by the role of entanglements and their effects on elasticity from different points of view[2,4-6]. The long dispute in the literature has now probably been solved by computer simulation where networks with and without entanglements can be generated very simply[7]. These results seem to favor a strong contribution of entanglements to the modulus.

We do not go in such details here, rather we pick up a new direction in the theory of networks, which can be studied first along the lines of phantom type models, which are considered as networks composed of phantom chains, i.e. chains which do not have topological restrictions. Therefore, the problem of entanglements is irrelevant. The only fact which should matter is the topology of crosslinking. These networks can be viewed as simplest first order models, from which more complicated and realistic considerations can be started.

It has been reported recently that the topology of crosslinkage influences deeply the behaviour of networks. Model systems, such as fractal networks, have been considered[8]. It has been suggested that such fractal "left overs" from a vulcanisation or gelation process create heterogeneities within the rubber sample which will effect the macroscopic and microscopic behaviour measured in rubbers. This will only be the case if no significant amout of entanglements are present between two crosslinks, since trapped entanglements will introduce a new relevant length scale, i.e. the tube diameter, and the chemical distance between crosslinks will be of less importance. The modulus is now determined by the distance between two entanglements rather than by the distance between crosslinks.[5,6]

Synthesis, Characterization, and Theory of Polymeric Networks and Gels
Edited by S.M. Aharoni, Plenum Press, New York, 1992

13

A simple mean field theory of random heterogeneous phantom rubbers has been considered by the present author[9] and independently by Heinrich and Schimmel[10]. It was found that such inhomogeneities do not effect the deformation behaviour but strongly influence the neutron scattering[9]. Moreover, inhomogeneities can reduce the effective modulus under certain circumstances[10,11].

So far we have spoken about the real topological structure. Here it is introduced a "thermodynamic topology", i.e. non trivial thermodynamic interactions, as present in blends, strongly influence the phase behaviour and elasticity of crosslinked blends. It is clear[12] that crosslinkage of partially miscible polymers prevent complete phase separation. Crosslinked blends cannot phase separate macroscopically but form microdomain structures as they undergo a microphase separation. The statistical mechanics of such materials is unknown yet and only first attempts have been made to solve this question[13].

The paper is organised as follows. First, the classical theories will be reviewed on their structure and topology relationship. We then turn to fractal networks and fractal containing networks. This will lead to the aspect of random heterogeneous networks and first results will be reviewed briefly. Finally, we discuss the case of crosslinked blends and semi IPNs, in more detail.

II. Classical models of crosslinked polymers

The classical models are divided in two limiting cases. The first is the Kuhn model, i.e. a single chain approximation where it turns out that the free energy is a sum over all contributions from each individual chain, neglecting all structural influences.

$$\beta F = \frac{1}{2} N \sum_{i=1}^{3} \lambda_i^2 \qquad (2.1)$$

where N is the number of strand and λ_i the deformation ratio in the i–th cartesian direction. The above model is quoted as affine model since the network deforms the same way on all length scales.

A more sophisticated analysis has been carried out by James and Guth[2] for four functional crosslinks. Their basic assumption was that several crosslinks of the rubber are fixed on the walls of the sample, forming the surface, whereas the internal crosslinks are free to move. Their result for the free energy is

$$\beta F = \frac{1}{4} N \sum_{i=1}^{3} \lambda_i^2 \qquad (2.2)$$

which is half of the free energy of the Kuhn model (2.1). This reduction of the free energy is due to additional motion of (entropy) crosslinks. For arbitrary functionalities f this result has been generalised by Graessley[14] for an exact calculation for networks of tree like structure. The macroscopic free energy in this case is given by

$$\beta F = \frac{1}{2} N (1 - 2/f) \sum_{i=1}^{3} \lambda_i^2 \qquad (2.3)$$

which is the basic result for phantom networks. Another proof of eq. (2.3) has been given recently by Higgs and Ball[15] using the elasticity–circuit analogy. It has to be noted that eq. (2.3) is only exact for tree like structures, corresponding to mean field solutions where no fluctuations of the structure, i.e. loops, are present. This can be seen as an elementary example (see ref. 15). On a cubic lattice ($f = 6$) one finds $\beta F = \frac{1}{3} N \sum \lambda_i^2$ ($\neq 1 - 2/f$) whereas on a tetrahedral lattice one finds $\beta F = 1/2 N \sum \lambda_i^2 (= 1 - 2/f)$.

Duiser and Staverman[16a] and Flory[16b] have generalised eq. (2.3) to

$$\beta F = \frac{1}{2} \xi \sum_{i=1}^{3} \lambda_i^2 \qquad (2.4)$$

where ξ is the cycle rank , i.e. the number of independent loops in the network which determines the elastically active network strands and the elastically active crosslinks in the following topological manner

$$\xi = N + 1 - M \simeq N - M \qquad (2.5)$$

M is the number of crosslinks. For perfect networks one has

$$f M = 2 N \qquad (2.6)$$

and eq. (2.4) reduces to eq. (2.3) immediately. Hence eq. (2.3) is exact for perfect phantom networks. The reduction of the modulus by $(2/f)$ $k_B TN$ is due to fluctuations of the crosslink points.[15–17]

Deam and Edwards have put forward another model[17]. It starts from the chain variables in continuous notation and uses the replica technique of quenched systems. The problem is the use of the crosslink constraint (see eq. (5.4) later in the paper), but it has been proposed to model it by a harmonic potential and carry out a variational calculation. The result for the free energy is given by:

$$\beta F = \frac{1}{2} N (\sum_i \lambda_i^2 + \log q_0 + \frac{\ell L}{12} \frac{q_0}{M}) \qquad (2.7)$$

15

for four functional crosslinks. The generalisation for $f \neq 4$ will be given later. The stationary condition for the free energy produces the well–known result for the localisation parameter

$$q_0 = \frac{6M}{\ell L} \qquad (2.8)$$

This result provides the physical interpretation of the localisation parameter. Its inverse defines the mean distance in which the crosslinks are localised; i.e. it is proportional to the mean square radius of gyration of a chain segment between two crosslinks, i.e. the mesh size of the system. $q_0^{-3/2}$ describes the volume which is explored by the crosslink upon diffusion. If this value is put back in the variational free energy it is found:

$$\beta F = \frac{1}{2} M \left\{ \sum_{i=1}^{3} \lambda_i{}^2 - \log\left(\frac{6M}{\ell L}\right) + 1 \right\} \qquad (2.9)$$

This is just the famous James and Guth result. For arbitrary functionalities f it can be shown that

$$q_0 = \frac{6}{\ell L} M \left(\frac{f}{L} - 1\right) \qquad (2.10)$$

This is shown easily for f even and assumed to be continued to odd values for f. Then one has

$$\beta F = \frac{1}{2}(\frac{f}{2} - 1) M \left\{ \sum_{i=1}^{3} \lambda_i{}^2 - \log\left(\frac{6M}{\ell L}(\frac{f}{2} - 1)\right) + 1 \right\} \qquad (2.11)$$

this can be transformed to the usual form by the perfect network equation, i.e. $fM = 2N$, but for later purpose it is kept in the present form. So far we have looked at the full energy which determines the macroscopic properties. In the following small angle neutron scattering (SANS) properties for phantom networks will be summarised.

The scattered intensity of a deuterated chain in a network is proportional to form factor, which contains the microscopic details of the chain in the network, i.e. the chain or the labeled path between two or several crosslinks. In general the form factor is defined as

$$S(k) = \frac{1}{L^2} \sum_{n,m}{}' \left\langle \exp(-ik(r_n - r_m)) \right\rangle \qquad (2.12)$$

where L is the number of scattering elements, r_n and r_m are the positions of scattering atoms on the chain, and k is the wave vector. The average $\langle \ \rangle$ has to be carried out with all conformations accessible to the segments. The procedure $\langle \ \rangle$ requires all information discussed in the case of the free energy, i.e. the positions of the cross—links and the location of the chains. The general result will be the structure factor of the Deam—Edwards model, i.e. the harmonic localization model. This structure factor contains the affine model (Kuhn) and the phantom model (James and Guth) as limiting cases, so we do not discuss them separately.

The calculation of the SANS form factor was carried out by Warner and Edwards.[18]

$$S(k) = \frac{1}{L^2} \int_0^L \int_0^L ds \ ds' \ exp \left[-\frac{\ell^2}{6} k_i^2 \lambda_i^2 |s-s'| - k_i^2(1-\lambda_i^2) \frac{1}{2q_0}(1-e^{\ell^2 q_0 |s-s'|/3}) \right] \tag{2.13}$$

where k_i is a Cartesian direction of the wave vector and λ_i is the Cartesian stretching ration. q_0 is the localisation parameter as before and enters naturally as the term accounting for fluctuations of the crosslinks around some mean position. Note that this fluctuation is determined by the environment, i.e. the crosslink density in the sample. Here it is assumed to be uniform, and q_0 is a number. If q_0 is small, i.e. in the weak cross—linking limit, the exponent containing q_0 may be expanded, and the classical expression for the James and Guth phantom network is recovered, i.e.

$$S(k) = \frac{1}{L^2} \int_0^L \int_0^L ds \ ds' \ exp \left[-\frac{\ell^2}{6} k_i^2 |s-s'| - \frac{\ell^2}{6} k_i^2(\lambda_i^2-1) \frac{1}{2} \frac{6M}{\ell L} |s-s'|^2 \right] \tag{2.14}$$

It has been shown that the affine model (Kuhn model), where the crosslinks are supposed to be fixed, gives a factor of 2 more in the second term in the exponential.

III. Networks of non classical connectivity

Fractals are objects with non classical connectivity, i.e. they represent spaces with a non integer dimension. Typical examples are the infinite percolation cluster, as an example for a random fractal or the Sierpinski gasket as a regular fractal. The fractals are normally treated as rigid with the bonds between the junction points being rigid. If these rigid bonds are replaced by flexible Gaussian chains, one has polymeric fractals.[19] The properties of such polymeric fractals are in general different from those of rigid lattice fractals. In rigid fractals the elasticity is enthalpic, whereas in polymeric fractals one has pure entropic elasticity.

The Sierpinski network[20a] is the most well–known example of a network with non classical connectivity. It has functionality four but there are holes on all scales, i.e. large fluctuations in structure. The other example the hierarchical fractal network is based on the idea of modeling the backbone of the percolation cluster.[20b]

We have to define the size and the topological behaviour of such (non entangled) fractal networks first before we discuss typical network properties of them. The basic quantities we need for later discussions are the relations between size and topology, i.e. connectivity of the network which determines the swelling behaviour of such networks in their own melts or in other fractal environment. We need first a relation between the connectivity (spectral dimension) and the ideal Gaussian fractal dimension of the polymeric fractal. The Gaussian fractal dimension can be easily shown to be

$$d_f = \frac{2\ d_s}{2 - d_s} \tag{3.1}$$

where d_s is the spectral dimension of the lattice fractal which is the same as that of the polymeric fractal since the topology is preserved. F_t (3.1) suggests that the Gaussian dimension of the network is

$$R_0^{d_f} \sim m \tag{3.2}$$

where m is the total mass

If this object is put in ordinary solvent, it swells and a smaller fractal dimension is obtained. This phenomenon is well known in linear chains, which are a special case of a polymeric fractal with $d_s = 1$ (see eq. (3.1)). Here, we are interested in more general situations, i.e. we put the fractal or in more complex solvent such as linear chains or another fractals with Gaussian dimension δ_f[21]. This will be relevant for later considerations in elasticity. Excluded volume becomes screened on the scale of the size of the different fractals and a generalised Flory free energy can be obtained[21]

$$F = k_B T\ \frac{R^2}{R_0^2} + \frac{v}{m^{\delta_f/d_f}}\ \frac{m^2}{R^d} \tag{3.3}$$

The first term is the usual elastic free energy ($\sim R^2$ since the fractal is Brownian) and the second term the screened excluded volume. Note that if $\delta_f = 0$ we obtain the ordinary Flory free energy of swelling in point like (m=1) and if $\delta_f = d_f$ we recover the classical expression for a melt of fractals. Minimisation of the free energy predicts

$$D_f = \frac{d + 2}{2 - \frac{1}{d_f}\ (\delta_f - 2)} \tag{3.4}$$

where d is the Euclidian space dimension. Eq. (3.4) contains all well known cases[21]. Most remarkably is the result that if any fractal with d_f is mixed in linear chains

$\delta_f = 2$ a universal size exponent $D_f = (d + 2)/2$ is predicted, independent of the topology d_s. Another limit which has to be noticed is the case if formally $D_f > d$ which is indeed unphysical. For such cases the fractals become saturated and takes $D_f = d$! For melts this condition is given by $d_f > d$, i.e. if that Gaussian fractal dimension is larger than the space dimension. In terms of the topological connectivity this means

$$d_s > \frac{2d}{2+d} = d_s^{max} \tag{3.5}$$

Therefore in $d = 3$ melts with $d_s > 6/5$ are always saturated and the fractal nature does not matter since $D_f = 3$, i.e. $R \sim m^{1/3}$.

We mention an experiment which can be discussed with the results. Antonietti et al.[22] synthesized microgels which are probably of fractal nature, and their connectivities are close to percolation cluster. In good solvent they obey $R^2 \sim m$ as isolated branched molecules (or percolation networks) with $d_s = 4/3$ or $d_f = 4$. Deuterated microgels in protonated microgels behave as $R^3 \sim m$, i.e. they are saturated whereas microgels solved in linear chains behave as $R^{2.6} \sim m$, whereas eq. (3.4) suggests $R^{2.5} \sim m$ which is in very good agreement with neutron scattering data by Antonietti et al.

This fact has now implications on networks containing fractal heterogeneities. Such fractal heterogeneities can be "left overs" from the vulcanisation or gelation process. Let us first look at a model example proposed by Boué and Bastide[8,25,24] which can be viewed as connected Sierpinski gaskets.

The structure represents a crosslinked melt of Sierpinski networks. Since the spectral dimension of the d–dimensional Sierpinski gasket is always larger than $2d/(2+d)$, i.e.

$$d_s = 2 \frac{\log(d+1)}{\log(d+3)} > d_s^{max} \tag{3.6}$$

Thus the polymeric gaskets are always saturated, i.e. their fractal nature does not matter and they form balls with $R^d \sim m_g$ where m_g is the mass in a single gasket. Therefore one may conclude that the elastic properties of a network are determined by the number of crosslinks which join the gasket rather than by the total number of crosslinks including those within the individual gaskets. Then the modulus is not determined by the total number of crosslinks but by a number far less.

More generally this applies for more complicated situations. Imagine again a network with some fractal "left overs", i.e. parts of percolation clusters or vulcanisation clusters at at a scale d_ξ which are crosslinked to other chains and other clusters. Therefore the size and physical behaviour of such systems is given by eq. (3.4) where the clusters with $d_f(d_s \simeq 4/3)$ are surrounded by linear chains with $\delta_f = 2$.

In $d = 3$ we have $D_f = 5/2$, i.e. the clusters are swollen and their internal crosslinks would contribute to elasticity.

Such considerations are only simple models and are not applicable for realistic systems but such models give some feeling into the behaviour of phantom networks containing fractal heterogeneities. Such conclusions become invalid if entanglements are present. Entanglements introduce a new length scale, the entanglement distance which determines the physical behaviour of such networks[5]. The effect of heterogeneities becomes blurred, due to entanglement sliding and the tube geometry.

We are now in position to discuss the elastic properties pure fractal polymer networks. Consider first a Sierpinski network. The topology of the Sierpinski network is given by the number of chains and the number of crosslinks as a function of its generation p. The number of chains in its p—th generation is given by

$$N_p = 3^p \tag{3.7}$$

The number of crosslinks can be counted also. In the first generation these are 3 crosslinks, i.e. the edges. In $p = 2$ there are $3 \cdot 3$—3 crosslinks, in $p = 3$ there are $3(3 \cdot 3$—3)—3 etc. For arbitrary p we have

$$M_p = 3^p - \sum_{n=1}^{p-1} 3^n = \frac{3^p + 3}{2} \tag{3.8}$$

crosslinks. Therefore the cycle rank of the structure in its p—th generation is given by (see eq. (2.5))

$$\xi_p = N_p + 1 - M_p = 1 + \sum_{n=1}^{p-1} 3^n = \sum_{n=0}^{p-1} 3^n \equiv \tfrac{1}{2}(3^p - 1) \tag{3.9}$$

According to Eq. (2.4) the free energy is then

$$F_p = \tfrac{1}{2}\, \xi_p\, k_B T \sum_{i=1}^{3} \lambda_i^2 \tag{3.10}$$

and the modulus is given by $G_p = \dfrac{\xi_p}{\Omega_p} k_B T$ where Ω_p is the volume in the p—th generation. If we assume a closed packed structure we assume $\Omega_p \sim N_p\, v_0$ where v_0 is the volume of the chain we predict for the modulus

$$G_p = \left(\frac{1}{2} - \frac{1}{2} 3^{-p} \right) k_B T \frac{1}{v_0} \qquad (3.11)$$

Taking the limit of $p \longrightarrow \infty$, the modulus becomes

$$G_\infty = \frac{1}{2} k_B T \frac{1}{v_0} \qquad (3.12)$$

which is just the modulus of the James and Guth—network for functionality $f = 4$. The Kuhn modulus would be $k_B T / v_0$. Note that if we calculate the ratio of the number of chains and the number of crosslinks we find

$$\frac{N_p}{M_p} = 2 \cdot \frac{1}{1 + \frac{3}{3^p}} \xrightarrow[p \to \infty]{} 2 \equiv \frac{f}{2} \qquad (3.13)$$

since the functionality $f = 4$ for the Sierpinski network. Therefore this method predicts a modulus which is in accordance with the James and Guth model $G \sim \frac{kT}{v_0} \left(1 - \frac{2}{f} \right)$.

This observation is in contradiction with calculations by Bastide and Boué[24] who predict an ultraweak modulus for the Sierpinski micro network. Their argument can be summarised as follows. The starting point is a Sierpinski network, where the three summits are fixed. Thus the Kuhn modulus is given by $G_p = N_K k_B T / \Omega_p$ where N_K is the number of chains attached to the fixed crosslinks, i.e. the edges. This is just $N_K = 3$, i.e. only the outer chains count. (This has been concluded from application of the star—triangle equivalence[24]). Therefore the volume Ω_p is increasing as $3^p v_0$ and the modulus is given by

$$G_p = 3 \, (k_B T / v_0) \, 3^{-p} \xrightarrow[p \to \infty]{} 0 \qquad (3.14)$$

Therefore the micro network would represent an "ultraweak solid".

This is because only three points are fixed which are the edges. This number remains constant for each generation, whereas the number of network elements (chains) is increasing exponentially. Bastide's and Boué's calculation is equivalent to Graessley's theory[14] where the free energy of a tree micro network in p—th generation is given by

21

$$F_p = \frac{1}{2} \nu\, k_B T\ \frac{N_{kp}}{N_p} \sum_{i=1}^{3} \lambda_i^2 \tag{3.15}$$

where N_{kp} is the number of fixed crosslinks and N_p the number of chains in total. For treelike structures this term goes as $(f-2)/(f-1)$, which is different from the macroscopic network. ν is the number of strands per unit volume, $\nu \simeq N_p/V$. A detailed statistical mechanical theory on the modulus of a fractal micro network of the Sierpinski type is also an open problem and is left to a separate publication.

IV. Random heterogeneous networks

We now turn to the case of networks containing random rather than fractal heterogeneities. This seems to be more realistic for randomly crosslinked networks. Here a oversimplified model is offered, and we start from the free energy of the Deam–Edwards model eq. (2.11). This model has an extra parameter, i.e. $q_1 = \frac{6M}{\ell L}(\frac{f}{L}$ $- 1)$ which can be interpreted in a different way. The localization parameter q_0 depends on the environment of a crosslink, since the cross–link density around a given cross–link determined the strength of the localization. That means around a given crosslink the crosslink density is high, the localization will be strong, whereas it is weak, the localization will be also weak. In order to make a model we assume now that the localization depends on a given distribution of crosslinks. Thus we label the localization parameter with the spatial coordinate r, and write $q_0(r)$. We find

$$q_0(r) = 6M(r)/(\ell L) \tag{4.1}$$

The free energy is now a functional of the crosslink density $M(r)$, i.e. it has the same functional dependence as before but the cross–link density is now a function of the spatial coordinate

$$\beta\, F(\{M(r)\}) = \frac{1}{2} M(r)\Big(\sum_i \lambda_i^2 - \log\Big\{\frac{M(r)}{\ell L}\Big\} + 1\Big) \tag{4.2}$$

Now we have to average over a distribution of $M(r)$

$$F = \int \mathscr{D} M(r)\, F(\{M(r)\})\, P(\{M(r)\}) \tag{4.3}$$

where $P(\{M(r)\})$ is the normalised distribution of all realisations of the crosslink density. This is a quenched average of the free energy functional. To have a simple model we difine for $M(r)$

$$M(r) = M + n(r) \qquad (4.4)$$

where M is the total cross–link number and n(r) is the fluctuation, which is a random variable. We set

$$P(\{M(r)\}) = \mathscr{N}\exp\left\{-\frac{1}{2\Delta\Omega}\int dr\,\{M(r) - M\}^2\right\} \qquad (4.5)$$

where Ω is a measure of the strength of the network disorder. Therefore the average of the free energy is very simple and the result is

$$\beta F = \frac{1}{2}M\left[\sum_i \lambda_i{}^2 - \log\left\{\frac{6M}{\ell L}\right\} + 1 + \frac{3}{2}\frac{\Delta}{M} + 0(\Delta^2)\right] \qquad (4.6)$$

The free energy is only marginally changed. An extra term proportional to the fluctuation of the crosslink density increases the free energy, but it does not affect the modulus or the deformation dependence. The modulus and the deformation dependence of the free energy are not altered. We conclude that the deformation dependence and the free energy are less sensitive to inhomogeneities. This agrees with the physical intuition: The free energy is a macroscopic quantity and this is less sensitive to microscopic details. This is not the case in small angle neutron scattering, where all the microscopic details matter. as we will see below.

We take a given realisation of the network, formulate the same theory as that described above, and calculate the structure factor and then we average over all possible realizations of the crosslink densities. Thus we write for the SANS form factor

$$S(k) = \frac{1}{L^2}\int_0^L\int_0^L ds\,ds'\exp\left[-\frac{\ell^2}{6}k_i{}^2|s-s'| - \frac{\ell^2}{6}k_i{}^2(\lambda_i{}^2-1)\frac{1}{2}\frac{6M}{\ell L}|s-s'|^2 - \right.$$
$$\left.\frac{\ell^2}{6}k_i{}^2(\lambda_i{}^2-1)\frac{1}{2}\frac{n(r)}{\ell L}|s-s'|^2\right] \qquad (4.7)$$

and perform the average over the Gaussian distribution of $\{n(r)\}$, and we find for the neutron scattering form factor

$$S(k) = \frac{1}{L^2}\int_0^L\int_0^L ds\,ds'\exp\left[-\frac{\ell}{6}k_i{}^2|s-s'|\right.$$
$$\left.\left\{1 + \frac{6M}{L}(\lambda_i{}^2-1)|s-s'|\left\{1 - \frac{\Delta}{ML}\frac{1}{6}k_i{}^2(\lambda_i{}^2-1)|s-s'|^2\right\}\right\}\right] \qquad (4.8)$$

Thus we find additional terms in the exponent of the form factor. The fluctuation of the crosslink density introduces a new wave vector and a new deformation dependence.

V. Rigid rod networks

Another example of the effect of the influence of structure is the rigid rod network. Here the topology of crosslinking can be the same as in flexible chain networks, but the flexible chains have been replaced by rigid rods. There are two different possibilities. The rods can be connected rigidly or flexibly.

The first one is an example for a entalpic network[25] and the latter one is purely entropic[26]. The rigid entalpic network has been discussed in ref. (25) on the basis of scaling arguments. One essential difference of both types of networks is that the scaling of the modulus with the volume fraction of the rods is significantly different, i.e.

$$G \sim \phi^2 \qquad \text{frozen crosslinks}$$

$$(5.1)$$

$$G \sim \phi^{3/2} \qquad \text{flexible crosslinks}$$

The flexibly hinged network gave rise to a field theory formulation for networks[26] which enables a general description for networks with arbitrary elements. We do not go into mathematical details.

The first question for the flexible rod network is: Does it deform anyhow? The answer is given by counting the number of the degrees of freedom[27]. If the degrees of freedom are greater than zero the network can be deformed. To count the degrees of freedom consider N rods in d dimensions. These free rods have $N(2d-1)$ degrees of freedom. Now there are M crosslinks which remove $M f$ degrees of freedom, where f is the functionality, but give $M d$ translational degrees of freedom. Thus one has the condition

$$N(2d-1) - M f d + M d \geq 0 \qquad (5.2)$$

For a perfect network eq. (2.6) can be employed and we find a limiting functionality

$$f_{max} = 2d \qquad (5.3)$$

Thus if $f < f_{max}$ the networks can be deformed entropically whereas for $f > f_{max}$ the network becomes entalpic, i.e. the crosslinks localise. Another argument has been put forward in ref. (25), i.e. M free crosslinks have Md degrees of freedom but there are $M f/2$ rigid rod constraints in the sample (for a perfectly linked network), so that $f_{max} = 2d$ as before.

The field theoretic Hamiltonian for flexible rigid rod networks is given by

$$H = \mu \int d^3r \int d^3r' \; \phi(r) \, g(r - r') \, \phi(r') + \nu \int d^3r \, (\phi^*(r))^f - \int d^3r \, \phi(r) \, \phi^*(r) \quad (5.4)$$

and the partition function is given by

$$Z = \oint \oint \frac{d\mu}{\mu^{N+1}} \frac{N!}{\nu^{M+1}} \frac{d\nu}{\nu} \frac{M!}{} \int \delta \phi \int \delta \phi^* \, e^{-H} \quad (5.5)$$

μ and ν are fugacities which encounter for the addition of rods and crosslinks. The propagator $g(r)$ has for rods the form

$$g(r) = \frac{1}{4\pi} \delta(|r| - L) \quad (5.6)$$

where L is the rod length. Eqs. (5.5) and (5.4) describe a non hermitian field theory for <u>endlinking</u> of rods, and all correlation functions are non zero if eq. (2.6) is satisfied.

The partition function can be evaluated in tree approximation (saddle point) and we obtain for small deformation the free energy

$$F = \tfrac{1}{2} N \, kT \left(1 - \tfrac{2}{f} \right) \sum \lambda_i^2$$

(5.7)

That is the James and Guth result. This is not surprising since there is enough entropy in the crosslinks to encount for eq. (5.7). The typical form of the modulus is an artefact of the tree approximation. Note that eq. (5.7) is exact for treelike networks. If loops are present the degrees of freedom become reduced and eq. (5.7) becomes invalid. A tree can always be deformed even at $f \to \infty$ (see 5.7) whereas a full loopy structure cannot only for $f < f_{max}$. Thus fluctuations in the structure (loops) play an important role. The exact calculation for modulus for $f < f_{max}$ beyond the tree approximation is an open problem.

VI. Crosslinked blends and semi IPNs

Crosslinked blends are technically very important and have a wide range of applications. The physics of such materials, i.e. their statistical mechanics is unknown yet and is an open problem. Here we want to introduce a new concept to this problem. Imagine a partially miscible blend of two polymers, say A and B. These polymers interact with an excluded volume interaction

$$\int\limits_0^{L_\sigma} ds \int\limits_0^{L_\tau} ds' \ V_{\sigma\tau}(\mathbf{R}_\sigma(s) - \mathbf{R}_\tau(s')) \qquad (6.1)$$

where σ, τ = A, B and L_σ is the length of the polymer in species σ. Usually $V_{\sigma\tau}(\mathbf{r})$ is taken to be short ranged, i.e. $V_{\sigma\tau}(\mathbf{r}) = V_{\sigma\tau}\delta(\mathbf{r})$. The phase behaviour is ruled by de Gennes famous RPA equation.[28]

$$\frac{1}{S_A^0(\mathbf{k})} + \frac{1}{S_B^0(\mathbf{k})} - 2\chi_F > 0 \qquad (6.2)$$

where the Flory χ–parameter $2\chi_F = 2V_{AB} - (V_{AA} + V_{BB})$ and $S_\sigma^0(\mathbf{k})$ are the bare structure factors of the species σ. A similar equation can be derived for block polymer melts[29,30]

$$\frac{S_A^0(\mathbf{k}) + S_B^0(\mathbf{k}) + S_{AB}^0(\mathbf{k})}{S_A^0(\mathbf{k})\, S_B^0(\mathbf{k}) - (S_{AB}^0(\mathbf{k}))^2} - 2\chi_F > 0 \qquad (6.3)$$

for stable melts. S_A^0, S_B^0 are the structure factors of the A and B block, whereas S_{AB}^0 is the structure factor of the AB correlations. If χ_F becomes larger a phase separation takes place. For ordinary blend this is a macroscopic phase separation whereas for block copolymer melts this happens at finite k–values which indicates a micro phase separation.

Crosslinked blends can be utilized by a partially miscible blend, which is homogeneous at high temperatures and is then crosslinked in the one phase region. Therefore we get different types of crosslinks $M_{\sigma\tau}$ which will behave differently. If now the χ_F–parameter is increased the uncrosslinked system phase separates but the crosslinked system cannot, since $M_{AB} \neq 0$. Therefore we expect a microphase separation where the result is A–rich regions and B–rich regions, separated by surfaces formed by the AB crosslinks. Note that AA and BB crosslinks can move cooperatively into the A–rich or B–rich phases. But we expect a renormalisation of the χ_F parameter due to the presence of crosslinks. The question is how can such a problem be quantified with the simplest version? Here the theory is only outlined and we leave the technical details for a separate discussion.

The first what we need is an appropriate Hamiltonian of the one component network. The partition function for N chains with M crosslinks can be written as[31]

$$Z = \int \prod_{\alpha=1}^{N} \mathscr{D}\, R_\alpha(s)\, e^{-\beta H} \prod_{e=1}^{M} \delta(R_{\alpha_e}(s_e) - R_{\beta_e}(s_{e'})) \tag{6.4}$$

here the Hamiltonian of the uncrosslinked chains is given by

$$\beta H = \sum_{\alpha=1}^{N} \int_0^L \left(\frac{\partial R_\alpha}{\partial s}\right)^2 + \sum_{\alpha\beta} \int ds \int ds'\, V(R_\alpha(s) - R_\beta(s')) \tag{6.5}$$

where $V(r - r')$ is the excluded volume interaction. The δ–function in eq. (5.4) represents the crosslink constraint. Now one has to perform an average over all crosslink constraints and since the crosslinks are quenched degrees of freedom the replica method has to be employed[5].

$$\langle Z^n \rangle = \int \prod_{a=1}^{n} \prod_{\alpha=1}^{N} \mathscr{D}\, R_\alpha^a(s)\, e^{-\beta H_{eff}} \tag{6.6}$$

with

$$\beta H_{eff} = \sum_{a=1}^{n} \sum_{\alpha=1}^{N} \int_0^L ds \left(\frac{\partial R_\alpha^a}{\partial s}\right)^2 + \sum_{a=1}^{n} \sum_{\alpha,\beta} \int_0^L ds \int_0^L ds'\, V(R_\alpha^a(s) - R_\beta^a(s')) - \tag{6.7}$$

$$- \mu \sum_{\alpha\beta} \int_0^L ds \int_0^L ds' \prod_{a=1}^{n} \delta(R_\alpha^a(s) - R_\beta^a(s'))$$

where n is the number of replicas and μ is proportional to the mean number of crosslinks. The crosslink term gives rise to a definition of new collective variables $\Omega_{\hat{k}}$ which depends on a super wave vector $\hat{k} = (k^1, k^2, ..., k^n)$

$$\Omega_{\hat{k}} = \sum_{\alpha=1}^{N} \int_0^L ds\, e^{i\hat{k}\cdot\hat{R}_\alpha(s)} \tag{6.8}$$

where $\hat{R}(s)$ is a super vector $\hat{R}_\alpha(s) = (R_\alpha^1(s), R_\alpha^2(s), ..., R_\alpha^n(s))$ and the scalar product $\hat{k}\cdot\hat{R}$ is defined in the usual way

$$\hat{\mathbf{k}} \cdot \hat{\mathbf{R}}_\alpha = \sum_{a=1}^{n} \mathbf{k}^a \cdot \mathbf{R}_\alpha^a \qquad (6.9)$$

Together with the ordinary density variables $\Omega_{(0,0,..\mathbf{k}^a,0..0)}$ the Hamiltonian can be rewritten

$$\beta H = \sum_{a=1}^{n} \sum_{\alpha=1}^{N} \int_{0}^{L} \left(\frac{\partial R_\alpha^a}{\partial s} \right)^2 + (V - \mu) \sum_{\mathbf{k}_a} \Omega_{(0,...\mathbf{k}_a..0)} \, \Omega_{(0,..-\mathbf{k}_a..0)}$$

$$- \mu \sum_{\hat{\mathbf{k}}}{}' \Omega_{\hat{\mathbf{k}}} \, \Omega_{-\hat{\mathbf{k}}} \qquad (6.10)$$

This Hamiltonian can be readily generalised to crosslinked blends. Note that in the one replica sector the potential becomes renormalised by the crosslink term μ. The one replica sector, i.e. these replicas which are not coupled describe now the thermodynamics of the system, whereas the coupled replicas describe the elastic properties and the rigidity of the system. The prime on the sum over $\hat{\mathbf{k}}$ means that the one replica sector is excluded.

With this information the RPA calculation can be done[32] which involves quite an amount of technicality so that only the main results are quoted. The Hamiltonian for the incompressible–crosslinked blend is given by

$$H = \sum_{\hat{\mathbf{k}}} \left\{ F(\mathbf{k}^a) - (2\chi_F - 2\xi_x) \left| \Omega_{(0,..\mathbf{k}_a..0)} \right|^2 + \right.$$

$$\left. + 2\zeta_x \left| \Omega_{\hat{\mathbf{k}}} \right|^2 + \text{higher orders} \right\} \qquad (6.11)$$

where $F(\mathbf{k}^a)$ is a complicated function of the wavevector $\mathbf{k}^a$ which contains information of the bare and the cross form factor of the network (similar to block copolymers, see eq. (5.3)) and the structure factor of the chain origins. The definition of the second ζ_x parameter is given by

$$2\zeta_x \sim (M_{AB} - (M_{AA} + M_{BB})) \, \mu_0 \qquad (6.12)$$

There will be a microphase separation at a finite wavevector k* given by

$$F(k^*) - 2\,(\chi_F - \zeta_x) = 0 \tag{6.13}$$

which depends on the number of crosslinks M_{AA}, M_{BB}, and M_{AB}, the chain length L and the number of chains $N = N_A + N_B$. $1/|k^*|$ determines the size of the microdomain structure.

Let us discuss a few special cases:

a) $M_{AB} = 0$

$$\frac{1}{S(k)} = F_0(k) - 2\,\chi_F - \mu\,(M_{AA} + M_{BB}) \tag{6.14}$$

If the last term is large the system will always phase separate into two unentangled networks. This is, of course, unrealistic, since an IPN lives from the effect of the entanglements.

b) $M_{AB} \gg M_{AA} + M_{BB}$

$$\frac{1}{S(k)} = F_1(k) - 2\,\chi_F + \mu\,M_{AB} > 0 \tag{6.15}$$

is always positive. The system is always stable at $k = 0$.

c) $M_{AB} = 0, \; M_{AA} > 0 \;\; M_{BB} = 0$

This defines a semi IPN. The structure factor is given by

$$\frac{1}{S(k)} = \frac{1}{G_{AA}^0(k)\,S_{AA}^0(k)} + \frac{1}{S_{BB}^0(k)} - 2\,\chi_F - \mu\,M_{AA} \tag{6.16}$$

where $G_{AA}^0(k)$ is the structure factor of the A–chain ends. The system will phase separate at $k = 0$ at

$$\frac{1}{\phi\,N_A} + \frac{1}{(1-\phi)\,N_B} - 2\,\chi_F - \mu\,M_{AA} = 0 \tag{6.17}$$

Then there will be a critical crosslink density of the network where the free chains will phase separate even at $\chi_F = 0$. Since $1/\phi\,N_A$ is the total amount of network monomers it can be neglected $\phi\,N_A \gg 1$ and we find

$$\mu\, M_{AA}^{crit} = \frac{1}{(1-\phi)\, N_B} - 2\, \chi_F \qquad (6.18)$$

For longer number of crosslinks the free chains will separate from the network, whereas for $M_{AA} < M_{AA}^{crit}$ the free chains will be solved in the network.

References

1. L.R.G. Treloar "The physics of rubber elasticity", Oxford Clarendon Press (1975)
2. J.E. Mark, B. Erman, "Rubberlike Elasticity — a molecular primer", John Wiley, New York 1988
3. T.A. Vilgis in "Comprehensive polymer science", Vol. 6 ed. G.F. Eastmond et al., Pergamon Press, Oxford (1989)
4. G. Heinrich, E. Straube, G. Helmis, Adv. Pol. Sci. 85, 33 (1988)
5. S.F. Edwards, T.A. Vilgis, Rep. Prog. Phys. 51, 243 (1988)
6. S.F. Edwards, T.A. Vilgis, Polymer 27, 483 (1987)
7. E. Duering, K. Kremer, G.S. Grest, to be published
8. F. Boué, J. Bastide, M. Buzier, C. Collette, A. Lapp, J. Herz, Prog. Coll. & Pol. Sci. 75, 152 (1987)
9. T.A. Vilgis, Macromolecules, accepted (1991)
10. K.H. Schimmel, G. Heinrich, Coll. Pol. Sci. (1991) to appear
11. G. Heinrich, T.A. Vilgis, to be published
12. M.G. Brereton, T.A. Vilgis, Macromolecules 23, 2044 (1990) and refs. therein
13. K. Binder, H.L. Frisch, J. Chem. Phys. 81, 2126 (1984)
14. W.W. Graessley, Macromolecules 8, 186 (1975)
15. P.G. Higgs, R. Ball, J. Phys. France 49, 1785 (1988)
16a. J. Duiser, A. Staverman, Physics of Non—crystalline Solids, North Holland, Amsterdam (1965)
16b. P.J. Flory, Proc. R. Soc. Lond. A 351, 351 (1976)
17. R. Deam, S.F. Edwards, Phil. Trans. R. Soc. A 280, 317 (1976)
18. M. Warner, S.F. Edwards, J. Phys. A 11, 1649 (1978)
19. M.E. Cates, J. Phys. France 46, 1059 (1985)
20a. B. Mandelbrot, "The Fractal Geometry of Nature", Freeman, San Francisco, (1975)
20b. A. Coniglio, Physica A 140, 51 (1986)
21. T.A. Vilgis, Physica A 153, 341 (1988)
22. M. Antonietti, D. Ehlich, K.J. Fölsch, H. Sillescu, M. Schmidt, P. Lindner Macromolecules 2, 2802 (1989)
23. J. Bastide, F. Boué, preprint
24. J. Bastide, F. Boué, Physica 140A, 251 (1986)
25. J.L. Jones, C.M. Marquez, J. Phys. France 51, 1113 (1990)
26. F. Boué, S.F. Edwards, T.A. Vilgis, J. Phys. France 49, 1635 (1988)
27. T.A. Vilgis, F. Boué, S.F. Edwards in "Molecular basis of polymer networks", Springer Proc. in Physics 42, p. 170 (1989)
28. P.G. de Gennes, "Scaling concepts in polymer physics", Cornell University Press, Ithaca (1979)
29. L.Leibler, Macromolecules 13, 1602 (1980)
30. T.A. Vilgis, R. Borsali, Macromolecules 23, 3172 (1990)
31. P. Goldbart, N. Goldenfeld, Phys. Rev. Lett. 58, 2676 (1987)
32. T.A. Vilgis, to be published

THE FRACTAL NATURE OF ONE-STEP RIGID NETWORKS AND THEIR GELS*

Shaul M. Aharoni

Polymer Science Laboratory, Research & Technology
Allied-Signal Inc.
P.O. Box 1021, Morristown, New Jersey 07962-1021

INTRODUCTION

Recently we prepared gelled, covalently linked rigid polyamide networks by polymerizing in solution in a single step the appropriate monomer mixture. By the correct choice of monomer size and functionality, f, we were able to prepare networks with stiff segments having <u>average</u> length, ℓ_0, as short as 6.5Å and as long as 208Å, and rigid branchpoint (crosslink joint) functionality of 3 to 6.[1,2,3] X-ray studies revealed that the dry or swollen networks and their pre-gel precursors were always amorphous. This work will be limited to systems with stiff segments 32Å and 38.5Å in length and with f = 3 branchpoints. Typical stiff segments with fully reacted branchpoints are shown in structures I and II below:

I

and

---

* Presented at the Networks-91 Conference in Moscow on April 21-26, 1991.

Synthesis, Characterization, and Theory of Polymeric Networks and Gels
Edited by S.M. Aharoni, Plenum Press, New York, 1992

31

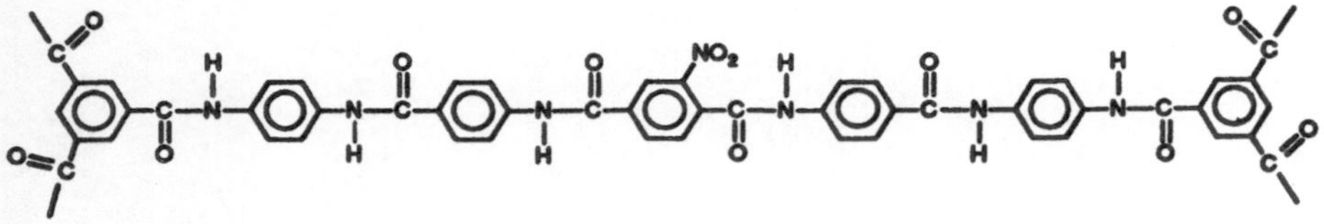

II

We strongly emphasize, however, that when built into the "infinite" network the segments may be as straight as in I and II, or contain one or more bends as will be described below. Furthermore, not all the branchpoints are fully reacted, resulting in network imperfections, especially in the regions where the pre-gel highly-branched polymeric species come in contact with one another in the process of forming the "infinite" network. We have previously shown that in the pre-gel stage the highly-branched rigid polyamides follow the expectations of the fractal model.[2,3] Therefore, we call them fractal polymers (FPs).

It is well-accepted that in aromatic polyamides the amide groups are present in the planar trans configuration.[4] In such chains, the amide groups on both sides of each phenylene ring can adopt either anti or syn configuration:

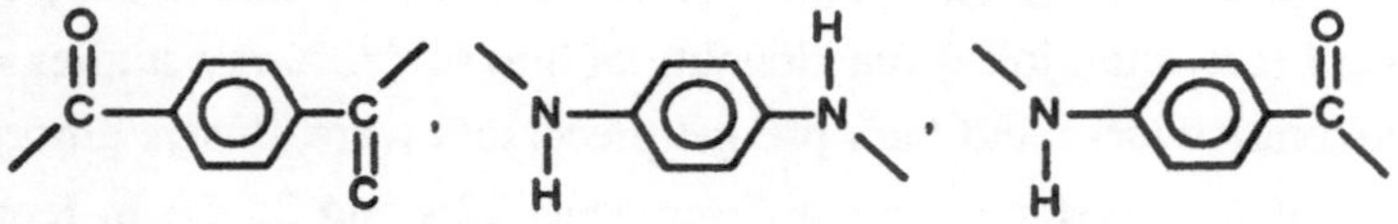

anti

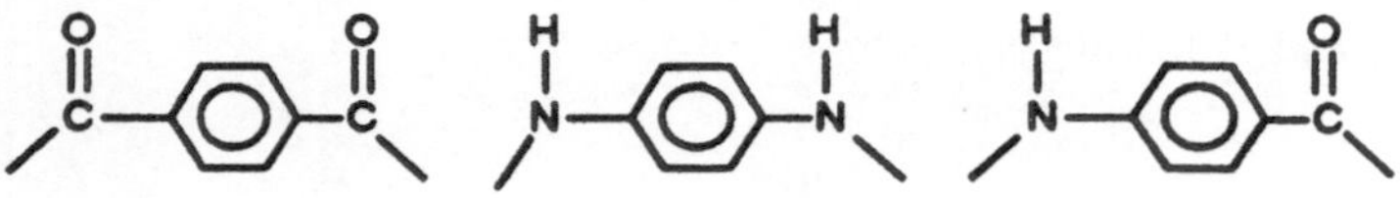

syn

The energetic difference between these two configurations was shown by Hummel and Flory[5] to be only about 0.3 kcal/mol in favor of the anti placements, which is about half RT at ambient temperatures. This means that as the monomers are added to the growing segments during polymerization, there is only a very slight preference to the anti placements and both anti and syn configurations are to be found in the stiff segments of the growing pre-gel FPs and the "infinite" networks created from them. We have previously shown[4] that a likely locus for interconversions from anti to syn or syn to anti is a 180° torsional rotation around the bond connecting the aromatic ring to the amide carbon or nitrogen. The activation energy for such a rotation is in the neighborhood of only 5 kcal/mol, while other allowed rotations are associated with substantially higher activation energies.[4] Solid state NMR relaxation studies[6] showed that in the network the carbonyl carbon is highly decoupled from the ring carbons, reflecting a high degree of rotational freedom around the ring-to-carbonyl bond. In addition to chain bending by anti-syn and syn-anti interconversions, there exist several other mechanisms of stiff polyamide chain bending. Prominent among these are incremental small departures from torsional angle energy minima, and $60°$ ring flipping from $\pm 30°$ to $\mp 30°$ relative to the amide groups plane. The properties of the "infinite" rigid networks and gels will be described below first, followed by a description of the fractal polymers network precursors.

"INFINITE" RIGID POLYAMIDE NETWORKS AND GELS

The highly branched rigid polyamide networks and precursor FPs were prepared in a single step Yamazaki-type reaction[1-3, 7-9] from mixtures of 1,3,5-benzenetricarboxylic acid (BTCA)/4,4'-diaminobenzanilide (DABA)/p-aminobenzoic acid in 2:1:2 molar ratio for structure I and from BTCA/DABA/nitroterephthalic acid (NTPA) in 2:2:1 molar ratio for structure II. The precursor species containing NTPA are slightly more soluble than those not containing the nitro group, but otherwise both substances are rather similar. All dry and solvent-swollen networks and FPs were found by x-ray techniques to be fully amorphous and isotropic. The polymeric gels were prepared at concentrations of $C_0 \leq$ 10% and then equilibrated in the good solvent N,N-dimethylacetamide (DMAc) until they were, first, purified of reagents and reaction by-products, and finally swelled no more. As a rule, no "infinite" network could be prepared at concentrations $C_0 < 2\%$. For a given ℓ_0 and f, gels prepared in the interval $2 < C_0 \leq 10\%$ swelled at equilibrium in inverse proportion to C_0; i.e., the higher is C_0 the lower the swelling. The degree of swelling is inversely dependent on f and directly on ℓ_0. The swelling dependence of several network families on ℓ_0 is shown in Figure 1.

Importantly, once swelled to equilibrium, the rigid networks with low levels of imperfections, typified by relatively short ℓ_0 and high C_0, are insensitive to changes in temperature and measurably change neither their volume nor their linear compressive modulus upon changes in temperature.[1]

The one-step reaction could be aborted at various times in the pre-gel stage, the polymeric species purified from monomers and oligomeric species, and then reacted again to create "infinite" network gels. The modulus, G, of the gels was found to be inversely proportional to the hydrodynamic radius, R_H, of the precursor FPs and to their intrinsic viscosity $[\eta]$ obtained from dilute solution viscosity measurements (Figure 2).

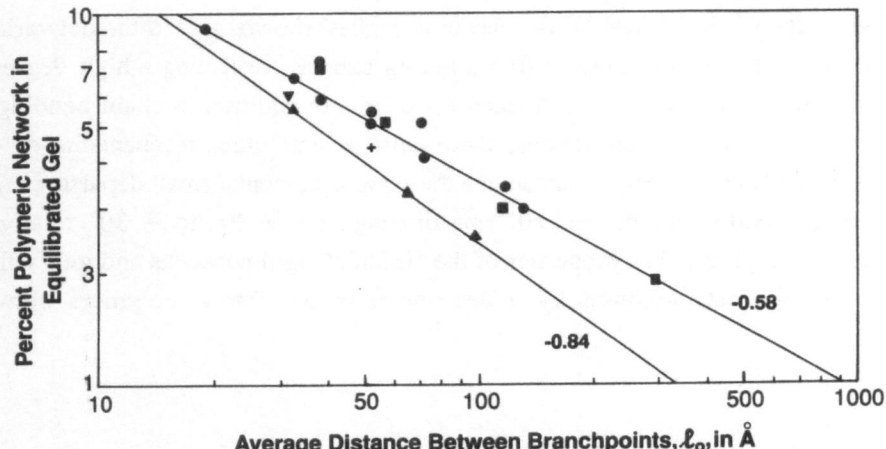

Figure 1.   Swelling of rigid polyamide networks as
function of ℓ_0. All networks were prepared
at $C_0 = 10\%$. All circles and squares for
$f = 3$. All triangles for $f = 4$.
+ for average $f = 3.5$ obtained by mixing
of branchpoint functionalities during the
polymerization.

The observed proportionalities were

$$G \propto R_H^{-3.3} \tag{1}$$

and

$$G \propto [\eta]^{-3.8} \tag{2}$$

An important observation of the study[3] was that the modulus of gelled networks prepared
from the smallest FPs ($M_w \cong 9600$) was about the same as the modulus of gelled
networks prepared directly from the same monomers but without the intermittent step of
aborting the reaction and removing all residual monomers and low-M oligomers.

Gels of well-formed networks showed limited ability to deform under stress and
to swell in good solvents. Their total strain before failure was in the range of several
percent, and their swelling was limited to about the same low volume percent. We
believe that the limited swelling of the well-formed rigid network gels is made possible
by the bent stiff segments straightening out and reaching or approaching their fully
extended state. Once this state is reached, the segments cannot stretch further and the
swelling stops. This explains the rather low level of swelling of well-formed rigid
polyamide networks. During swelling, the increase in the population of straight segments

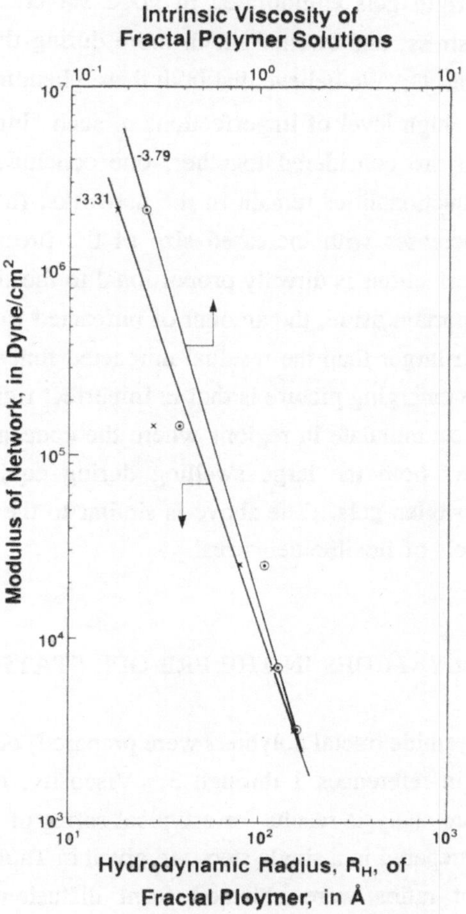

Intrinsic Viscosity of Fractal Polymer Solutions

Figure 2. Modulus of completed network gels against R_H and $[\eta]$ of the precursor FPs. All are at $C_0 = 5\%$, $f=3$, $\ell_0 = 38.5\text{Å}$.

at the expense of bent segments is randomly distributed in space and direction. In the case of deformation under stress, the changes in populations of straight and bent stiff segments are not directionally random. Under linear compressive (negative) strain, straight segments in the compression direction may bend and bent segments in perpendicular directions will straighten out to accommodate the deformation. When these can stretch no more, the sample deformation will either cease or the sample fail. Under conditions of linear positive strain, the above description is reversed.

Poorly-formed rigid gels equilibrated in good solvent supported much larger strains under external stress, and swelled much more during the equilibration from C_0 to the final concentration, C. We believe that both these phenomena are inter-dependent and connected with the high level of imperfections in such "infinite" networks. When references 1, 2, and 3 are considered together, one concludes that (a) a substantial amount of unreacted functionalities remain in the networks, (b) the relative amount of these functionalities increases with increased size of the precursor FPs, and (c), the degree of network imperfections is directly proportional to the size of the precursor pre-gel species. As an important aside, the amount of unreacted functionalities in the rigid one-step networks is far larger than the residual unreacted functionalities in comparable flexible networks. The emerging picture is that in imperfect rigid gels large amounts of solvent preferentially accummulate in regions where the concentration of imperfections is largest. This allows both the large swelling during equilibration and the large deformations of such swollen gels. The above is similar to the observations of Bastide and associates[10,11] on gels of flexible networks.

RIGID FRACTAL POLYAMIDES IN THE PRE-GEL STATE

The pre-gel polyamide fractal polymers were prepared, purified and characterized as described in detail in references 1 through 3. Viscosity, dynamic and static light scattering and end-group analysis results for a typical series of pre-gel highly branched polyamide precursors prepared in a single step, are given in Table I below. In it, R_H and both molecular weight ratios were obtained from diffusion coefficients and their distributions as measured by dynamic light scattering. M_w was measured by static light scattering, N_{ends} was obtained by end-group titrations and iodine-decoration. N_{calc} was calculated from model structures such as in Figure 3 below. The results in the table are very similar to other such series of branched polyamide FPs. They follow the proportionalities

$$[\eta] \propto M_w^{0.419} \tag{3}$$

$$R_H \propto M_w^{0.487} \tag{4}$$

and

$$M_w/M_n \propto M_w^{0.35} \tag{5}$$

all characteristic of highly branched polymers.

Table I

Dilute Solution Characteristics of Pre-Gel One-Step Rigid Polyamides

Code	$[\eta]^a$	M_w	M_w/M_n	M_n	N_w	N_n	N_{ends}	N_{calc}	M_z/M_w	$R_H, \text{Å}$
45XE	0.21	7800	(2.0)	3900	11	5.5	4.1	4-5	1.7	17±2
45XF	0.24	10700	2.2	4860	15	6.8	5.8	6	1.5	22.0
45XG	0.30	16700	2.7	6200	23.4	8.7	6.9	7	1.6	26.6
45XB	1.26	410000	8.6	47650	574	67	37	35	1.8	120
45XH	1.61	700000^b	-	-	-	-	-	-	-	155^b
59E	0.26	9600	2.5	3850	13.4	5.4	4.6	4-5	1.6	18.1

(a) In dL/g.   (b) Extrapolated from log-log plots.

Small-angle x-ray scattering (SAXS) intensities, I, were obtained from pre-gel members of series 45X and of the "infinite" network progeny in the dry and solvent-swollen states. They were plotted on log-log paper against the scattering vector q (in Å^{-1}). The data points for each polymer fell on a rather straight line whose slope gave the scattering exponent

$$I \propto R^{Df}/q^{Df} \tag{6}$$

or

$$I \propto R^{Ds}/q^{2D-Ds} \tag{7}$$

where R is the radius of the scattering species, D_f is the mass fractal dimensionality, D_s is the surface dimensionality and D is the geometric dimensions, i.e., $D = 3.$[12,13] As can be seen in Table II, in our case all scattering exponents were larger than 3.0. This immediately excludes the mass fractal dimensionality from consideration, since $D_f \leq$ 3.0. We are, therefore, left with the initially unexpected result that our pre-gel polymers and the post-gel network all conform with the surface fractal model. The values of D_s obtained from proportionality 7 are given in Table II. They fall in the correct interval for highly corrugated surfaces. The fact that upon swelling, the values of D_s increase toward 3.0, indicates, we believe, that in this case we see by SAXS the FPs with the solvent they imbibed, making their surfaces appear less and less ramified. The fact that we altogether see surface fractality and not mass fractality is, we believe, due to the fact that the highly branched polymers were made from monomers in a one-step reaction

Table II

Scattering Exponents and D_s Values for Several Members of the 45X Series

Code	Form	Dry Samples		Solvent-Swollen Samples	
		Scattering Exponents	D_s	Scattering Exponents	D_s
45XE	pre-gel	3.75	2.25	3.5	2.5
45XF	pre-gel	3.5	2.5		
45XG	pre-gel	4.0	2.0	3.3	2.7
45XB	pre-gel	3.9	2.1	3.35	2.65
45XD	network	3.3	2.7	3.0	3.0

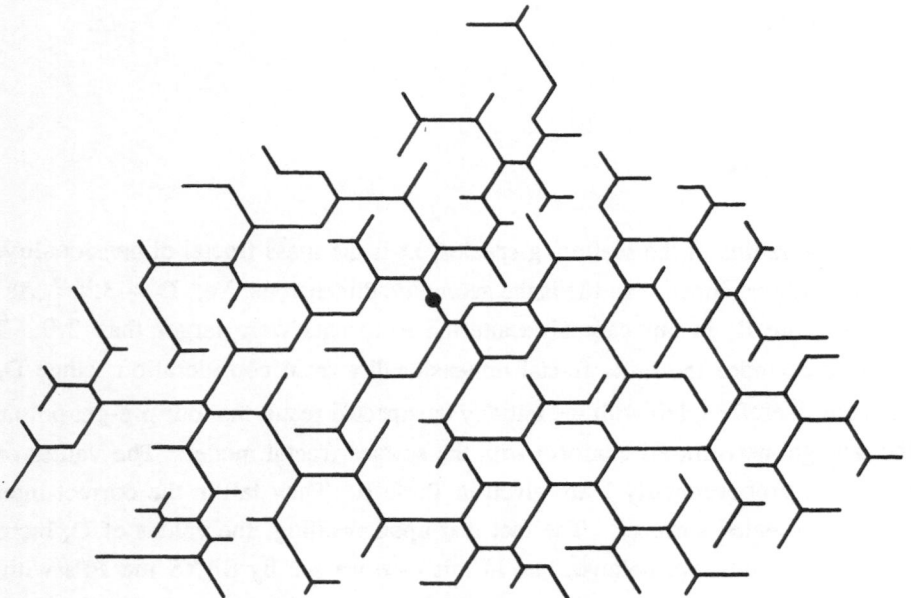

Figure 3.   Two-dimensional schematic representation of FP with $M_n = 65000$, $\ell_0 = 38.5$ Å and $f = 3$. Heavy dot is the nucleus of the FP.

resulting in the segments between branchpoints being not of uniform length but of an average length, ℓ_0, characteristic of each polymer composition. The branchpoint is, however, of constant functionality, f = 3, in this work. A two-dimensional schematic representation of the 3-dimensional FPs with segments described by structure II above, is shown in Figure 3. Here ℓ_0 = 38.5 Å and f = 3. In this Figure, unlike reality, all stiff segments are drawn straight; i.e., fully extended. Because the segments are not of uniform length, they can grow without crowding and frustration, unlike the case of uniform length segments[14] where the polymer concentration in each particle increases upon going from the nucleus outwards. On the other hand, the polymer concentration appears not to drop greatly upon progressing away from the nucleus, as happens in cases of diffusion limited aggregates.[15] The reason for this is not clear to us at present. Because of their rather uniform internal density and highly ramified surfaces, the pre-gel particles conform with the surface fractal model and not with the mass fractal one.

When present in the appropriate concentrations in solution under the correct conditions for reaction, the pre-gel species will grow in size and number until they finally form a single "infinite" cluster filling the whole solution volume. At that point, gelation takes place. As was shown in references 1 and 2, the gel point for systems polymerized at $7.5 \le C_0 \le 10.0\%$ was reached at a viscosity of about 200 centipoises, while the viscosity of the system increased extremely fast in a pattern best described by percolation theory. This is characteristic[16] of highly-branched pre-gel polymeric particles.

REFERENCES

1) S.M. Aharoni and S.F. Edwards, Macromolecules **22**, 3361 (1989).

2) S.M. Aharoni, N.S. Murthy, K. Zero and S.F. Edwards, Macromolecules **23**, 2533 (1990).

3) S.M. Aharoni, Macromolecules **24**, 235 (1991).

4) S.M. Aharoni, G.R. Hatfield and K.P. O'Brien, Macromolecules, **23**, 1330 (1990) and references therein.

5) J.P. Hummel and P.J. Flory, Macromolecules **13**, 479 (1980).

6) S.A. Curran, C.P. LaClair and S.M. Aharoni, Macromolecules **24**, 5903 (1991).

7) N. Yamazaki, M. Matsumoto and F. Higashi, J. Polymer Sci. Polym. Chem. Ed. **13**, 1373 (1975).

8) S.M. Aharoni, Macromolecules **15**, 1311 (1982).

9) S.M. Aharoni and D.H. Wertz, J. Macromol. Sci.-Phys. **B22**, 129 (1983).

10) L. Leibler and J. Bastide, Macromolecules **21**, 2647 (1988).

11) J. Bastide, E. Mendes Jr., F. Boue, M. Buzier and P. Lindner, <u>Makromol. Chem. Macromol. Symp.</u> **40**, 81 (1990).

12) J.E. Martin and A.J. Hurd, <u>J. Appl. Cryst.</u> **20**, 61 (1987).

13) P.W. Schmidt, in "The Fractal Approach to Heterogeneous Chemistry", Ed. D. Avnir; Wiley, Chichester, 1989; pp. 67-79.

14) D.A. Tomalia, A.M. Naylor and W.A. Goddard III, <u>Angew. Chem. Int. Ed. Engl.</u> **29**, 138 (1990).

15) P. Meakin, in "Encyclopedia of Polymer Science and Engineering" 1990 Suppl. Vol.; Wiley, New York, 1990; pp. 323-342.

16) E.J.A. Pope and J.D. Mackenzie, <u>J. Non-Cryst. Solids</u> **101**, 198 (1988).

STATIC AND DYNAMIC LIGHT SCATTERING

FROM SOLUTIONS AND GELS OF RF PARTICLES

Patricia M. Cotts

IBM Research Division
Almaden Research Center
San Jose, California 95120

INTRODUCTION

The name *aerogel* is used to describe a gel from which the solvent has been removed under supercritical conditions.[1,2] This results in little or no surface tension, so that there is minimal shrinkage of the gel during drying. The extremely light, high surface area foams are called *aerogels*. These materials have applications as catalyst supports, thermal and acoustic insulators, and most commonly Cerenkov detectors for high energy physics experiments. The recent increase of studies on aerogels has been largely devoted to silica aerogels, those formed from silicon dioxide.

Gels formed from the reaction of resorcinol and formaldehyde with sodium carbonate (Na_2CO_3) as catalyst may also be dried under supercritical conditions, producing organic aerogels.[3,4] These organic RF aerogels are the focus of this study. Like their silicon dioxide counterparts, the RF aerogels comprise linear chains of small (less than 20 nm diameter) particles. These chains form the network of the dried gel and may be seen using transmission electron microscopy (TEM). Previous studies have shown that the primary particle size in the dried RF aerogel is dependent on the ratio of the resorcinol to catalyst (R/C ratio).[3] The compressive modulus of the dried gel as a function of the density has also been reported.[4] In this study, dynamic light scattering is used to investigate the sol to gel transition in the aqueous solution of RF particles. Recently, power-law behavior at the gel point has been reported in the shear modulus for poly(dimethylsiloxane) gels.[5,6] Since the intensity autocorrelation function measured by dynamic light scattering also reflects molecular relaxations, it is expected that similar power-law behavior might be observed in the autocorrelation func-

Synthesis, Characterization, and Theory of Polymeric Networks and Gels
Edited by S.M. Aharoni, Plenum Press, New York, 1992

tion, and it has been reported for silica gels[7], polyurethane gels[8], polysaccharide gels[9], and poly(methylmethacrylate) gels[10]. In the present study, we report near power-law behavior of the intensity autocorrelation function for the RF gels, which occurs in the region of the gel point.

EXPERIMENTAL TECHNIQUES

Preparation of Samples

Stock solutions of 1 M resorcinol, 2 M formaldehyde, and 0.1 M Na_2CO_3 in distilled deionized water were prepared. These solutions were held in an oven at 70°C prior to reaction at 85°C to minimize the time required to obtain thermal equilibrium. For each measurement, the requisite amounts of the solutions were combined and filtered (0.5 μm Fluoropore filter, Millipore Corp.) into cleaned 16 mm diameter vials which were sealed with caps. Solutions were prepared at a fixed molar ratio of 2:1 formaldehyde to resorcinol with recorcinol/catalyst (R/C) ratios from 50 to 300. Transparent gels were obtained for all formulations. The vials were placed immediately into the thermostatted light scattering instrument containing filtered silicone oil as an index matching fluid, at 85°C.

Dynamic Light Scattering

Dynamic light scattering measurements were made using a Brookhaven variable angle photogoniometer (BI200-SM), with a Brookhaven correlator (BI2030AT). The light source was a Spectra-Physics 120B HeNe laser (632.8 nm). The oxidation products from the resorcinol result in reddish-brown gels at the final stages of gelation, so that red light is necessary to minimize absorption. The experimental time correlation fuction, $C(t)$, may be expressed in terms of the relaxation function $\phi(t)$:

$$C(t) = <n>^2[1 + f(A)\phi^2(t)], \tag{1a}$$

and was analyzed by the method of cummulants:[11]

$$\ell n\left(\frac{C(t)}{B} - 1\right)^{1/2} = \ell n\, f(A)^{1/2} - \Gamma t + \mu_2 t^2/2 + \cdots \tag{1b}$$

with B the baseline of the correlation function ($<n>^2$), and $f(A)$ a spatial coherence factor which is dependent upon the optics (e.g., the number of coherence areas viewed) as well as the properties of the scattering centers. This expression is valid for *homodyne* conditions. The measured baseline (average of 4 delay channels 1029-1032 times the sample time Δt) was used for B, however, the calculated baseline obtained from the count rate was within 0.5% of the measured baseline. Measurements were made at a number of scattering vectors q where:

$$q = \frac{4\pi n \sin(\Theta/2)}{\lambda} \tag{2}$$

with Θ the scattering angle, n the refractive index of the solution, and λ the wavelength of the incident light in vacuum. For dilute solutions, at small q, $qR_g < 1$, the mutual diffusion coefficient at each concentration c may be obtained from the extrapolation of the reduced first cummulant, $D_c(q) \equiv \Gamma_{c,q}/q^2$ to $q = 0$. The normalized second cumulant, μ_2/Γ^2, reflects the width of the molecular weight distribution. The limiting diffusion coefficient at infinite dilution, D_0, is then used to obtain the hydrodynamic radius R_H through the Stokes equation:

$$D_0 = \frac{k_b T}{6\pi\eta R_H} \tag{3}$$

with k_b the Boltzmann constant, T the absolute temperature, and η the viscosity of the medium. For polydisperse samples, the R_H measured in this manner is given by:

$$R_H \equiv <1/R_H>_z^{-1} \tag{4}$$

where the z-average is taken over the *inverse* hydrodynamic radius, since the experimentally measured parameter is the diffusion coeffcient.

For gels in which the effective length between crosslinks, the mesh size ξ, is small, i.e. $q\xi < 1$, a continuum model such as that proposed by Tanaka and co-workers[12] may be used to interpret the dynamic light scattering. These cooperative motions of the gel network (an elastic network in a viscous medium) are diffusive and exhibit a q^2 dependence. The measured time correlation function reflects the modulus of the gel and approaches a single exponential. Under suitable conditions, the cooperative diffusion coefficient, D_{coop} may be related to the effective mesh size with a Stokes equation like Equation 3 above:[13]

$$D_{coop} = \frac{k_b T}{6\pi\eta\xi} \tag{5}$$

The above analysis requires a well-defined measurement of D_{coop} from the dynamic light scattering. For gels, there is frequently a substantial *static* contribution to the correlation function:

$$C(t) = <n>^2[1 + f(A)[2r(1-r)\phi(t) + r^2\phi^2(t)]] \tag{6}$$

where r is the fraction of intensity associated with the time dependent light scattering. For scattering dominated by the staic component, $r << 1$:

$$C(t) = <n>^2[1 + 2f(A)r\phi(t)] \, , \tag{7}$$

so that:

$$\frac{\Gamma}{2q^2} = D \tag{8}$$

Measurements for which Equations 7 and 8 are valid are *heterodyne* conditions. Under these conditions the apparent $f(A)$ will be much smaller than the true $f(A)$ measured for a solution. Recent work by Pusey and van Megen[14] suggests that even the heterodyne analysis is not adequate for many gelled systems because they are *non-ergodic*. In many gels, *inhomogeneities* exist such that both the scattered intensity as well as the decay of the correlation function depend significantly on the specific coherence area viewed. The average over these areas for the sample is not equivalent to the time average observed for a specific area.

RESULTS

Initial reaction

In the first hour of reaction at 85°C, the measured R_H remains relatively constant, although the intensity of the scattered light is increasing rapidly (5-10 times the initial intensity after 1-2 hours). The correlation functions obtained are not single exponential, but may be fit adequately with a 3rd order cummulants analysis. The normalized second cumulant, μ_2/Γ^2, is approximately 0.3 in this initial region of the reaction. An example of a typical correlation function is shown in Figure 1 (upper curve), obtained for R/C = 200, 15 minutes after preparation of the mixture as described above. For values of the scattering vector $qR_H << 1$, measured values of D_c were independent of q, and essentially independent of concentration c, over the range investigated. The small concentration dependence is expected for small globular particles, in contrast to the more pronounced dependence observed for long chain polymers. Thus, values of R_H may be calculated from the measured D_c using Equation 3 above. These values are shown in Figure 2 as a function of R/C ratio. Also shown are the estimated bead radii of the dried aerogels obtained from TEM measurements.

Aggregation and Gel Formation

The aggregation of the primary particles into linear chains forming the gel network was followed using dynamic light scattering. 15-20 hours were usually required for the disappearance of the long-time tranlational diffusion modes and the dominance of the fast cooperative diffusion mode in the time correlation function. During this time the correlation functions display a very wide range of relaxation times, and a single delay time is inadequate. For this reason, measurements were done using the Brookhaven 8000 correlator, which permits a geometric spacing of delay times, and 4 decades in relaxation times were covered. The main observations during the sol-to-gel transition may be summarized:

1. The effective hydrodynamic radius estimated from a cumulant fit to the measured correlation function increases with time, approaching values of several hundred nm in the last minutes before gelation. An example of the growth of the effective R_H for R/C = 50 is shown in Figure 3.

2. The scattered intensity also increases with time; at first rapidly, then either levelling off (for R/C = 50 or 100), or continuing to increase exponentially (for R/C = 200 or 300), as shown in Figure 4.

3. The normalized second cumulant, μ_2/Γ^2 increases as the polydispersity of the particle size increases. After gelation, when only the cooperative diffusion coefficient of the gel network is seen, μ_2/Γ^2 decreases again as the correlation function approaches a single exponential.

4. The amplitude of the correlation function, as measured by the apparent $f(A)$, decreases continuously as the particles aggregate into linear chains and form a network.

5. The correlation functions can be fit with a stretched exponential prior to the gel point, and approach power-law behavior in the region of the gel point.

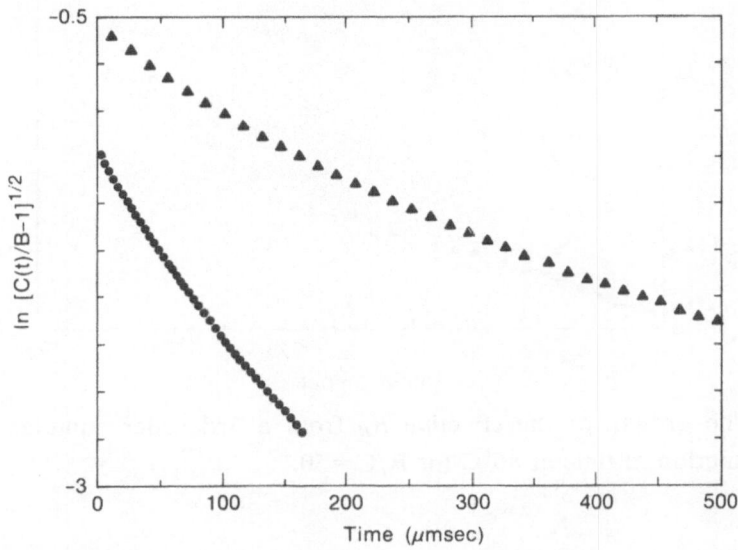

Figure 1. Normalized time correlation functions obtained for R/C = 200 at 85°C. Top curve: 15 minutes after preparation of the sample, bottom curve: 855 minutes.

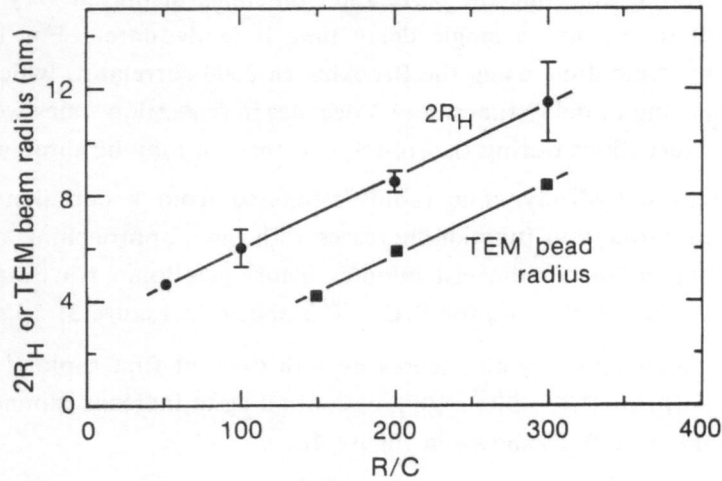

Figure 2. Values of $2R_H$ obtained in the first hour of reaction at 85°C, as a function of R/C ratio. Also shown are the estimated bead radii from TEM on the dried gels.

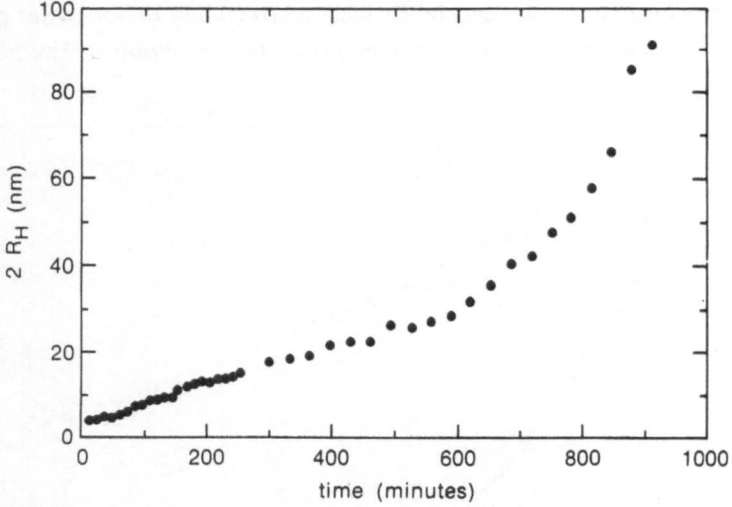

Figure 3. The growth of the effective R_H from a 3rd order cumulant fit as a function of time at 85°C for R/C = 50.

Figures 5 and 6 show logarithmic correlation functions obtained as a function of reaction time for two different RF gel preparations. Figure 6 shows longer time behavior of $g_2(t)$ obtained by neglecting the intitial fast relaxations. Power-law behavior over 6 or more decades has been reported in several gelling systems.[5-10] As suggested by Lang and Burchard[9], the gel point may be seen even more dramatically by considering the slope of the time correlation functions shown in Figures 5 and 6. These slopes are plotted as a function of time of reaction in Figure 7, and show a dramatic change in the region of the gel point.

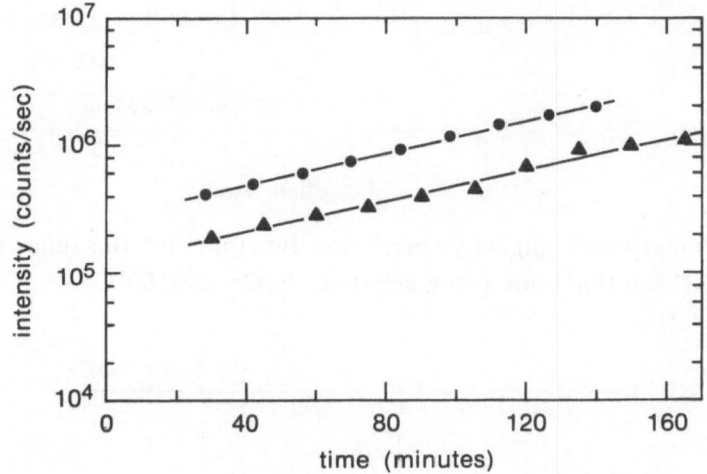

Figure 4. Exponential growth in intensity observed for RF samples undergoing gelation.

Gelled samples

After times ranging from 600-1500 minutes, the fast cooperatve diffusion of the gel network is dominant and nearly single exponential behavior is observed. The lower curve in Figure 1 is the correlation function obtained after 855 minutes for $R/C = 200$, and the semilog plot is nearly linear. Figure 8 shows that this cooperative mode is diffusive, with the expected q^2 dependence. However, as pointed out by Pusey and van Megen[14], a time-averaged measurement of a single scat-

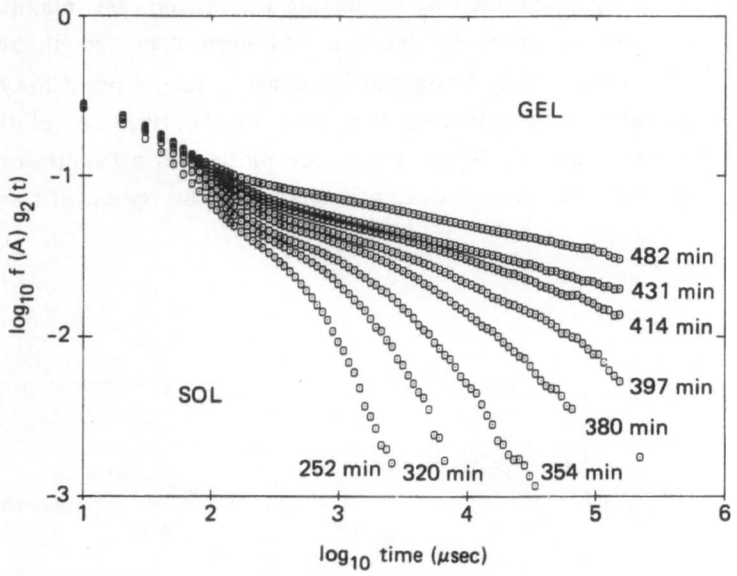

Figure 5. Logarithmic intensity correlation functions for the times listed for an RF solution undergoing gelation. R/C = 200, 80°C.

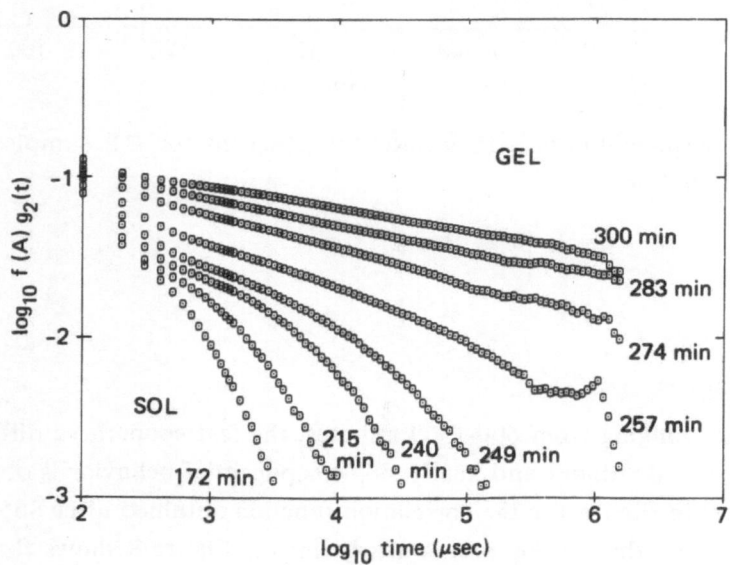

Figure 6. Logarithmic intensity correlation functions for the times listed for a gelling RF solution, neglecting intitial fast relaxations. R/C = 200, 75°C.

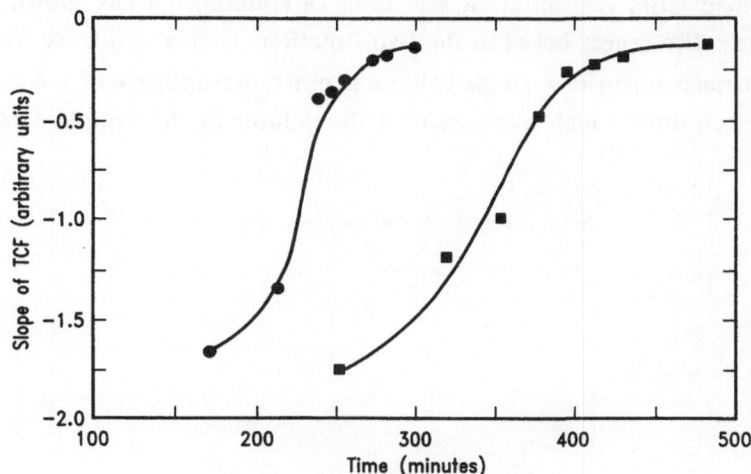

Figure 7. Slopes of the TCF's shown in Figures 5 and 6 as a function of reaction time, showing the large change near the gel points of 225 and 360 minutes, respectively.

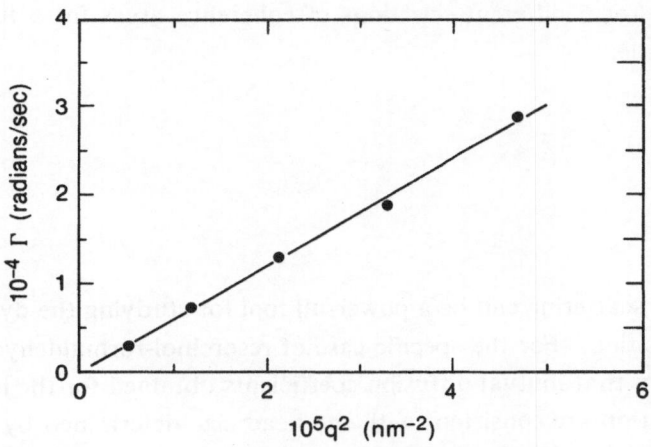

Figure 8. Γ versus q^2 for a gelled sample (R/C = 200, after 855 minutes at 85°C), showing the linear q^2 dependence.

tering volume is insufficient for gels which are non-ergodic. The dramatic loss in amplitude observed for the gelled RF samples suggests that the non-ergodic effects are significant. This may also be seen by comparison of correlation functions obtained from two different selections of coherence areas shown in Figure 9. The large differences between the two functions is clear evidence that a time-averaged measurement of a single volume is not representative of the gelled sample, and a technique which averages over the volume of the sample is required.

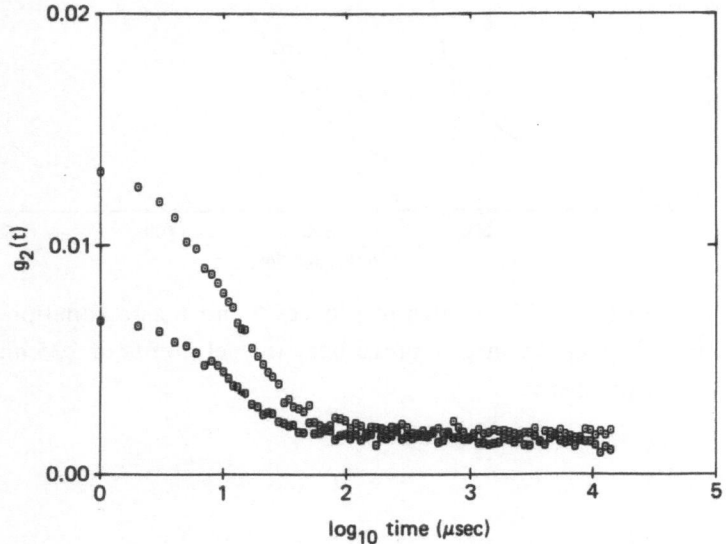

Figure 9. $g_2(t)$ for 2 different selections of coherence areas for a fully gelled RF sample.

SUMMARY

Dynamic light scattering can be a powerfull tool for studying the dynamics of the sol-to-gel transition. For the specific case of resorcinol-formaldehyde aerogels, it has been shown that mutual diffusion coefficients obtained for the intial particles formed in solution are consistent with the bead size determined by TEM on the final dried gels. At the sol-gel transition, near power-law behavior is observed in the intensity correlation function, with an exponent near 1/3. This is very similar to the power-law behavior reported for a variety of gelling systems observed with either rheological or optical techniques.[5-10] In the fully gelled systems, non-ergodic behavior is observed, and a measurement of a single coherence area is not sufficient to characterize the sample.

REFERENCES

1. R. Pool, *Science*, **247**, 807 (1990).

2. J. Fricke, *J. Non-Crystalline Solids*, **100**, 169 (1988).

3. R. Pekala, *Journal of Materials Science*, **24**, 3221, (1989).

4. R. Pekala, in *Polymer Based Molecular Composites*, D. W. Schaefer and J. E. Mark, Eds., MRS Symposium #172, (1990), p.285.

5. H. H. Winter and F. Chambon, *J. Rheol.*, **30**, 367 (1986).

6. F. Chambon and H. H. Winter, *J. Rheol.*, **31**, 683 (1987).

7. J, E. Martin and J. P. Wilcoxon, *Phys. Rev. Lett.*, **61**, 373 (1988).

8. M. Adam, M. Delsanti, J. P. Much, and D. Durand, *Phys. Rev. Lett.*, **61**, 706 (1988).

9. P. Lang and W. Burchard, *Macromolecules*, **24**, 8142 (1991).

10. L. Fang, W. Brown and C. Konak, *Macromolecules*, **24**, 6839 (1991).

11. D. E. Koppel, *J. Chem. Phys.*, **57**, 4814 (1972).

12. T. Tanaka, L. O. Hocker, and G. B. Benedek, *J. Chem. Phys.* **59**, 5151 (1973).

13. P. deGennes, *Scaling Concepts in Polymer Physics*, Cornell University Press: Ithaca, 1979 New York, 1961, p.391.

14. P. N. Pusey and W. van Megen, *Physica A*, **157**, 705 (1989).

SUPPRESSION OF FLUCTUATION-DOMINATED KINETICS BY MIXING

I.M. Sokolov[*][S] and A.Blumen[*]

[*]Theoretische Polymerphysik, Universität Freiburg, Rheinstraße 12, W-7800 Freiburg i. Br., Germany

[S]P.N.Lebedev Physical Institute of the Academy of Sciences of Russia, Leninsky prosp. 53, Moscow 117924, Russia

1. INTRODUCTION

Fluctuation effects in diffusion-controlled reactions have attracted much attention since the pioneering works of Ovchinnikov and Zeldovich [1] and of Toussaint and Wilczek [2]. Thus even the simplest bimolecular reaction of $A + B \to 0$ type does not in general obey classical kinetics, since the long-time decay of the reactants' concentrations is dominated by fluctuations. It is believed that stirring destroys large fluctuations and therefore restores classical kinetics. Some simple models support this point of view, see Ref.3 for a review. The purpose of the present Chapter will be to show the origin of fluctuation effects, to present a theoretical approach allowing to treat diffusion and stirring on an equal footing and to elucidate the influence of different mixing procedures on the course of the reactions.

2. THE CLASSICAL REACTION SCHEME

Let us first consider the classical reaction scheme for the $A + B \to 0$ reaction. The basic equations for this reaction are:

$$\begin{cases} \dfrac{dA(t)}{dt} = -\kappa A(t)B(t) \\[2mm] \dfrac{dB(t)}{dt} = -\kappa A(t)B(t) \end{cases} \tag{1}$$

Here $A(t)$ and $B(t)$ are the average concentrations of A and B reactants at time t and κ is the reaction-rate coefficient (assumed constant). The conservation of the difference in the overall concentrations $Q = B(t) - A(t) \equiv B(0) - A(0)$ makes it possible to rewrite Eq.(1) in the form:

Synthesis, Characterization, and Theory of Polymeric Networks and Gels
Edited by S.M. Aharoni, Plenum Press, New York, 1992

$$\frac{dA(t)}{dt} = -\kappa A(t)[A(t)+Q] \tag{2}$$

Separation of variables yields then the exact solution of Eq.(2):

$$\frac{1 + Q/A(0)}{1 + Q/A(t)} = e^{-Qkt} \tag{3}$$

For nonstoichiometrical situations the long-time asymptotic decay of the minority species will approach an exponential at long times: $A(t) \sim exp(-Qkt)$. Here we focus on the stoichiometrical case $A(0) = B(0)$, i.e. $Q=0$. In this case Eq.(3) gives:

$$A(t) = B(t) = \frac{1}{\kappa t + 1/A(0)} \tag{4}$$

and hence $A(t) \sim 1/t$ for large t. However, an in-depth analysis of the many-body effects, which is also strongly supported by the numerical modelling of the situation shows that in general the decay does not follow $A(t) \sim 1/t$ [1,2,4-15]. The analysis and the numerical simulations of the reactions show that the decay obeys a more general power-law:

$$A(t) = \frac{1}{t^\alpha} \tag{5}$$

where the exponent α depends on the dimension d of the space and is $\alpha = d/4$ for $d < 4$. For a discussion of the nonstoichiometrical case see e.g. Ref.15.

3. DIFFUSION-CONTROLLED REACTION WITHOUT STIRRING

The main reason of the failure of the classical reaction scheme, Eq.(1), is that it implicitly assumes the system to be spatially homogeneous. The numerical work evidences clearly, however, that during the reaction clusters of similar particles emerge, the mean cluster size growing with time. These clusters develop from the initial Poisson fluctuations in the local distributions of the reactants. The occurence of large clusters slows the reaction down, since only particles near the cluster boundaries can react with each other. In order to deal with such structures one must take the dependence of the concentration on position into account. One deals hence with the local quantities $A(r,t)$ and $B(r,t)$ which in the simplest approximation are governed by the following set of equations:

$$\begin{cases} \dfrac{dA(r,t)}{dt} = D\Delta A(r,t) - \kappa A(r,t)B(r,t) \\[2mm] \dfrac{dB(r,t)}{dt} = D\Delta B(r,t) - \kappa A(r,t)B(r,t) \end{cases} \tag{6}$$

Here only diffusion is accounted for: D is the diffusion coefficient of the reactants (assumed equal, $D=D_A=D_B$) and Δ is the Laplace operator.

The problematic part of Eq.(6) is the reaction term; judicious, in-depth analyses [13,14] show that $- \kappa A(r,t)B(r,t)$ is only approximate, since it involves a decoupling of the many-body problem at the two-body stage and total disregard of the three-body correlations; furthermore, the term also involves preaveraging [16,17]. Nontheless, since our aim is to study the influence of *mixing* (not yet included in Eq.(6)) we will start from these, widely accepted approximate forms, Eq.(6).

Now the average concentrations $A(t)$ and $B(t)$ are connected with the position-dependent ones through $A(t) = <A(r,t)>$ and $B(t) = <B(r,t)>$, where the averaging procedure can be understood either as an ensemble or as a spatial average. A convenient way to proceed from Eq.(6) is by reverting to the local sum $s(r,t) = A(r,t) + B(r,t)$ and the local difference $q(r,t) = A(r,t) - B(r,t)$. In these variables Eq.(6) takes the form:

$$\frac{\partial q(r,t)}{\partial t} = D\Delta q(r,t) \tag{7}$$

and

$$\frac{\partial s(r,t)}{\partial t} = D\Delta s(r,t) - \frac{\kappa}{2}[s^2(r,t)-q^2(r,t)] \tag{8}$$

Note that Eq.(7) involves only the variable q and that it is linear; moreover Eq.(8) depends only parametrically on q. Now it is a simple matter to express q in terms of the Green's function $G(r,t)$ of the diffusion equation. The solution of Eq.(7) is

$$q(r,t) = \int G(r,r',t)q(r',0)dr' \tag{9}$$

where $q(r,0)$ corresponds to the initial particle distribution. The explicit form for G in d dimensions is:

$$G(r,r',t) = (4\pi Dt)^{-d/2}exp\left[-\frac{|r-r'|^2}{4Dt}\right] \tag{10}$$

Eqs.(9) and (10) make it possible to obtain all moments of the q-distribution. Starting from a random uncorrelated initial distribution of particles and from stoichiometrical conditions one has $<q(r,0)> = 0$ and $<q(r_1,t)q(r_2,t)> = 2c\delta(r_1-r_2)$ with $c = A(0) = B(0)$. Therefore:

$$<q(r,t)> = \int G(r,r',t)<q(r',0)>dr' = 0 \tag{11}$$

and

$$<q^2(r,t)> = \iint G(r,r',t)G(r,r'',t)<q(r',0)q(r'',0)>dr'\,dr''$$

$$= 2c\int G^2(r,r',t)dr' = c\sqrt{2}(2\pi Dt)^{-d/2} \tag{12}$$

Thus $q^2(t) \equiv <q^2(r,t)> = c\sqrt{2}(2\pi Dt)^{-d/2}$.

At longer times the distribution of q tends to a Gaussian with zero mean and dispersion given by Eq.(12) [2,13,18], which can be seen either from the analysis of higher moments or directly from Eq.(9), by writing $q(x,t)$ as a weighted sum of random, independent, identically distributed variables.

From the fact that q is Gaussian-distributed at longer times it follows:

$$\langle q^2(r,t)\rangle = \frac{\pi}{2}\langle|q(r,t)|\rangle^2 \tag{13}$$

Since one always has $s(r,t) > |q(r,t)|$, one sees that $s(t) \equiv \langle s(r,t)\rangle = 2A(t)$ cannot decay faster than $\langle|q(r,t)|\rangle \simeq \sqrt{\frac{2}{\pi}}\langle q^2(r,t)\rangle^{1/2} \sim (Dt)^{-d/4}$; now in dimensions d less then 4 this decay law is slower than the classical prediction $A(t) \sim t^{-1}$; this demonstrates that low dimensions and spontaneously occuring fluctuation effects slow the reaction down.

A more detailed analysis of Eq.(8) shows that asymptotically $\langle s(x,t)\rangle$ tends to $q(t) = \langle q(x,t)\rangle^{1/2}$. The numerical solutions of Eq.(7) and (8) support this conclusion, where, however the crossover to the long-time behavior strongly depends on the dimension; higher dimensions need considerably longer times [13,14,23,24].

Up to now only the very-long-time asymptotic domain for $s(t)$ was discussed. To obtain the behavior of $s(t)$ in a more extended time domain one can start from Eq.(8) and notice that after ensemble averaging, terms involving spatial derivatives in Eq.(8) vanish. To obtain a closed form for $s(t)$ one can approximately set $\langle s^2(r,t)\rangle = s^2(t)$, see Ref. [3] for a discussion. The equation for $s(t)$ then reads:

$$\frac{ds(t)}{dt} = -\frac{1}{2}\kappa[s^2(t) - q^2(t)] \tag{14}$$

For $q^2(t)=C/t^\alpha$ Eq.(14) can be solved analytically, see Ref.[3]. The qualitative behavior of $s(t)$ can be inferred without explicitly solving the equation, by simply comparing $q(t)$ with the solution of Eq.(7) for $q=0$, namely, $s(t) = [2/(\kappa t) + s(0)^{-1}]^{-1}$. For long times we obtain that

$$s(t) \sim \begin{cases} \dfrac{2}{\kappa t} & \text{for } q(t) \ll \dfrac{2}{\kappa t} \\[4mm] q(t) & \text{for } q(t) \gg \dfrac{2}{\kappa t} \end{cases} \tag{15}$$

The characteristic crossover time t_c between the two regimes follows then from the comparison of the two forms; implicitly one has $2/(\kappa t_c) \sim q(t_c)$.

4. MIXING PROCEDURES

Now we proceed by incorporating stirring aspects into the diffusion-reaction scheme. For this we focus on two basic models for mixing: on the one side the baker's transformation, which mixes strongly, and, on the other side, shear-flow mixing, which is less effective. These rather simple-looking procedures are, however, related to industrially used mixing devices; for an overview one may consider Refs.20 and 21.

Baker's transformation is one of the simplest theoretical models for mixing, see Ref.22. Fig.1 exemplifies the model, as applied to the unit square. Each step of baker's transformation (which requires, say, a time τ for completion) consists of three substeps: (i) squeezing the square to half

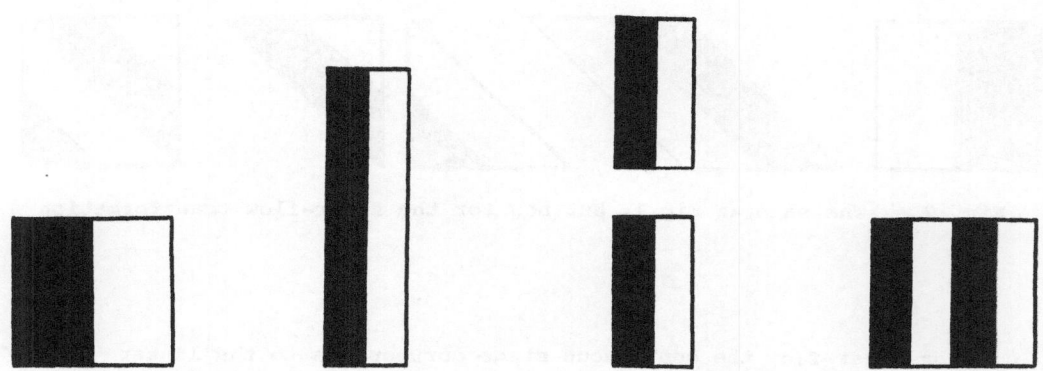

Fig. 1.   From left to right are shown the initial configuration of the
          system and the three stages of baker's transformation, see text
          for details.

its initial width and double height, (ii) cutting the obtained object into
two parts and (iii) pasting the upper part to the right of the lower one. In
Fig.1 in order to render evident the mixing the system is shown as
consisting of white and black regions. We assume that the time required for
the two last stages of the transformation is negligible compared to the
first stage.

        The second mixing model (shear flow) can be visualized as consisting
in: (i) shearing a square into a rhombus with an acute angle of $\pi/4$, (ii)
cutting the rhombus into two equal rectangular triangles and (iii) pasting
the right triangle to the left of the left one (see Fig.2). Here we again
assume the last two stages of the transformation to be extremely fast.

        For the mathematical description we focus on these transformations in
two dimensions. Stage (i) of the baker's transformation can be described by
the linear mapping:

$$r(t) = \hat{L}(t-n\tau)r(n\tau+0) \qquad (16)$$

Here $n\tau+0$ corresponds to the moment of time just after completing the
cutting and pasting stages of the n-th step of the transformation; the
relation between the time t and the number of steps is $n=[t/\tau]$, where $[x]$
denotes the whole part of the number x. The linear operator $\hat{L}$ is given by:

$$\hat{L}(\theta) = \begin{pmatrix} \eta^{-1}(\theta) & 0 \\ 0 & \eta(\theta) \end{pmatrix} \qquad (17)$$

with $\theta = t - n\tau$ and $\eta(t) = 2^{t/\tau}$. The cutting and pasting stages correspond
to the discontinuous mapping:

$$\begin{cases} x(n\tau+0) = (x(n\tau-0) + [y(n\tau-0)]/2) \\ y(n\tau+0) = \{y(n\tau-0)\} \end{cases} \qquad (18)$$

where $x(n\tau-0)$ and $y(n\tau-0)$ are the coordinates of the particle just before
the cutting and pasting stages of the transformation, and $\{y\}$ denotes the
fractional part of y.

Fig. 2.   The same as Fig.1, but now for the shear-flow transformation

For shear-flow the continuous stage corresponds to the linear transformation

$$r(t) = r(n\tau+0) + v\Delta t \qquad (19)$$

where v denotes the liquid's velocity, $v = (y/\tau, 0)^T$. The discontinuous stage of the shear-flow model is

$$\begin{cases} x(n\tau+0) = \{x(n\tau-0)\} \\ \\ y(n\tau+0) = y(n\tau-0) \end{cases} \qquad (20)$$

Already from these equations, and also using Fig.1 and 2 one sees that the baker's transformation mixes effectively: both the number of layers and also most distances between two initially very close points of the system grow exponentially with time. Shear-flow mixes less effectively, since these quantities generally grow linearly.

The analytical procedure for mixing can be further simplified by using appropriate boundary conditions, which correspond to attaching to the initial system copies of itself along the x-direction. In fact one may even go a step further and use *statistical* copies of the system, as shown by us in Refs.16 and 17. Under such conditions one can dispense with the discontinuous step: The infinite domain undergoes now continuous squeezing for baker's transformation and continuous shearing for shear-flow. In a liquid in motion both cases can be described by introducing position-dependent velocity fields. We can obtain these fields using the $r(t)$ forms given by Eqs.(16) and (19). In the case of continuous mixing this information suffices, since $v(t)$ follows by simple differentiation. In both cases we find for v the structure $v_i = \sum_j \alpha_{ij} r_j$, where i and j denote the coordinates and the α_{ij} are position- and time-independent. Moreover in both cases $\sum_j \alpha_{ii}=0$ is obeyed, i.e. the liquid is incompressible, $\nabla \cdot v = 0$.

In the two-dimensional case discussed above one finds for baker's transformation that the matrix α_{ij} is diagonal with $\alpha_{xx}=-ln2/\tau$ and $\alpha_{yy}=ln2/\tau$. For shear-flow only one element α_{ij} is nonzero; one has namely $\alpha_{xy}=1/\tau$. In three-dimensions the baker's transformation has again a diagonal matrix_with $\alpha_{xx}=-\alpha$ and $\alpha_{yy}=\alpha_{zz}=\alpha/2$ ($\alpha>0$). For shear-flow the matrix α_{ij} again has only one nonzero element, namely $\alpha_{xy}=1/\tau$.

5. REACTIONS UNDER FLOW

Putting together the results of the last sections it becomes obvious that a diffusion-controlled reaction in a moving, incompressible liquid is described by the following pair of differential equations [3,16,17,21]:

$$\begin{cases} \dfrac{\partial A}{\partial t} + v\cdot\nabla A = D\Delta A - \kappa AB \\[2mm] \dfrac{\partial B}{\partial t} + v\cdot\nabla B = D\Delta B - \kappa AB \end{cases} \tag{21}$$

Here A and B depend both on position and on time. Eq.(21) extends Eqs.(6) through the inclusion of velocity-dependent (drift) terms. As before, we revert now to the difference q and the sum s of the local concentrations and obtain:

$$\frac{\partial q}{\partial t} + v\cdot\nabla q = D\Delta q \tag{22}$$

and

$$\frac{\partial s}{\partial t} + v\cdot\nabla s = D\Delta s - \frac{1}{2}\kappa(s^2 - q^2) \tag{23}$$

Thus we can proceed by parallelling the treatment of Eqs.(7) and (8) above, the only difference being now that we need to have the Green's functions for diffusion under drift.

For baker's transformation the matrix (α_{ij}) is diagonal and Eq.(22) decouples in cartesian coordinates into independent equations for each coordinate. One has thus $G(r,r_0,t) = \prod_i \gamma_i(r_i,r_{0i},t)$, where i runs (in 3d) over x,y and z, r_i is the i-th component of r and $\gamma_i(r_i,r_{0i},t)$ is the solution of the one-dimensional equation

$$\frac{\partial\gamma}{\partial t} + \alpha_{ii}r_i\frac{\partial\gamma}{\partial r_i} = D\frac{\partial^2\gamma}{\partial r_i^2} \tag{24}$$

corresponding to the initial condition $\gamma_i(r_i,r_{0i},t) = \delta(r_i-r_{0i})$. A simple way to solve Eq.(24) is to look for its solution in the form

$$\gamma_i(r_i,r_{0i},t) = (2\pi Df(t))^{-1/2}exp\left[-\frac{(r_i-m(t))^2}{2Dg(t)}\right] \tag{25}$$

Inserting this form into Eq.(24) and comparing powers of r_i one obtains the following equations for the functions f, g and m:

$$\dot{f} = 2f/g$$
$$\dot{g} = 2(\alpha_{ii}g + 1) \tag{26}$$
$$\dot{m} = \alpha_{ii}m$$

from which the expressions for the functions f, g and m follow:
$f(t) = [1-exp(-2\alpha_{ii}t)]/\alpha_{ii}$, $g(t) = [exp(2\alpha_{ii}t)-1]/\alpha_{ii}$ and
$m(t) = r_{0i}exp(\alpha_{ii}t)$. To obtain these expressions we used as additional
condition the fact that for $\alpha_{ii}=0$ the function γ corresponds to the Green's
function for simple, one dimensional diffusion, Eq.(10) for $d=1$. Hence the
explicit form of the Green's function for baker's transformation is:

$$G(r,r_0,t)=(2\pi D)^{-d/2}\prod_i\sqrt{\frac{\alpha_{ii}}{1-exp(-2\alpha_{ii}t)}}exp\left(-\sum_i\frac{\alpha_{ii}(r_i-r_{0i}e^{\alpha_{ii}t})^2}{2D[exp(2\alpha_{ii}t)-1]}\right) \tag{27}$$

Furthermore one can check by straightforward integration that
$\int_{-\infty}^{\infty}\gamma(r_i,r_{0i},t)dr_i = e^{\alpha_{ii}t}$ and $\int_{-\infty}^{\infty}\gamma(r_i,r_{0i},t)dr_{0i} = 1$. Thus Green's
function $G(r,r_0,t)$ satisfies the normalization conditions:

$$\int_{-\infty}^{\infty}G(r,r_0,t)dr = exp\left(\sum_i\alpha_{ii}t\right) = 1$$

and $\tag{28}$

$$\int_{-\infty}^{\infty}G(r,r_0,t)dr_0 = 1$$

The case of shear-flow is somewhat more complicated. The Green's
function $G(x,y,x_0,y_0,t)$ for two-dimensional shear-flow is governed by the
equation:

$$\frac{\partial G}{\partial t} + \alpha y\frac{\partial G}{\partial x} = D\left[\frac{\partial^2 G}{\partial x^2} + \frac{\partial^2 G}{\partial y^2}\right] \tag{29}$$

where the initial condition is $G(x,y,x_0,y_0,0) = \delta(x-x_0)\delta(y-y_0)$. The
expression does not separate anymore with respect to the two variables x and
y. For the derivation of the solution we sketch some heuristic arguments
which allow to guess its form, and then display the final result, which can
be verified by substitution in Eq.(29). First we note that the Galilean
transformation

$$\begin{cases}x' = x-x_0-\alpha ty_0 \\ \\ y' = y-y_0\end{cases} \tag{30}$$

60

reduces the problem to finding a Green's function which depends only on three variables. Namely, by an explicit change of variables it follows that

$$G(x,y,x_0,y_0,t) = G_0(x',y',t) \qquad (31)$$

is the desired solution, if the function $G_0(x,y,t)$ is the solution of the same Eq.(29) for the initial condition $G_0(x,y,0) = \delta(x)\delta(y)$.

To find the form of the auxiliary function G_0 we consider the random process underlying Eq.(29). This process corresponds to a simple diffusion in the y-direction, while the motion in the x-direction has a superimposed drift term. This motivates the ansatz:

$$G_0(x,y,t) = \int_{-\infty}^{\infty} \mathcal{G}(x',y,t)G_d(x-x',t)\,dx' \qquad (32)$$

with $G_d(x,t)$ being the Green's function of a one-dimensional diffusion equation, see Eq.(10). From Eqs.(29), (31) and (32) it follows after partial integration that $\mathcal{G}(x,y,t)$ must obey the equation:

$$\frac{\partial \mathcal{G}}{\partial t} + \alpha y \frac{\partial \mathcal{G}}{\partial x} = D \frac{\partial^2 \mathcal{G}}{\partial y^2} \qquad (33)$$

Equation (33) describes a diffusion motion in the y-direction and a drift motion in the x-direction.

In a discrete picture one may view the movement in the y-direction as a random walk $y_j = \sum_{i=1}^{j} s_i$ with a mean-square displacement of $<s_i^2> = a^2$ and a characteristic step time of τ. The displacement in the x-direction depends on y. During one step its magnitude is $\alpha\tau y$, and therefore, since $y_j = \sum_{i=1}^{j} s_i$ one has $x_i = \alpha\tau \sum_{j=1}^{N} \sum_{i=1}^{j} s_i = \alpha\tau \sum_{i=1}^{N} (N+1-i)s_i$, N being the number of steps.

Because the s_i are statistically independent one can view for N large the random process $\{x_N, y_N\}$ as being Gaussian. This process has zero mean $<x_N> = <y_N> = 0$; its second moments are $<y_N^2> = \sum_{i=1}^{N} <s_i^2> = Na^2$;

$<x_N^2> = \alpha^2\tau^2 \sum_{i=1}^{N} (N+1-i)^2 <s_i^2> \sim N^3\alpha^2 a^2\tau^2$ and $<x_N y_N> = \alpha\tau \sum_{i=1}^{N} (N+1-i)<s_i^2> \sim N^2\alpha a^2\tau$.

Using the fact that a^2, τ and the diffusion coefficient $D \sim a^2/\tau$ are related, one can rewrite these moments as functions of time, by remembering that $N \propto t/\tau$. One obtains: $<x^2> \sim D\alpha^2 t^3$, $<y^2> \sim Dt$ and $<xy> \sim D\alpha t^2$. We expect therefore that $\mathcal{G}$ has the form:

$$\mathcal{G}(x,y,t) = \frac{A}{D\alpha t^2} \, exp\left[-c_1 \, \frac{x^2}{D\alpha^2 t^3} -c_2 \, \frac{y^2}{Dt} + c_3 \, \frac{xy}{D\alpha t^2}\right] \tag{34}$$

with the coefficients A, c_1, c_2 and c_3 to be determined by substituting Eq.(34) into Eq.(33). These procedure leads to the following values: $A=\sqrt{3}/2\pi$, $c_1=c_3=3$ and $c_2=1$. Using Eqs.(34),(32) and (31) one obtains now the explicit expression for the Green's function $G(r,r_0,t)$ for the diffusion equation under shear-flow in 2d:

$$G(r,r_0,t) = \frac{\sqrt{3}}{2\pi Dt\sqrt{\alpha^2 t^2 + 12}} \, exp\left[-\frac{3[x-x_0-\frac{\alpha t}{2}(y+y_0)]^2}{Dt(\alpha^2 t^2 + 12)} - \frac{(y-y_0)^2}{4Dt}\right] \tag{35}$$

For $\alpha \to 0$ the expression reverts back to Eq.(10). For the three-dimensional case one notes that the z-variable separates from the x- and y-ones; the equation in z-direction reads as an ordinary diffusion equation. The full Green function in this case has the form:

$$G(r,r_0,t) = \left[\frac{3}{16\pi^3 D^3 t^3 (\alpha^2 t^2 + 12)}\right]^{1/2} \times$$

$$\times \, exp\left[-\frac{3[x-x_0-\frac{\alpha t}{2}(y+y_0)]^2}{Dt(\alpha^2 t^2 + 12)} - \frac{(y-y_0)^2}{4Dt} - \frac{(z-z_0)^2}{4Dt}\right] \tag{36}$$

These forms will be used in the following analysis. Note that for both functions

$$\int_{-\infty}^{\infty} G(r,r_0,t)\,dr = \int_{-\infty}^{\infty} G(r,r_0,t)\,dr_0 = 1 \tag{37}$$

6. REACTION BEHAVIOR UNDER MIXING

Parallelling closely Sec.3, especially Eq.(12), we are now able to obtain the mean square value of the difference variable $q(r,t)$, namely $q^2(t)=<q^2(r,t)>$. Due to the Gaussian form of the Green's functions, Eqs. (27),(35) and (36), the integration in Eq.(12) is very simple. We use that in all cases considered the structure of the Green's function is $G(r,r_0,t) = F(t)exp(-Q_t(r,r_0))$, with Q_t being a quadratic form of r and r_0. Therefore $G^2(r,r_0,t) = F^2(t)exp(-Q_t(r\sqrt{2},r_0\sqrt{2})) = F(t)G(r\sqrt{2},r_0\sqrt{2},t)$, as may also be verified by simple inspection. This leads to the following expression:

$$q^2(t)=2c\int G^2(r,r',t)\,dr' =\sqrt{2}cF(t)\int G(r\sqrt{2},r'\sqrt{2},t)\,dr'\sqrt{2} = \sqrt{2}cF(t) \tag{38}$$

as the last integral is equal to unity according to Eqs.(28) and (37).

62

According to Eq.(38) we obtain for baker's transformation (for times long enough so that $e^{\alpha_{xx}t} \ll 1$ and both $e^{\alpha_{yy}t}, e^{\alpha_{zz}t} \gg 1$):

$$q^2(t) = \sqrt{2}c(4\pi D)^{-d/2} \left[\prod_i |\alpha_{ii}| \right]^{1/2} exp(\alpha_{xx}t) \tag{39}$$

(note that α_{xx} is negative). For shear-flow one has for $\alpha t \gg 1$ from Eq.(35):

$$q^2(t) = \frac{\sqrt{3}c}{\sqrt{2\pi D\alpha}} t^{-2} \tag{40}$$

in $d=2$ and from Eq.(36):

$$q^2(t) = \frac{\sqrt{3}c}{2\sqrt{2\pi^3 D^3 \alpha}} t^{-5/2} \tag{41}$$

in $d=3$.

According to Eq.(39) the decay of $q^2(t)$ for baker's transformation is exponential: $q^2(t) = Cexp(-\alpha t)$. The solution of Eq.(14) for this form of $q(t)$ is given in Refs.23 and 24 and reads

$$s(t) = \frac{\kappa\sqrt{C}exp(-\alpha t/2) \ K_1(\kappa\sqrt{C}exp(-\alpha t/2)/\alpha)}{\alpha \ K_0(\kappa\sqrt{C}exp(-\alpha t/2)/\alpha)} \tag{42}$$

where K_0 and K_1 are modified Bessel functions. Note that this form does not describe the very early stages of the reaction. The asymptotical analysis of this form shows that for intermediate times one obtains $s(t) = q(t)$ if the reaction rate coefficient is large enough; for longer times one has $s(t) = 2/\kappa t$, which reproduces the classical kinetics result. To estimate the crossover time T_C we use the parameters that correspond to concentrated ionic solutions in water ($c \approx 10^{22} cm^{-3}$, $D \approx 10^{-5} cm^2/s$) and to quite fast reactions ($\kappa \approx 10^{-12} cm^3/s$). As caracteristic values we take the α's in the range of 1 s^{-1}. With these parameters we obtain $T_C \simeq$ 10s for the crossover between the exponential and the classical regimes (at this time the concentration is around 10^{-10} of its initial value).

For shear-flow in $3d$ one can evaluate the crossover time T_C according to Eqs (15) and (41). The calculation shows that T_C is extremely long, being around 10^{11}s, which is of course far from the experimentally attainable range. For the values of parameters used one has thus $s(t) = q(t) \sim t^{-5/4}$. In both cases a decrease in the initial concentrations (smaller c) or slower elementary reaction acts (smaller κ) render the crossover time T_C shorter and the mixing-controlled stage less pronounced.

The two-dimensional shear-flow situation is special. In this case Eqs.(14) and (40) lead to

$$s(t) \sim (Kt)^{-1} \tag{43}$$

with an effective reaction rate coefficient

$$K = \frac{P^2}{\kappa}\left[\sqrt{1+\frac{\kappa^2}{P^2}} - 1\right] \qquad (44)$$

where $P = (2/3)^{1/4}\sqrt{\pi\alpha D/c}$. The value of K is determined by the smaller of the two parameters κ and P. For our values of parameters $P \simeq 10^{-13}cm^3/s$, being smaller than κ. Therefore the reaction is mainly controlled by slow mixing, i.e. $K \simeq P$. The dilution of the system up to $c \sim 10^{18}cm^{-3}$ renders, however, P large, so that the reaction aspect becomes more prominent, i.e. we have $K \simeq \kappa/2$.

7. CONCLUSIONS

In this chapter we have investigated the influence of mixing on the kinetics of the stoichiometrical $A + B \rightarrow 0$ reaction, and have studied two special mixing schemes in detail. As a main result we find that in general the important initial stages of the decay (these are experimentally of main interest) are controlled by stirring. However, the duration of these stages varies, depending on the type of mixing considered and on the dimension of space in which the reactants move. On the other hand, one often recovers asymptotically, at long times, the classical kinetic behavior. Our calculations allow also to conclude that the classical kinetic scheme is obeyed only in the limiting case of very effective mixing and very diluted solutions.

ACKNOWLEDGEMENTS

Thanks are due to Profs. J.Klafter and G.Zumofen for discussions and to D.Loomans and S.Luding for technical help in analytical programming. This work was supported by NATO research grant RG 0115/89, by the SFB 60 of the DFG and by the Fonds der Chemischen Industrie.

REFERENCES

1.   A. A. Ovchinnikov and Ya.B.Zeldovich, Chem.Phys. 28:215 (1978)
2.   D. Toussaint and F. Wilczek, J.Chem.Phys. 78:2642 (1983)
3.   I.M.Sokolov and A.Blumen, Int.J.Mod.Phys.B 5:3127 (1991)
4.   K. Kang and S. Redner, Phys.Rev.Lett. 52:955 (1984)
5.   K. Kang and S. Redner, Phys.Rev.A 32:435 (1985)
6.   G. Zumofen, A. Blumen and J. Klafter, J.Chem.Phys. 82:3198 (1985)
7.   I. M. Sokolov, Pis'ma Zh.Eksp.Teor.Fiz. 44:53 (1986) [JETP Lett. 44:67 (1986)]
8.   A. Blumen, J. Klafter and G. Zumofen, in "Optical Spectroscopy of Glasses", I. Zschokke, ed., Riedel, Dordrecht (1986) pp. 199-265 9
9.   A. G. Vitukhnovsky, B. L. Pyttel and I. M. Sokolov, Phys.Lett.A 128:161 (1988)
10.   R. Kopelman, Science 241:1620 (1988)
11.   V. Kuzovkov and E. Kotomin, Rep.Progr.Phys. 51:1479 (1988)
12.   A. S. Mikhailov, Phys.Repts. 184:307 (1989)
13.   G. Zumofen, J. Klafter and A. Blumen, J.Stat.Phys. 65:1015 (1991)
14.   G. Zumofen, J. Klafter and A. Blumen, Phys.Rev.A 44:8390 (1991)
15.   H. Schnörer, I. M. Sokolov and A. Blumen, Phys.Rev.A 42:7075 (1990)
16.   H. Schnörer, V. Kuzovkov and A. Blumen, J.Chem.Phys. 93:7148 (1990)
17.   E. Clément, L. M. Sander and R. Kopelman, Phys.Rev.A 39:6455 (1989)
18.   I. M. Sokolov and A. Blumen, Phys.Rev.A 43:2714 (1991)

19. M. Abramovitz and I. A. Stegun, eds., "Handbook of Mathematical Functions", Dover, N.Y. (1971)

20   S. Middleman, "Fundamentals of Polymer Processing", McGraw-Hill, N.Y. (1977)

21. J. M. Ottino, "The Kinematics of Mixing: Stretching, Chaos and Transport", Cambridge Univ.Press, Cambridge (1989)

22. R. S. Spencer and R. M. Wiley, J.Coll.Sci. 6:133 (1951)

23. I. M. Sokolov and A. Blumen, J.Phys.A: Math. and Gen. 24:3687 (1991)

24. I. M. Sokolov and A. Blumen, Phys.Rev.Lett. 66, 1942 (1991)

PHASE TRANSITIONS IN LIQUID CRYSTAL POLYMERS

S F Edwards
Cavendish Laboratory
Cambridge CB3 OHE

1. Introduction

Liquid crystal polymers have locally a rather rigid structure which for short molecules would be classic nematics. When the molecule is long some analogue of the nematic transition can be expected to occur, and there is already a substantial literature on this matter (1-4). The complication that occurs with long lcp's and quite generally with long and short lcp networks, is that the ability of the molecules to lie parallel is frustrated by the topological conditions imposed by the entanglement following from their length, and the impossibility of making the members of a network, in general, to all lie parallel. The comparison of concentrated polymer solutions or melts with spaghetti is not a good one for flexible polymers which are much more kinky that spaghetti can be, but it is not unreasonable for long lcp's which can take up worm-like configurations. The difficulty over a nematic state is thereby easily visualised, for polymers can lie parallel for a while, but then they wander off into the general background, and thermodynamic forces are inadequate to remove the topological constraints thus imposed unless the molecules are rather short, ie have a length not greatly in excess of the length between entanglements.

It is interesting to contrast the situation with flexible polymers. With these, thermodynamic forces which would like to crystalise the polymer, can suck the polymer along its enveloping tube, and by reptation bring it up to other parts of the same chain and hence permit crystallisation. It is like a diner on spaghetti who sucks the pasta out of a bowl into his mouth, where (if it is flexible enough) he can rearrange it with his tongue. However, if the spaghetti is not sufficiently flexible this will not work.

In this paper a mathematical framework will be given for a self-consistent calculation of the properties of the system. It is believed that the approach is simpler than others in the literature, and also uncovers some difficulties which have not previously been addressed.

Synthesis, Characterization, and Theory of Polymeric Networks and Gels
Edited by S.M. Aharoni, Plenum Press, New York, 1992

2. Mathematical description

For a flexible polymer of Kuhn length b, for large distances a convenient description is by the Weiner type of integral, ie a weight for a single self interacting chain

$$\exp(-\frac{3}{2b}\int_0^L \underline{R}'^2(s)ds - \underline{w}\int_0^L\int_0^L \delta(\underline{R}(s_1) - \underline{R}(s_2))ds_1 ds_2) \qquad (2.1)$$

is attached to each configuration $\underline{R}(s)$ where s is the monomer label, ie the arc length along the locus $\underline{R}(s)$ of the polymer (1). There is an obvious extension to many chains. This weight represents a continuous curve $\underline{R}(s)$ whose tangent is continuously changing abruptly in direction and corresponds to the limiting mathematics whereby a random walk of step length b is described by the diffusion equation. The interaction term acknowledges that the large scale behaviour of the polymer is governed solely by the integral of the detailed potential of interaction so only one parameter w needs to be introduced.

A calculation of the free energy associated with the δ function interaction diverges, but all physical quantities involve differences and these all converge. Thus in the sense in which the word was introduced into quantum field theory, this theory is renormalisable.

When one turns to the worm-like chain, the first change will be to ensure that the locus $\underline{R}(s)$ no longer changes direction abruptly, but its second derivative, ie its curvature can do so. Once the curve is differentiable, one can (and must) use the normal formula of differential geometry which comes from Pythagoras' theorem

$$dR^2 = ds^2 \quad \text{ie} \quad R'^2 = 1 . \qquad (2.2)$$

Thus the new weight without interaction is

$$\exp(-\varepsilon\int R''^2 ds)\prod_s \delta(R'^2 - 1) \qquad (2.3)$$

and now the interaction can involve R' as well as R. Thus in addition to a contact term there will be a term like

$$(R'(s_1) \times R'(s_2))^2 \delta(R(s_1) - R(s_2)) \qquad (2.4)$$

where the square is needed because there is no physical reason for the tangent to be $\underline{R}'$ or $-\underline{R}'$; this is a standard form for liquid crystals. For lcp's one needs a shape and again a contact form is taken.

The previous contact term $\delta(R_1 - R_2)$ now causes trouble because it is no longer renormalisable, ie the self interaction $s_1 \to s_2$ is now so strong that even differences are divergent. Physically of course the molecule cannot self interact at all because of its stiffness, but bringing this in with too much detail produces needless complexity. The simplest solution is to modify the δ function to

$$w\left\{\delta(R(s_1) - R(s_2)) - \delta(\tfrac{1}{2}(R'(s_1) + R'(s_2))(s_1 - s_2))\right\} \tag{2.5}$$

This again is a one parameter interaction, and it is just this one parameter which will survive in long distance effects if a detailed potential were to be employed. The detail of this problem will not be pursued in this paper and the simple δ interaction used.

The mathematical problem of the $\delta(R'^2 - 1)$ is substantial. the standard theorem of the Weiner integral is that if the weight factor is

$$\exp\left[-\frac{b_n}{2} \int_0^L R^{(n)2}(s)ds + \int f(R^{(n-1)}...,R) \right], \tag{2.6}$$

this is equivalent to the differential equation [9]

$$\left(\frac{\partial}{\partial s} - \frac{1}{2b_n} \frac{\partial^2}{\partial R^{(n-1)2}} - f(R^{(n-1)}...,R) + \sum_{m=1}^{n-1} R^m \frac{\partial}{\partial R^{m-1}} \right) P(R^{(n-1)}...,R) = 0 . \tag{2.7}$$

For example, when s is the time variable of a Brownian particle $\dot{R} = v$, and the physical problem of diffusion of the velocity due to collisions gives the Fokker Planck equation

$$\left(\frac{\partial}{\partial t} + v.\frac{\partial}{\partial r} - \frac{D}{2} \frac{\partial^2}{\partial v^2} - f(v,r) \right) P = 0 . \tag{2.8}$$

(It usually appears in a slightly different structure of

$$\left(\frac{\partial}{\partial t} + v.\frac{\partial}{\partial r} - \frac{D}{2} \frac{\partial}{\partial v}\left(\frac{\partial}{\partial v} + \frac{v}{kt} \right) \right) P = 0 \right) . \tag{2.9}$$

If one has the Ehrenfest wood wind problem, the velocity is conserved on scattering with fixed obstacles, v^2 is a constant, so wlg $v^2 = 1$.

This is then expressed as

$$\underline{v} = \underline{n}, \text{ a unit vector}$$
$$= (\sin\theta\sin\phi, \sin\theta\cos\phi, \cos\theta) \tag{2.10}$$

and

$$\left(\frac{\partial}{\partial t} + \mathbf{n}\cdot\frac{\partial}{\partial r} - \frac{D}{2}\,\mathcal{L}(\theta,\phi)\right)P=0 \tag{2.11}$$

where $\mathcal{L}$ is Legendres operator, ie the angular part of the Laplacean whose eigensolutions are the Legendre functions. Although familiar, eqn (2.11) is not a good starting point for a calculation and (2.8) ie Hermites equation is much simpler.

Thus returning to the polymer problem, if instead of using $R'^2 = 1$, one could use

$$<R'^2> = 1 \tag{2.12}$$

calculations become much simpler. If one took

$$\exp\left(-\int\frac{\varepsilon R''^2}{2} - \int\frac{3}{2b}R'^2\right) \tag{2.13}$$

as the weight factor, at large distances the R'^2 term dominates the calculation and

$$\left\langle(R(s_1) - R(s_2))^2\right\rangle = b|s_1 - s_2| \tag{2.14}$$

so b is still the Kuhn length. But if one calculates

$$<R'^2> \text{ it gives } 3\pi^2\varepsilon\,/\,b\,. \quad (6) \tag{2.15}$$

Thus the Kuhn length is $(3\pi^2\varepsilon)^{-1}$ and the diffusion equation becomes

$$\left(\frac{\partial}{\partial s} + v\frac{\partial}{\partial r} - \frac{1}{2\varepsilon}\frac{\partial^2}{\partial v^2} - \frac{3}{2b}v^2\right)P(r,v,s) = 0 \tag{2.16}$$

where v is now used for $\dfrac{\partial R(s)}{\partial s}$.

In general it is not valid to replace the Kratky-Porod type equation by the Hermite equation but when one finds many other approximations required to build up a self consistent field theory of a nematic transition it is possible to argue that some version of the Hermite form is equally valid. To this one can note that

$$\delta(R'^2(s) - 1) = \frac{1}{2\pi}\int\limits_{-\infty}^{\infty} d\alpha(s)\exp\left[-i\alpha(s)(R'^2(s) - 1)\right] \tag{2.17}$$

So if one writes each $\alpha(s) = \lambda(s)\,ds$

$$\prod_s \delta(R'^2(s)-1)=N\int_{-\infty}^{\infty}\cdots\int_{-\infty}^{\infty}[d\lambda(s)]\exp\left[-i\int ds\ \lambda(s)[R'^2(s)-1]\right]$$

(2.18)

a functional integral over λ with N the normalisation. This produces an exponent

$$\exp\left(-\frac{\varepsilon}{2}\int R''^2(s)ds - i\int\lambda(s)R'^2(s)ds + i\int\lambda(s)ds\right)$$

(2.19)

This will now be built into the much more elaborate structure with interactions and a mean field value for λ deduced on a par with all the other mean field variables. The simple version above is $\lambda = 3/2bi$, but in general it will not take this value.

3. Interactions

For flexible polymers the central variable describing a solution is the concentration as a function of position, ie the density of polymer at a point

$$\rho(r) = \sum_{\alpha}\int\delta(r - R_{\alpha}(s_{\alpha}))ds_{\alpha}.$$

(3.1)

The interaction can be expressed in terms of ρ

$$w\sum_{\alpha\beta}\iint\delta(R_{\alpha}(s_{\alpha}) - R_{\beta}(s_{\beta}))ds_{\alpha}ds_{\beta} = w\int d^3r\rho^2(r)\ .$$

(3.2)

When one now turns to the nematic interaction (2.4) it is natural again to look for a collective variable, and this can be easily effected in a well known way:

$$\sum_{\alpha\beta}\iint\left(\underline{R}'_{\alpha}(s_{\alpha})\times\underline{R}'_{\beta}(s_{\beta})\right)^2\delta\left(R_{\alpha}(s_{\alpha}) - R_{\beta}(s_{\beta})\right)ds_{\alpha}ds_{\beta}$$

(3.3)

$$= \sum\iint R'^i_{\alpha}(s_{\alpha})R'^a_{\alpha}(s_{\alpha})R'^j_{\beta}(s_{\beta})R'^b_{\beta}(s_{\beta})(\delta^{ia}\delta^{jb} - \delta^{ib}\delta^{ja})\delta\left(R_{\alpha}(s_{\alpha}) - R_{\beta}(s_{\beta})\right)$$

$$= \int\sigma^{ij}(r)\sigma^{ab}(r)(\delta^{ia}\delta^{jb} - \delta^{ib}\delta^{ja})d^3r$$

(3.4)

$$= \int\left(\sigma^{ij}\sigma^{ij} - \sigma^{ii}\sigma^{jj}\right)d^3r,$$

(3.5)

where

$$\sigma^{ij}(r) = \sum_{\alpha}\int ds_{\alpha}R'^i_{\alpha}(s_{\alpha})R'^j_{\alpha}(s_{\alpha})\delta(r - R_{\alpha}(s_{\alpha}))\ .$$

(3.6)

From the definition $\sigma^{ii}(r) = \rho(r)$

(3.7)

so that the second term can be incorporated into w. In thermal physics it is customary to define such quantities relative to ρ, eg the velocity of a fluid is

$$\sum_{\alpha} \underline{\dot{R}}_{\alpha} \delta(r - R_{\alpha}) = \underline{v}(r)\rho(r) \tag{3.8}$$

and $\underline{v}$ is used in say the Stokes fluid equations rather than the momentum

$$\underline{P}(r) = \sum_{\alpha} m_{\alpha} \underline{\dot{R}}_{\alpha} \delta(r - R_{\alpha}) \tag{3.9}$$

In the present case it seems much simpler to use the analogue of momentum, so that the definition of σ above will be used.

For flexible polymers $kT\sigma$ is the stress tensor and also gives the birefringence[1]. When

$$\left\langle \sigma^{ij} \right\rangle \neq \delta^{ij}\rho \tag{3.10}$$

the material is nematic. It will here be called the orientation tensor of the material.

One may note that given a system of polymers, one can define $\rho(r)$, but given $\rho(r)$ one cannot get back to the polymers. To get back to the polymers from the collective coordinates one needs an infinite set of collective coordinates of which the density and orientation tensor are the first two. The density of curvature

$$\underline{K}(r) = \int \underline{R''}(s)\delta(r - R(s))ds \tag{3.11}$$

is even in the arc length, so is physically significant, but to be a sensible collective coordinate, the torsion of the curve must exist, and this does not seem required for the simplest model of lcp's. Nevertheless one may note that when two molecules fit together in such a way that their normals are parallel, if there is an energetic advantage, an interaction like

$$\sum_{\alpha\beta} \iint \underline{R}_{\alpha}''(s_{\alpha}) \bullet \underline{R}_{\beta}''(s_{\beta}) W(R_{\alpha}(s_{\alpha}) - R_{\beta}(s_{\beta}))ds_{\alpha}ds_{\beta} \tag{3.12}$$

describes. The particular case of the local self contribution of a single chain, $\varepsilon R''^2/2$ is contained in this formula. This implies that if ε contains kT, it will contribute to the free energy in a non-trivial way, so that one has to go to the level where torsion exists, or equivalently, use cut offs in the definitions. In this paper ε is treated as a constant and the torsion is not invoked.

4. Conjugate Fields

So far the description of the lcp is via R(s) which has $\underline{R}(s)$ and $\underline{R}'(s)$ continuous. The constraint $\underline{R}'^2(s) = 1$ can be expressed by a $\lambda(s)$ in the exponent. Likewise the density and orientation tensor can be usefully expressed by conjugate fields which decouple the polymers, ie the polymers interact with a field and the elimination of this field produces the interaction, just as the Coulomb interaction in charged particles can be replaced by each particle interacting with an electric field whose equations of motion when solved reproduce the Coulomb interaction.

Thus

$$\sum_{\alpha\beta} \iint \delta(R_\alpha - R_\beta) ds_\alpha ds_\beta = \frac{1}{(2\pi)^3} \int d^3k \sum_{\alpha,\beta} \int ds_\alpha \int ds_\beta e^{ik(R_\alpha - R_\beta)} \tag{4.1}$$

$$= \int \rho^2(r) d^3r \tag{4.2}$$

$$= \frac{1}{(2\pi)^3} \int \rho_k \rho_k^* d^3k \tag{4.3}$$

where

$$\rho_k = \sum_\alpha \int ds_\alpha e^{ikR_\alpha(s_\alpha)} . \tag{4.4}$$

If one writes

$$\exp\left(-\frac{w}{(2\pi)^3}\int \rho_k \rho_k^* d^3k\right) = N \int \prod d\phi d\phi^* \exp\left(-\frac{1}{(2\pi)^3}\int \frac{\phi_k \phi_k^* d^3k}{2w} - \frac{i}{(2\pi)^3}\int \phi_k \rho_k\right) \tag{4.5}$$

or

$$\exp\left(-w\int \rho^2(r) d^3r\right) = N \int \prod d\phi(r) \exp\left(-i\int \phi(r)\rho(r) d^3r - \frac{1}{2w}\int \phi^2(r) d^3r\right) \tag{4.6}$$

$$= N \int \prod d\phi \exp\left(-i\sum_\alpha \int ds_\alpha \phi(R_\alpha(s_\alpha)) - \frac{1}{2w}\int \phi^2(r) d^3r\right),$$

where

$$N^{-1} = \int \prod d\phi \exp\left(-\frac{1}{2w}\int \phi^2(r) d^3r\right) . \tag{4.7}$$

This now leaves a path integral over the $R_\alpha(s_\alpha)$ which separates

$$\exp\left(-\sum_\alpha \int ds_\alpha \left(R_\alpha''^2(s) + i\underline{\lambda}_\alpha(s_\alpha)R'^2(s_\alpha) + i\phi(R_\alpha(s_\alpha))\right)\right) . \tag{4.8}$$

A similar transformation is possible for the orientation via a tensor field ψ_{ij} or $\underline{\underline{\psi}}$

$$\int (\mathcal{D}\underline{\psi}) \exp\left(-\frac{1}{2u}\int \underline{\underline{\psi}}\cdot\underline{\underline{\psi}}\ d^3r -i\sum_\alpha\int \underline{R}'_\alpha(s_\alpha)\underline{\underline{\psi}}(R(s_\alpha))\underline{R}'_\alpha(s_\alpha)ds_\alpha\right)$$

$$=\exp\left(-u\sum_{\alpha\beta}\int\int ds_\alpha ds_\beta (R'_\alpha R'_\alpha)(R'_\beta R'_\beta)\delta(R_\alpha-R_\beta)\right).$$

$$(4.9)$$

Thus finally the free energy of our system is given by

$$\exp(-A/kT)=N\int \mathcal{D}(\lambda)\mathcal{D}(\underline{\psi})\mathcal{D}(\phi)\exp(-Q(R))\exp(-P(\phi,\lambda,\psi))\mathcal{D}(R)$$

$$(4.10)$$

where Q is

$$\int\sum_\alpha\left(\frac{\varepsilon R''^2}{2}+i\lambda(s_\alpha)R'^2(s_\alpha)+iR'_\alpha\underline{\underline{\psi}}R'_\alpha-i\phi(R_\alpha)\right)$$

$$(4.11)$$

and

$$P=-\sum_\alpha i\int\lambda_\alpha(s_\alpha)ds_\alpha -i\int\rho\phi d^3r-i\int\underline{\underline{\sigma}}\underline{\underline{\psi}}\ d^3r.$$

$$(4.12)$$

The exponent Q gives rise to the differential equation

$$\left(\frac{\partial}{\partial s}-\frac{1}{2\varepsilon}\frac{\partial^2}{\partial v^2}+i\lambda(s)\underline{v}^2+i\underline{v}\underline{\underline{\psi}}(r)\underline{v}+i\phi(r)\right)P=0$$

$$(4.13)$$

or the same operator on the Green function

$$G(r_1 r_2 v_1 v_2 s_1 s_2)\quad\text{giving a r.h.s. }\delta(r_1-r_2)\delta(v_1-v_2)\delta(s_1-s_2)\ .$$

(Had one gone to the next order of accuracy with a R'''², it would be like a diffusing acceleration a

$$\left(\frac{\partial}{\partial s}-\frac{1}{2v}\frac{\partial^2}{\partial a^2}-\frac{\varepsilon}{2}a^2+i\lambda v^2+iv\psi v+i\phi+i\theta a\right)$$

$$(4.14)$$

where θ is now the field conjugate to the interaction (3.12)).

5. Evaluation

The simplest mathematics obtains when the polymer is long, for then one can describe $R(s)$ by a Fourier decomposition, ie the Rouse modes are a continuum and one can write

$$R_q=\frac{1}{2\pi}\int_{-L/2}^{L/2}R(s)e^{-iqs}\ ds$$

$$(5.1)$$

as if L were infinite.

The exponent Q becomes a simple polynomial in q provided that the mean field values of ϕ, ψ, λ are constants and these fields are expanded about these constant values. For simplicity use the same symbols for the mean fields. Then Q is

$$\int (\mathcal{D}R) \exp\left[-\frac{L}{2\pi} \int R_q(q^4 + i\lambda q^2 + iq^2\underline{\underline{\psi}} + i\phi) R_q dq\right] \tag{5.2}$$

which will have the usual Plemelj form

$$= V \exp\left(-\frac{L}{2\pi} \sum_\ell \int dq \log(q^4 + iq\psi_\ell + i\lambda + i\phi)\right) \tag{5.3}$$

$$= V \exp\Lambda \quad \text{(say)}$$

where L is the polymer length and ℓ is a symbol labelling the three eigenvalues of the matrix in (5.2). For N polymers in the box, one has $\mathcal{L} = NL$ replacing L in (5.3). The V^N is just the perfect gas term, one freedom per chain, and does not affect the argument so will be omitted.

Without loss of generality assume the lcp's have the same length. Then if the Green function

$$G(r_1 r_2 v_1 v_2 L[\phi, \lambda, \psi]) \tag{5.4}$$

can be derived (in some approximation) one can obtain the free energy from

$$\exp(-A/kT) = \int (\mathcal{D}\phi\, \mathcal{D}\lambda\, \mathcal{D}\psi) \exp\left(-N \log \int G(r_1 r_2 v_1 v_2 L[\phi\lambda\psi])\right) \tag{5.5}$$

The approximation which immediately suggests itself is that of mean fields, and the basic argument is that there will be a mean field approach for ϕ, λ, ψ. The particular phase change which is to be expected is that of a nematic type, and it is this aspect which will be the focus of the rest of the paper. The following analysis is based on the argument that a consistent mean field theory treats all the fields on the same basis and does not for example use the exact but very awkward form (2.11). By handling all at the same level, it enables one to proceed to a consistent theory of fluctuations, although there is no space for that in this paper.

Although 5.3 is a divergent expression, the divergence is a constant, and does not enter when the free energy is minimised. Thus A can be expressed as

$$A/kT = \Lambda \tag{5.6}$$

or as

$$\frac{A}{kT} = \Lambda - i\lambda \mathcal{L} - \frac{i\phi^2 V}{2w} - \frac{i\psi^2 V}{2u} \tag{5.7}$$

so if the concentration $\mathcal{L}/V$ is denoted by c, ie c is ϵ/ρ :

$$\sigma_\ell = \frac{c}{\sqrt{\psi_\ell + \lambda}} \tag{5.8}$$

$$c = \sum_\ell \frac{c}{\sqrt{\psi_\ell + \lambda}} \tag{5.9}$$

$$wc = i(\lambda + \phi) \tag{5.10}$$

$$u\left(\sigma_\ell - \sum_\ell \sigma_\ell\right) = i\psi_\ell \tag{5.11}$$

The ϕ equation does not affect the orientation behaviour, and so is discussed no further (but is vital in that it gives the fluctuations which give screening at the next order).

Note that if $u = 0$, $\psi_\ell = 0$ and $\sigma_\ell = 4/3$. In the general case since $\sum_\ell \sigma_\ell = c$, one has

$$\sigma_\ell = c\left[(\sigma_\ell - \tfrac{1}{3})u + \lambda\right]^{-\frac{1}{2}} . \tag{5.12}$$

This is of course in the standard form expected for transitions of this kind. It gives a second order transition for two dimensions and a first order transition for three dimensions. The algebra is slightly more complicated than that of liquid crystals because of the root appearing.

As the correlations appear they will be badly described by mean fields which normally only work well past the critical region. Since the transition is first order however it should be possible to handle the evolution of ordered regions. The topological constraints on chains does not affect the final state of aligned polymer provided that time is given for the polymer to find that state, and that it is accessible. For networks this final state is not attainable for the topological constraints are permanent. One then has a picture of a thermodynamically favored ordered state being frustrated by frozen topological constraints. Thus one can expect the system to show birefringence after the transition but this can never reach, in a network, the level in free polymers [10-12].

6. General comments on the orientation tensor

It has been argued in this paper that the collective coordinate $\rho(r)$, the density or concentration is not adequate for lcp solutions or melts and the orientational tensor is also required.

$$\sigma^{ij} = \int R'^i R'^j \delta(r - R)$$

This is of course to be expected from lc theory and has been used by all studies of the problem. It is worth commenting however that the use of ρ in flexible polymer problems does not preclude σ from being of central importance there also as soon as one turns to dynamic problems. The point is that the success of tube models of the melt suggest that although σ varies wildly from point to point in a melt of flexible polymers, it is extremely stable in time, as it requires the long disengagement time τ_d to dissipate, ie a polymer has to reptate out of its tube to create a new tube and the amount of polymer left in the time of $t = 0$ is e^{-t/τ_d} at t [8]. Thus redefining σ to remove the constant value

$$\left\langle \sigma^{ij}(r,t)\sigma^{ij}(r',t)\right\rangle \sim e^{-|r-r'|/b} \tag{6.1}$$

but $$\left\langle \sigma^{ij}(r,t)\sigma^{ij}(r,t')\right\rangle \sim e^{-|t-t'|/\tau_d} \tag{6.2}$$

This effect will be enhanced for lcp's where again the tube is applicable, indeed rather more obvious than for flexible polymers. But it is easy to imagine the kind of motion required by slithering along tubes to align long lcp molecules; it will take many times the equivalent of τ_d for lcp's.

For networks difficult but visualisable problems emerge if the lcp's are long when part of the neighbourhood can align, essentially up to an entanglement distance. For short chains however the problem depends vitally on the history of creation of the network [10-12].

7. Acknowledgements

This work has been stimulated by the experimental work of Shaul Aharoni with whom the author has had much fruitful interaction. He would also like to thank Drs Griffin, Gupta and Warner for help in the work, and the AFRC for financial support.

8. References

1.   Doi M, Edwards SF. The Theory of Polymer Dynamics. OUP (1986).
2.   Abramchuk SS, Nyrkova IA, Khokhlov AR. Polymer Sci, 490 and 1759 (1989).
3.   Khodolenko AL. J Chem Phys, 95, 628 (1991).
4.   Lansac Y and A ten Bosch. Jn Chem Phys. 94, (3) 2168 (1991).
5.   Jolanta B, Lagoushi & Jaan Noolandi. J Chem Phys, 95, 1266 (1991).
6.   MG Bawendi & Karl F Freed J Chem Phys, 83, 2491 (1985).
7.   Warner M, Wang XJ. Macromolecules, 24, 4932 (1991).
8.   Ref 1, ch 6.
9.   Freed KF. Adv Chem Phy, 22, 1 (1972).
10.   Aharoni S, Edwards SF. In press (Macromolecules).
11.   Aharoni S, Murthy NS, Zero K, Edwards SF. Macromolecules, 23, 2532 (1990).
12.   Aharoni S, Edwards SF. Macromolecules, 22, 3361 (1989).

ONE-STEP AND TWO-STEP RIGID POLYAMIDE NETWORKS AND GELS: SIMILARITIES AND DIFFERENCES

Shaul M. Aharoni

Polymer Science Laboratory, Research & Technology
Allied-Signal Inc.
P.O. Box 1021, Morristown, New Jersey 07962-1021

INTRODUCTION

Condensation and addition are two fundamentally different polymerization processes.[1] In the case of addition polymerizations, typified by free radical polymerization, the growing chain-ends are highly reactive while the monomers are inert. When the reaction mixture is sampled before the reaction is complete and all monomers are consumed, one finds in it a large number of monomers together with a rather small number of fully grown polymer chains.[1] Because of the rapid growth of individual chains, the concentration of oligomeric species in the reaction bath is very small. In the case of condensation polymerizations the reactivity of every like functional group is the same independently of the size of the molecular species to which it is attached.[1a] This leads to random nucleation and the presence in the condensation reaction bath of a broad distribution of growing species. During chain growth, monomers and various size oligomers and chains participate together in the creation of larger oligomeric species and longer chains.[1] It was repeatedly demonstrated that this process of condensation polymerization is equally valid for linear and highly branched polymers.[1] In this chapter we shall discuss exclusively condensation polymers in which the products of the condensation reaction are aromatic amide groups.

Polymer networks are three-dimensional highly-branched structures. When they are swollen with liquid they are called gels. In this work we deal solely with covalently-bonded permanent gels and not with gels in which the interchain bonds may be reversibly formed or broken by changes in temperature or solvent quality. There are three fundamentally different methods to prepare covalently-linked polymeric network gels. One is to conduct the polycondensation in a single step starting with a solution of the appropriate monomers whose average functionality is higher than 2.0. This procedure

Synthesis, Characterization, and Theory of Polymeric Networks and Gels
Edited by S.M. Aharoni, Plenum Press, New York, 1992

is called by us a one-step method. In one-step polymerization, the growing polymeric entities are highly branched and in the pre-gel stage conform with the fractal model.[2-4] We therefore call these species fractal polymers (FPs). When the total polymer concentration in the reaction mixture, C_0, is above a characteristic critical threshold, C_0^*, the system eventually solidifies and a gel is formed. Some traits of the fractality of the pre-gel precursor species were found[3,4] to carry through to the "infinite" network progeny. The second preparative method requires the preparation of high molecular weight (M) linear polymer chains in a first step and then crosslinking them many times along the chains in a second, separate step. This is called by us a two-step method.[5] In this chapter we discuss only the case where the crosslinking leads to aromatic polyamide networks in which the species connecting the chains are chemically indistinguishable from the precursor chain segments between junction points. Because of the chemical similarity of all segments in the final network, we prefer to call the crosslink junctions branchpoints. The third method is similar to the two-step method, except that the cross-linking takes place only at reactive chain-ends. In this chapter, chain-end-linked networks will not be discussed.

The gels we describe in this chapter are rigid. They comprise rigid aromatic polyamide networks and swelling liquids. The rigid networks consist of stiff segments where the amide groups are placed in the para positions of the aromatic rings, and rigid aromatic branchpoints of functionality $f \geq 3$. The segments may all be of identical length ℓ or of an average length ℓ_0. Because of the segment shortness, the length distribution about ℓ_0 is not particularly broad. Because the stiff short segments can not adopt random coil configuration and conform with Gaussian statistics, the rigid networks and gels do not show the temperature dependence of the modulus, a defining feature of flexible networks. The one-step and two-step preparative methods produce two kinds of polymer networks.[5,6] This chapter is devoted to describing the similarities and differences between the two kinds of rigid networks.

This chapter is organized as follows. In the Introduction section, some definitions and the scope of the discussion were given. In the Syntheses section, the synthetic procedures we used to obtain gels of one-step and two-step rigid aromatic polyamide networks will be given in broad terms. The reader is referred to our previous papers[2-5] for specific details of the syntheses, work-up, characterization and shear modulus determination. In the Similarities section, the common denominators of the one-step and two-step network gels will be discussed, emphasizing likely mechanisms for the bending and straightening deformations of stiff polyamide segments in a network. In the Differences section, structural and performance differences will be described, emphasizing the effects of network defects and the fundamental difference in characteristic lengths: in one-step networks the radius of the fractal polymers at the point of gelation, and in two-step networks the average length of the precursor stiff chain.

SYNTHESES

Because their presumed melting point is higher than their decomposition temperature, rigid three-dimensional networks of aromatic polyamides can not be prepared in the absence of solvent. For the purpose of macro- and micro-homogeneity, they must be prepared in a single phase solution and care must be exercised to prevent, or, at least, minimize precipitation or the appearance of microsyneresis during polymerization. We found two convenient ways to achieve just that. Both use a 5 weight/volume percent anhydrous LiCl dissolved in N,N-dimethylacetamide (DMAc/LiCl) as the solvent in which the reaction is carried out. This is an excellent solvent for practically all monomers and for most of the aromatic polyamides. One polymerization is a Schotten-Baumann-type polycondensation, reacting aromatic diamines with diacid chlorides in the presence of a slight molar excess of pyridine. Depending on the solubilities of the species involved, the reaction may be carried from below room temperature up to about 85°C. When acid chlorides are not available or when AB monomers are used, a variation of the Yamazaki[7] procedure is employed to produce gels of rigid networks. In this case, in addition to the desired monomer mixture, the DMAc/LiCl contains slight molar excess of pyridine and of triphenylphosphite (TPP). This reaction is carried out in the temperature interval of $85 \leq T \leq 115°C$ and is propelled forward by the conversion of TPP to diphenylphosphite and phenol.[8] It is important to recognize that due to the presence of non-solvent reagents and reaction products in the reaction mixtures of both the Schotten-Baumann-type and Yamazaki-type procedures, the solvent quality of these media is far poorer than that of DMAc/LiCl or DMAc alone.

The following will serve to illustrate rigid networks with typical stiff segments and rigid branchpoints prepared by the one-step method. In the first case, the monomer tris(p-carboxyphenyl)-1,3,5-benzenetriamide is prepared separately from p-aminobenzoic acid and 1,3,5-benzenetricarboxylic acid chloride. This monomer is reacted under Yamazaki conditions with 4,4'-diaminobenzanilide (DABA) in 2:3 molar ratio to produce a network typified by the following fragment:

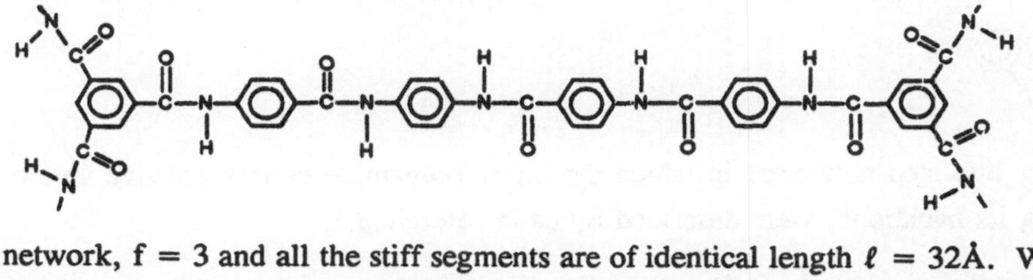

In this network, $f = 3$ and all the stiff segments are of identical length $\ell = 32$Å. When the network is prepared from 2:6:3 molar ratio of 1,3,5-benzenetricarboxylic acid (BTCA), p-aminobenzoic acid and DABA, one obtains a network chemically identical with the above but with the segment length being an average $\ell_0 = 32$Å instead of identical ℓ. Due to the unavailability of appropriate monomers, longer segments with identical ℓ are very hard to prepare and segments with average ℓ₀ are usually prepared instead. As an example, the typical segment with average $\ell_0 = 38.5$Å in the fragment

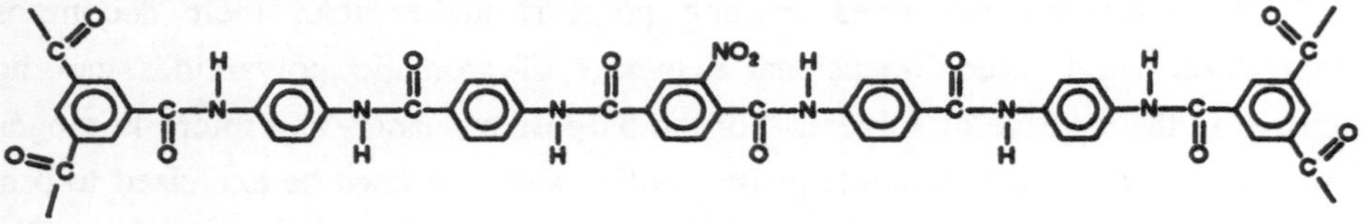

was prepared from 2:6:3 molar ratio of BTCA, DABA and nitroterephthalic acid (NTPA).

An example for the two-step procedure is shown below. Here, a linear polyamide with reactive sites is prepared first by Schotten-Baumann-type reaction from 2:1:1 molar ratio of terephthaloyl chloride, DABA and 3,5-diaminobenzoic acid:

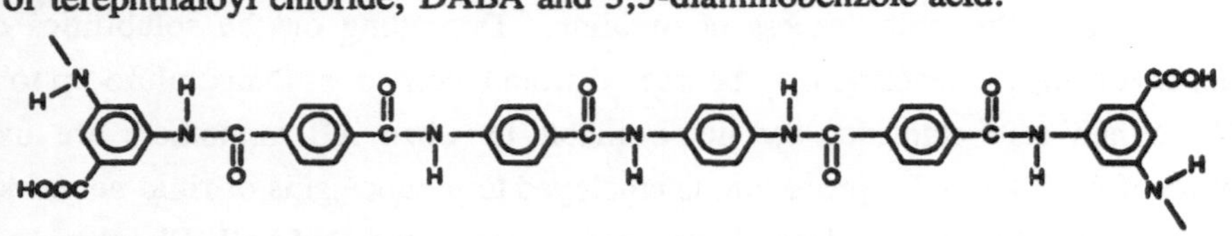

After purification and characterization, the linear polyamides are re-dissolved in DMAc/LiCl and reacted with DABA under Yamazaki conditions to obtain the following network:

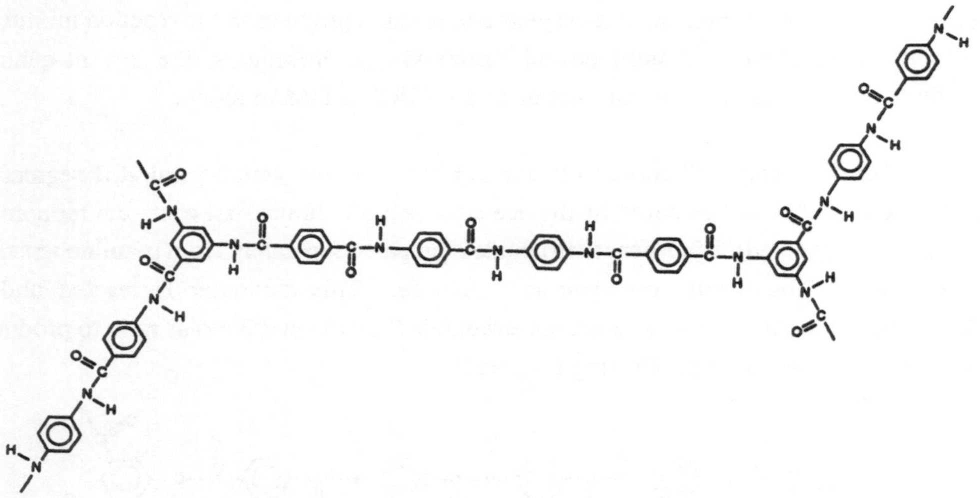

Other two-step networks, in which the linear polyamide carries reactive amine groups along its backbone, were described by us in reference 5.

In one-step polycondensations under Yamazaki conditions at $C_0 = 10\%$, an incubation period of about 10 minutes was generally observed. After the incubation, a rapid increase in viscosity ensued followed by the onset of gelation. Within the temperature interval of 85°C to 115°C, the higher the temperature, the shorter was the

82

incubation period and the faster the gelation. At any given temperature within that interval, the rate of viscosity increase and the time it took to reach the gel point were both strongly influenced by the closeness to stoichiometry of amine and carboxyl groups, the concentration C_0, the length of the segments between branchpoints and their stiffness, and the functionality, f, of the branchpoints. The shorter and stiffer the segments, the higher are f and C_0 and the closer the system to stoichiometry, the shorter was the incubation period and the more rapid the increase in viscosity and the onset of gelation. The one-step polycondensations were allowed to continue for 3 hrs after the gel-point before stopping. No change in properties was noted whenever the reaction was allowed to continue beyond this time. The level of network defects is easily estimable from the amount of swelling rigid networks experience upon equilibration in a good solvent such as DMAc. Correlations between swelling and structural features established[2-5] that deviations from stoichiometry, reduced C_0 and f, or increased ℓ_0, all contribute to the creation of more defective networks that swell in DMAc more than their "defect-free" counterparts. There are interesting differences in the definition of stoichiometry between one-step and two-step networks but here it suffices to state that the onset of gelation is remarkably rapid in the second step of the two-step procedure, sometimes in a matter of a few seconds, provided that the network is rigid and stoichiometry exists.

For the purpose of demonstrating the effects of changes in f, ℓ_0, and C_0 on the equilibrated concentration C and the shear modulus G at C_0 and at C, a series of one-step rigid network gels were prepared at a constant temperature from BTCA, DABA and NTPA. This series is typical of most of our previous one-step rigid gels.[2-4] The changed variables are shown in Table I below together with the resulting C and G:

Table I
Moduli of Rigid Network Gels Series 88 With Stiff DABA/NTPA Segments

Code	Functionality f	ℓ_0,Å	C_0,%	As-Prepared G, N/m²	Equilibrated C %	Equilibrated G, N/m²
88B	3	71	10.0	1.70×10^5	5.11	1.375×10^5
88A	3	52	10.0	2.37×10^5	5.48	1.81×10^5
88C	3	38.5	10.0	3.20×10^5	5.84	2.27×10^5
88D	4	52	10.0	4.535×10^5	7.67	3.54×10^5
88E	6	52	10.0	5.66×10^5	8.54	4.35×10^5
88F	3	52	7.5	0.399×10^5	2.82	0.296×10^5
88G	3	52	5.0	0.0087×10^5	Too soft to measure.	

Our previous work[2-5] is replete with additional examples demonstrating the same effects and confirming that the above dependencies hold true for both the one-step and two-step systems. They need not be shown here again.

SIMILARITIES

The one-step and two-step rigid polyamide gels are generally prepared in the concentration interval $C_0^* \simeq 2.0\% \leq C_0 \leq 12\%$. Below C_0^* no infinite network is formed. The highest concentration is a practical one, being the saturation point for many of the monomers employed in the polycondensations and of their polymeric products. The relatively low solubility of the monomers and polymers is caused by the non-solvent effects of the pyridine and TPP in the reaction mixtures. The reduced solvent quality is reflected by the fact that when the nascent gels are transferred to pure DMAc, they tend to swell significantly. We therefore conclude that during polymerization, the rigid polyamide gels are formed in a medium which is of modest solvent quality.

In gels prepared by both procedures, the stiff segments are short with the distance between branchpoints being, with few exceptions, an average one. The branchpoints in the networks are rigid. The overall rigidity of both types of gels is the most important factor affecting their properties. It leads to their modulus being far higher than that of comparable concentration gels of flexible networks[2,3] and to the modulus being generally temperature-insensitive.[2,3] At high concentration of network defects the modulus decreases and starts showing a very weak inverse dependence on temperature. The strains to failure of the rigid gels also depend on network defects: relatively perfect network gels fail at very low strains of about 2%. With increased defects the strain to failure increases to about 25%. The effects of network defects are clearly manifested in the swelling pattern of the gels during equilibration in the good solvent DMAc: the more defective the network the more the gel swells. The gelation time is also defect-dependent: at the same C_0 and T, gelation times for more defective networks are longer than for less defective networks.

In both one-step and two-step gels, stress is transmitted mostly along the stiff segments and through rigid branchpoints. A small fraction of the stress is propagated in the gelled network by means of excluded volume effects involving segments connected to the network at both ends or at only one end. This last feature is fundamentally different from the behavior of gelled flexible networks where dangling chain ends serve as diluent and in fact, reduce the modulus.

In para-substituted aromatic polyamide segments the amide groups are all coplanar. The planes of the aromatic rings are at ca. $\pm 30°$ relative to the plane of the amide groups. In that plane, the amide groups can be in anti or syn positions relative to one another. If all amide groups in a segment are of the same kind, i.e., all anti or all syn, then the segment will be completely straight. Because the two placements are almost identical in the free energy,[9] which is very small, both placements are incorporated into the network segments during the polymerization and the stiff segments may not be all straight. However, because the anti placements are slightly favored energetically[9] and because the stiff segments are rather short, the number of bends per segment is expected to be small. We believe that in a low-defect rigid network, an energy-efficient mechanism for segment bending or straightening is by anti-syn or syn-anti interconversions. The introduction of a single anti or a single syn placement in a segment of the opposite kind, creates a bend in the segment of about 20 degrees.[10] Two syn placements in a segment otherwise anti, or vice versa, creates in the segment a double-bend or a kink. In segments anchored into the gelled network at both ends by fully reacted branchpoints, the double-bends facilitate the displacement of one branchpoint relative to the other without either of them performing any "forbidden" torsional motions and with relatively small energy investment.[11] Small deviations from torsional angle energy minima may also participate in stiff polyamide segment bending and straightening. An important outcome of the above is that, in both one-step and two-step rigid network gels, stiff segments straight at rest can not stretch out further under stress. Segments containing bends may straighten out under stress but simple geometric calculations reveal that the straightening contributes very little to the macroscopic strain. We believe that these small strain contributions and the inability of straight segments to stretch further are the main reasons for the high modulus and very low ultimate strain of low-defect rigid network gels.[10,11]

To summarize: the behavior of rigid gels prepared by one-step or two-step procedure is dominated by the segments being short and stiff and the branchpoints rigid, by the presence of both straight and bent segments in the nascent network gels and the inability of straight segments to stretch under stress, and by the minor contributions to strain due to straightening of bent segments.

DIFFERENCES

We believe the most important difference between rigid one-step and two-step networks to be the size, spatial distribution and concentration of network defects. The difference is associated with the fact that the characteristic length in one-step networks is the diameter of the fractal polymers at the gel point while in the two-step networks the characteristic length is the length of the precursor chains used in the network-forming second step. To explain this statement, Figures 1 and 2 will be used to help us in describing the different morphologies we believe are present in the one-step and two-step pre-gel systems, how the "infinite" networks are formed, our perception of the final network morphologies and the difference in behavior between gels of low-defect one-step and two-step rigid networks. In both figures, the 3-dimensional reality was simplified by a 2-dimensional representation. In both the branchpoint functionality is 3. In Figure 1 the one-step network is described with the simplifying assumption that all the stiff segments are uniformly 38.5Å in length. This simplification does not change significantly the evolving structures and the final network morphology. In Figure 1A, relatively small highly branched polymeric species randomly nucleated in solution are shown. As the polycondensation reaction progresses, the polymeric species grow in size (Figure 1B) and in number until they finally connect. When the growing species throughout the reaction mixture connect with each other above a threshold concentration C_0^*, flow ceases and the system gels. Figure 1C represents the one-step system at the gel point. The polycondensation does not stop at this point. The network continues to grow until the reaction tapers off due to monomer depletion and reduced diffusivity of monomeric and oligomeric species. The final texture of the gelled network is shown schematically in Figure 1D. We have found that the growing rigid polymers are fractal in nature[3,4] and that the size of the highly branched rigid FPs at the gel point was not larger than about 300 to 400Å in diameter.[3] Network defects and small voids are expected to exist throughout the "infinite" network, some in, but most of them out, of the fractal polymers. The size and spatial distribution of the larger imperfections and voids, that can not be accommodated inside the FPs, are determined by the size and concentration of the FPs at the gel point: the larger the FPs or the more dilute the system, the larger and the more numerous are the network defects and voids, and the lower its modulus.[4] Because large oligomers and small FPs can hardly diffuse in the rigid gel, the continued network growth after the gel point is more uniform in nature and does not lead to preferential filling of the voids. The defects and voids, and their spatial distribution, are to a large extent trapped in the gelled rigid network.

When normal flexible networks swell, the imbibed solvent tends to preferentially concentrate in regions where the crosslinks concentration is the lowest[12,13] offering the least resistance to the swelling. This leads to enhanced network heterogeneity upon swelling. Recently, similar conclusions were theoretically obtained for swollen networks consisting of rigid rods connected by freely-jointed tetrafunctional branchpoints.[14] In the case of rigid networks, where the segments can not stretch and branchpoints can not deform in order to accommodate the imbibed solvent, the concentration of defects, imperfections and voids in the system determined the network swelling: the more imperfect the network, the more solvent it imbibed and the more it swells. On the molecular level, the solvent preferentially collects at and near voids, defects, etc., where the highest concentration is found of "empty space", partly reacted branchpoints and segments connected to the network at only one end. The distances between the larger pockets of high solvent concentration are dictated by the size of the FPs at the gel point and are of the same order of magnitude as the diameters of the FPs.

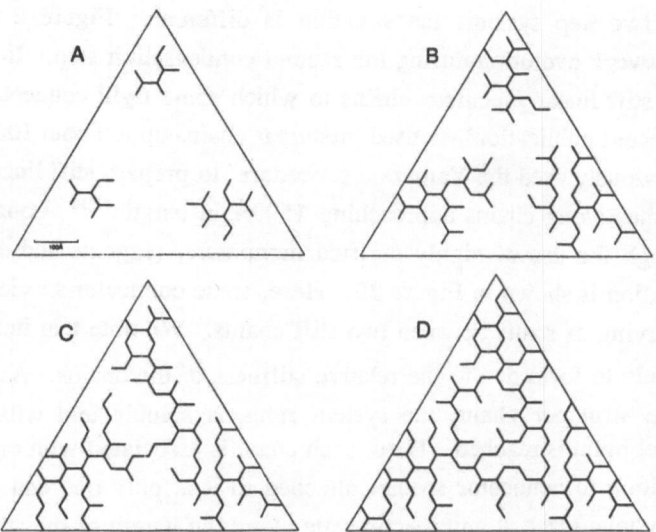

Figure 1.   Evolution of one-step rigid network in solution.   (A) Small fractal
             polymers after nucleation. (B) Larger FPs in solution prior to gelation.
             (C) An "infinite" network during formation at the gel point. (D) The
             mature network at the end of the reaction, monomers and oligomers
             depleted. Scale bar: about 100Å.

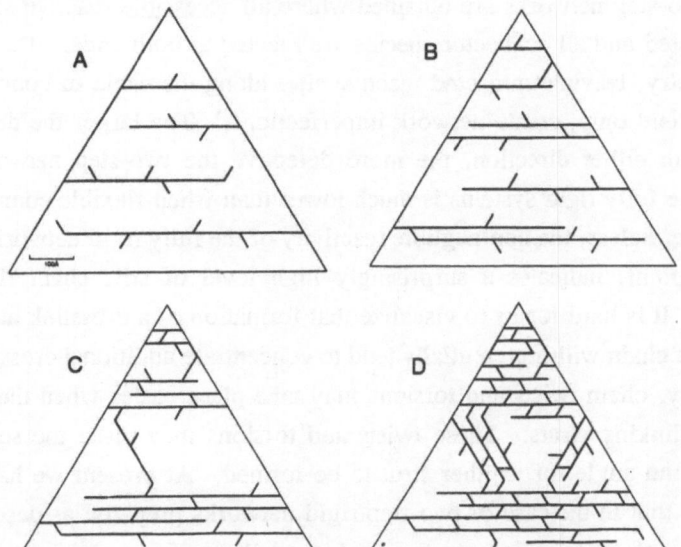

Figure 2.   The second step in the creation of two-step rigid network in solution.
             (A) Individual linear stiff chains with some connector species attached at
             one end. (B) Some linear chains are connected by struts, polymer still in
             solution. (C) An "infinite" network being formed at the gel point.
             (D) The mature network at the end of the second step reaction. Scale
             bar: about 100Å.

In rigid two-step systems the situation is different. Figure 2 schematically describes the network evolution during the second condensation step. In panel 2A we start with long, stiff linear precursor chains to which some rigid connector species are attached. In a recent publication[5] we used precursor chains up to about 1000Å in length, but we have previously used the Yamazaki procedure[7] to prepare stiff linear polyamides with reactable sites along chains approaching 1500Å in length.[15,16] Longer chains are obtainable through the use of highly purified monomers, reagents and solvents. The onset of aggregation is shown in Figure 2B. Here, some connector species are attached at both ends, serving as struts between two stiff chains. We note that intra-chain struts are highly unlikely to form due to the relative stiffness of the chains. As long as there is less than one strut per chain, the system remains soluble and will not gel. In Figure 2C the gel point is reached. Here, each chain is associated with one strut on the average, in addition to connector species attached to it at only one end. Because the crosslinking may take place at any reactive site along the length of the stiff chains, and because only one strut per chain is sufficient in order to reach the gel point, gelation is extremely fast under stoichiometry and concentration C_0 conditions comparable with one-step systems. After gelation, additional struts are formed in the network from connector species reacting with stiff chains, until the reaction mixture is depleted of reactable species. This stage is shown in Fig. 2D. It is important to recognize that the lowest defect rigid two-step networks are obtained where all accessible reactive sites along the chains are reacted and all connector species are reacted at both ends.[5] Deviations from this stoichiometry, leaving unreacted reactive sites along the chain or connector species reacted at one end only, create network imperfections.[5] The larger the deviation from stoichiometry in either direction, the more defective the two-step network is.[5] The reactivity of the fully rigid systems is much lower than when flexible connector species are used. Nevertheless, the non-neglible reactivity of the fully rigid networks, especially after the gel point, indicates a surprisingly high level of stiff chain flexibility and deformability. It is hard for us to visualize that formation of a crosslink at one position on a rather stiff chain will preferentially tend to concentrate additional crosslinks nearby. On the contrary, chain twists and torsions may take place easier when they are farther from the crosslinking struts. These twists and torsions may place the segment in the right position and angle for another strut to be formed. At present we have no proof, but we believe that in the case of two-step rigid networks prepared as described above, the distribution of crosslinking struts is substantially uniform along the stiff chains. Therefore, besides the length ℓ_0 which can be made similar in both systems, the characteristic length in two-step networks is the chain length, L, which is much larger than the diameter of the fractal polymers of one-step networks at the point of gelation. In one-step network gels the defects and voids are concentrated, not exclusively, at FP interfaces while in two-step systems the defects are distributed more uniformly and are likely to be smaller in size. They may be associated with misfit of chain reactive sites and connector position or angle, steric hindrance limiting site accessibility, and lack of stoichiometry.

When rigid networks prepared at C_0 swell in a good solvent to C, the shear modulus, G, changes. In highly-defective networks, the modulus of the gels decreases upon swelling, independently of whether the network was prepared by one-step or two-step procedures. When the networks are well-formed with only low levels of defects, then the preparation methods become important: the modulus of the rigid one-step gels gets smaller with swelling while the modulus of rigid two-step network gels increases upon swelling. This unique behavior was observed[5] only for low-defect networks of polyamides prepared by the two-step method. In low-defect networks, upon equilibration in DMAc, the amount of swelling itself was found to depend also on the preparative procedure. In one-step networks, the swelling from C_0 to C was generally larger than for two-step systems. Furthermore, the manner of failure was different: the one-step gels failed by highly brittle fracture while the two-step analogues were much more resilient and did not fail in such a brittle fashion. The above differences in swelling and in the response to stress reflect what we believe to be the fundamental difference between gels of rigid one-step and two-step networks: the one-step networks are much more defective than two-step networks of comparable C_0, f and ℓ_0. The differences in size, distribution and distances between defects in the two kinds of gelled networks, together with their possible causes, were mentioned above. In the case of two-step networks, we may approach network perfection by fully reacting stoichiometric amounts of accessible reactive chain sites with all ends of the strut-forming connector species; the defects are small and more or less uniformly distributed. The increased modulus upon swelling of low-defect gels of rigid two-step networks is due to the coexistence of two important factors: in such gels there are not enough defects and voids to disrupt the network to such an extent that a sufficient amount of weakening solvent pockets is created, and, at the same time, the bent stiff segments straighten out to their limit in order to accommodate the largest possible amount of swelling solvent. The straightening mechanism is probably by syn-anti or anti-syn interconversions or small deviations from torsional angle energy minima. In the swollen gel the population of straight segments increases at the expense of the bent segments. Because the straight segments can not stretch further, the resistance to deformation of the swollen gel is higher than that of the nascent gel, leading to higher shear modulus. As the number of defects in two-step networks increases, the increase in modulus with swelling fades away and is gradually replaced by the usual decrease in G with swelling, as was observed in the one-step rigid network gels and is the case with all gels of flexible networks.

One final point. There were recently described in the literature two novel one-step non-polyamide rigid networks.[17,18] They both show an unexpectedly large degree of swelling when immersed in their respective good solvent. A careful examination of the structures convinced us that the likeliest reason for most of the swelling can be traced to network defects created by incomplete reaction of the functionalities. A two-step network made in a rather complicated way from flexible linear chains and rigid struts[19,20] also shows an "unexpected" amount of swelling. We believe that in this case, network defects may be traced to connector species attached to a flexible chain at only one end. The likely reason for a major part of the swelling of all these networks is, hence, the same as for our rigid polyamide networks.

REFERENCES

1) (a) P.J. Flory, "Principles of Polymer Chemistry"; Cornell University Press, Ithaca, N.Y., 1953; chapters 3 and 4. (b) R.W. Lenz, "Organic Chemistry of Synthetic High Polymers"; Interscience, N.Y., 1967; Sections II and III. (c) B. Vollmert, "Polymer Chemistry"; Springer, New York, 1973; chapter 2. (d) G. Odian, "Principles of Polymerization"; Wiley-Interscience, N.Y., 1981; chapters 2 and 3.

2) S.M. Aharoni and S.F. Edwards, Macromolecules 22, 3361 (1989).

3) S.M. Aharoni, N.S. Murthy, K. Zero and S.F. Edwards, Macromolecules 23, 2533 (1990).

4) S.M. Aharoni, Macromolecules 24, 235 (1991).

5) S.M. Aharoni, Macromolecules 24, 4286 (1991).

6) M. Daoud, E. Bouchaud and G. Jannink, Macromolecules 19, 1955 (1986).

7) N. Yamazaki, M. Matsumoto and F. Higashi, J. Polym. Sci. Polym. Chem. Ed., 13, 1373 (1975).

8) S.M. Aharoni, W.B. Hammond, J.S. Szobota and D. Masilamani, J. Polym. Sci. Polym. Chem. Ed., 22, 2579 (1984).

9) J.P. Hummel and P.J. Flory, Macromolecules 13, 479 (1980).

10) S.M. Aharoni, G.R. Hatfield and K.P. O'Brien, Macromolecules 23, 1330 (1990).

11) S.M. Aharoni, Intern. J. Polym. Mater., in press (1991).

12) J. Bastide and L. Leibler, Macromolecules 21, 2647 (1988).

13) J. Bastide, L. Leibler and J. Prost, Macromolecules 23, 1821 (1990).

14) M. Rubinstein, L. Leibler and J. Bastide, preprint (1991).

15) S.M. Aharoni, Macromolecules 15, 1311 (1982).

16) S.M. Aharoni, Macromolecules 20, 2010 (1987).

17) T.M. Miller and T.X. Neenan, Chem. Mater., 2, 346 (1990).

18) O.W. Webster, F.P. Gentry, R.D. Farlee and B.E. Smart, <u>Polymer Preprints</u>, <u>32(1)</u>, 412 (1991).

19) V.A. Davankov and M.P. Tsyurupa, <u>Pure & Appl. Chem.</u>, <u>61</u>, 1881 (1989).

20) V.A. Davankov and M.P. Tsyurupa, <u>Reactive Polym.</u>, <u>13</u>, 27 (1990).

NEW INSIGHTS INTO AROMATIC POLYAMIDE NETWORKS FROM

MOLECULAR MODELING

W. B. Hammond, S. M. Aharoni and S. A. Curran

Allied-Signal, Inc.
Research and Technology
Morristown, N.J. 07962-1021

Introduction

Historically, the behavior of linear polymers was modeled on one of two extremes: either highly flexible coil-like chains mathematically describable by means of Gaussian statistics and perturbations[1-7], or stiff rod-like macromolecules incapable of bending and flexing[8-10]. Later, models were put forth describing relatively stiff chains in terms of either wormlike chains[11-13] with constant curvatures, or chains composed of rods connected by flexible spacers[13,14], in order to close the gap between the highly flexible and rod-like extremes. Because of ease of modeling, the para-aromatic polyamide liquid crystalline polymers were treated initially in terms of rod-like chains[15,16]. Among these polymers one may enumerate poly(p-benzamide)(p-PA), poly(p-phenylene terephthalamide)(p-PT), poly(p-benzanilide terephthalamide)(p-BT) and poly(p-benzanilide nitroterephthalamide)(p-BNT).

Experimental results have shown, however, that the aromatic para-substituted polyamides deviate substantially from the rodlike model and are far better described in the terms of the wormlike chain model. Primary among these results are chain persistence lengths[11,17-26], a, which appear to have settled recently in the interval of 150 < a <300Å for p-PT and about twice these values for p-BA[22]. Values at the lower end of these ranges were measured on aromatic polyamides containing occasional ring substituents such as nitro[23] or phenyl [26] groups. Considering the fact that the projected length of a monomeric unit of para-substituted aromatic polyamides

Synthesis, Characterization, and Theory of Polymeric Networks and Gels
Edited by S.M. Aharoni, Plenum Press, New York, 1992

93

is of the order of 6.45Å [27-29], a persistence length of 150 to 300Å corresponds to only 23 to 46 such units. This reflects a substantial backbone flexibility and a large deviation from the rodlike model. As will be shown below, the reasons for this flexibility are complex. In this paper, some local motions and structural features will be presented and discussed, which together may explain the flexibility of stiff aromatic polyamide linear chains and network segments.

Polymer Structure

The polyamides to be studied in this work are schematically described by the following structures:

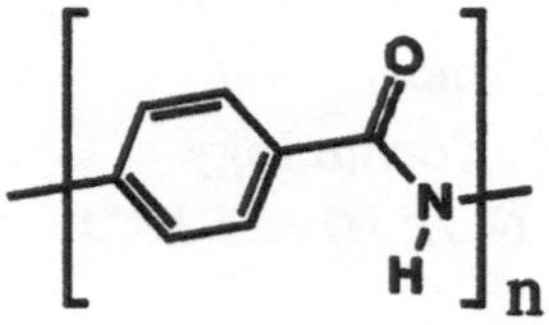

p-BA

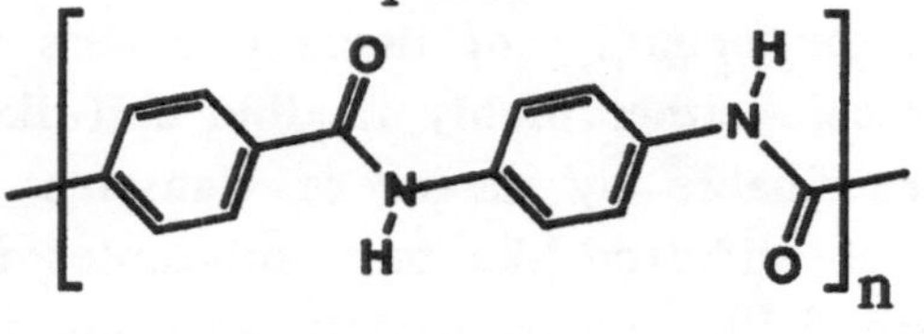

p-PT

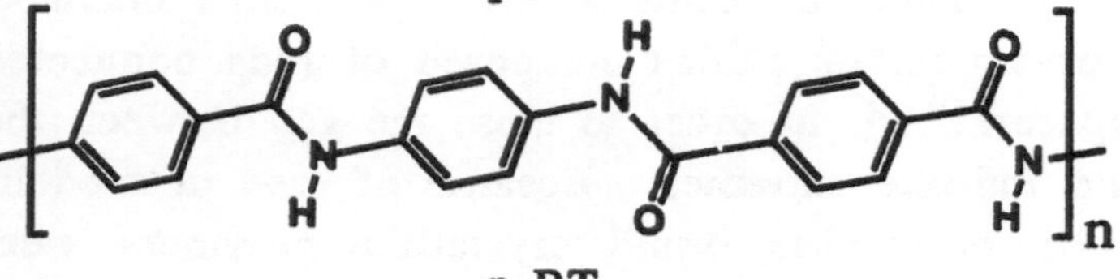

p-BT

The torsional and valence bond angles are defined in Figure 1 below:

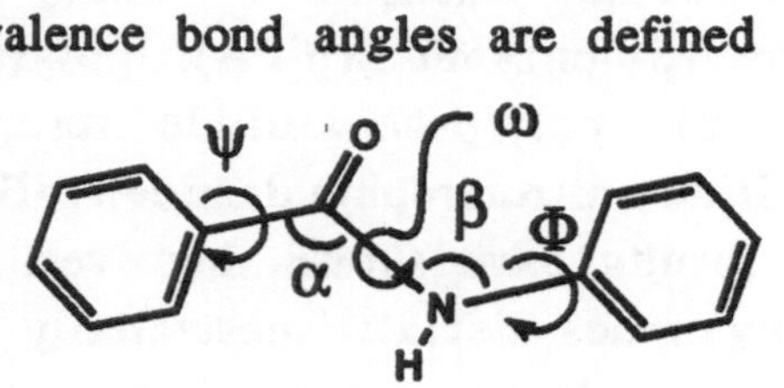

Figure 1

The directions of rotational axes in all the chain units are almost colinear and as a consequence, the entire chain acquires the shape of a crankshaft[30]. However, because neither angle α nor β equals 120° and $| \beta - \alpha | \neq 0$, *vide infra*, there is a curvature imparted to the overall chain direction, a curvature reflected in the value of the persistence length, **a**. The magnitude of **a** is inversely dependent, hence, on the difference between α and β. Using a difference of 10°, Flory and associates [31] calculated a = 410Å and 435Å for p-BA and p-PT, respectively. Because in their calculations they did not consider other bending modes these results should be regarded as the upper limit for the respective persistence length[18]. The experimental results cited above[19-26] support this assertion. The important conclusion from the above is that $|\beta - \alpha| \approx 10°$ and values of $|\beta - \alpha|$ significantly smaller or larger than ≈10° are in disagreement with experimental results [29,32] and should not be used for modeling of the aromatic polyamides.

In the crystalline state practically all the amide groups are hydrogen bonded (H-bonded) to one another. In order to optimize the bond strength, the H-bonded amide groups lie in a plane and the aromatic rings are twisted by the angles ψ and ϕ out of that plane. Crystallographic data teaches us that the values of both ψ and ϕ are about 30 degrees, and that these values are about the same for polymeric[27,28,29] as well as monomeric or oligomeric species[23,33,34]. It is important to recognize that in aromatic amide systems where $|\beta - \alpha| << 10°$, the aromatic rings are twisted out of the H-bond plane at angles substantially different from 30° [35]. In the common situation where $\psi \approx \phi \approx 30°$ the hydrogen bond energy is maximized at about 7 kcal/mol [36] and the distance between the axes of chains participating in H-bonds is about 5Å. If the angles between the aromatic rings and the H-bonded plane decrease below 30°, the length of the H-bonds increase and the bond energy decreases. When the aromatic rings are coplanar with the amide groups in the H-bond plane, then the distance between chain axes grows to about 7Å [37] and the H-bond energy is greatly reduced. From the above we conclude that in order to maximize the H-bond energy in the crystalline state, small deviations are allowed from the values of the angles α, β, ϕ and ψ in aromatic polyamide chains which are not H-bonded to one another. Such chains or chain-segments exist in dilute or semi-dilute solutions of long linear polyamides and in rigid isotropic network gels characterized by stiff segments connected by rigid branchpoints[38,39]. In both systems interchain or intersegmental H-bonds are replaced by chain-solvent or segment-solvent H-bonds.

It is well accepted[27-35] that in p-BA, p-PT and p-BT the amide group is present in a planar *trans* and not in a *cis* conformation. The placement of amide groups on both sides of each aromatic ring is not so clear cut. In the aromatic polyamide chains, the amide groups

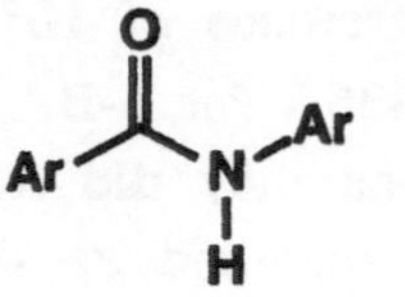

trans 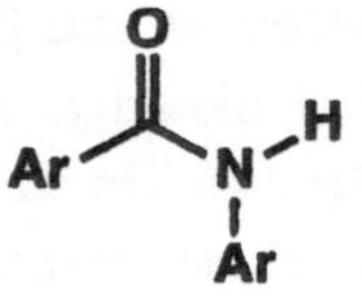   *cis*

can adopt either an *anti* or a *syn* conformation. However, even this is
an oversimplification. Since each substituent, carbonyl or amide
group, forms a ±30° angle with the benzene ring to which it is
attached, the para substituents can adopt two *syn* and two *anti*

anti

syn

conformations. We illustrate this for the terephthalamide group. For
the *anti* conformation, one group can be above the plane of the
phenyl group and the second group below the plane. We designate
this conformation *(-)anti.* (the dihedral angle between the amide
carbonyl groups is 180°). A second conformation *(+)anti*, places the

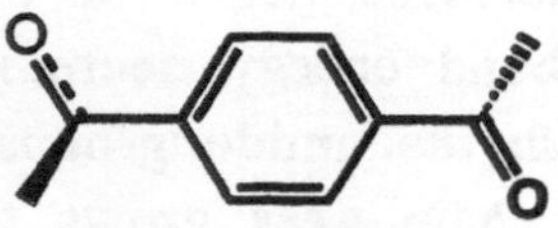

(-)anti (180°) 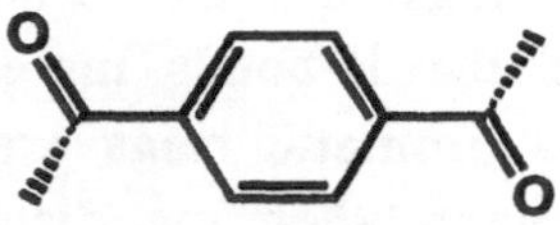   *(+)anti* (120°)

carbonyl oxygens on the same side of the phenyl plane (the dihedral
angle between the carbonyl groups is 120°). In the same manner we

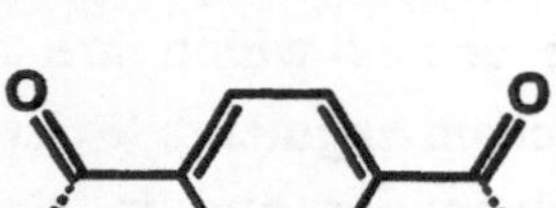

(+)syn (0°)    *(-)syn* (60°)

represent the *syn* conformations as *(+)syn* and *(-)syn* with dihedral
angles between the carbonyl groups of 0 and 60°, respectively. A
similar analysis can be applied to the p-benzamide and p-phenylene
diamine units where the conformations are defined between
carbonyl and NH and between NH and NH, respectively.

It has been shown by Tashiro et al.[29] that in the crystalline state, the placement of the amide groups in p-BA is exclusively

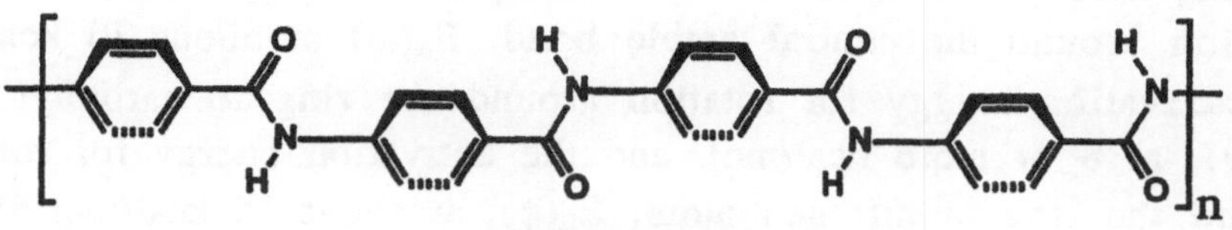

(+)syn -**p-BA**

(+)syn and in p-PT is exclusively *(+)anti* with respect to the para

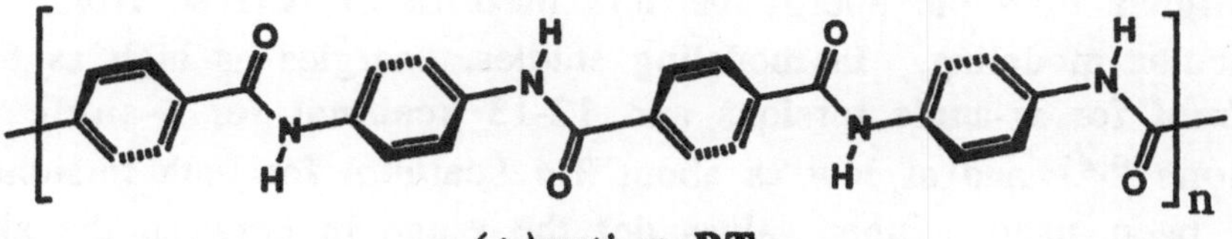

(+)anti -**p-PT**

substituents attached to each ring. The exact placements of the amide groups in p-BT is not known to us. However, because solutions of p-BT require much higher concentrations to become liquid crystalline than is required for p-BA[23], we can safely assume that the persistence length of p-BT is significantly shorter than that of p-BA. This may be brought about by having a mixture of both *(±)anti* and *(±)syn* placements along the p-BT chain.

An important observation[40,41] is that if a single *syn* placement is inserted into a chain or segment which is otherwise wholly *anti*, a bend in the chain of about 20 degrees results. The very same bend happens when a single *anti* placement is inserted in a fully *syn* chain. From our discussion above, this is an oversimplication, since one must consider *(±)syn* and *(±)anti* conformations and their interconversion. The existence of such bends can, and most likely does, contribute to a decrease of the chain persistence length from the upper limits mentioned above. Hummel and Flory[32] have shown that the difference in dipole-dipole interactions between the *anti* and *syn* placements in a polyamide chain is on the order of 0.3 kcal/mol. This indicates that there is practically no preference for *anti* over *syn* placements and, in the amorphous state or in solutions of isotropic network gel, such placements may be built at random into the chain during synthesis or thereafter. The existence of both such placements allows for some chain flexibility and a variety of deformation modes in crosslinked networks [41]. Conversions from one placement to another, which may be called amide-flips, require the amide groups to perform rotations between *syn* and *anti* conformers of 120° relative to an attached aromatic ring and to overcome energy barriers of substantial heights. Experimental studies, mostly NMR, IR and thermal, place the activation energy for rotation around the central amide bond, $E_a(\omega)$ at about 20 kcal/mol., the activation energy for rotation around the ring to carbonyl bond, $E_a(\psi)$, at 8 or more kcal/mol. and the activation energy for rotation around the ring to nitrogen atom, $E_a(\phi)$, at about 15 kcal/mol [40]. The high $E_a(\omega)$ values render *trans* to *cis* amide interconversions highly improbable. In general, the experimental values of $E_a(\psi)$ and $E_a(\phi)$ are higher than the energy barriers used in or derived from molecular modeling. In modeling studies, energies as high as 6 to 9 kcal/mol for ψ-angle torsions and 12-13 kcal/mol for ϕ-angle torsions[29,42] and as low as about 3-4 kcal/mol for both torsions[43,32] have been used. Other values dot the range in between the above extremes[32,44,45]. When torsional energies are considered, care should be taken to verify that the angles of $\psi \approx \phi \approx 30°$ emerge from the calculations and that barriers at 0° and ± 90° of substantial heights occur for both ψ and ϕ angle torsions.

As will be shown below, 120° amide flips (*anti-syn* or *syn-anti* interconversions) are rare but do occur on occasion. Their frequency appears to be higher in stiff polyamide segments in rigid networks, whose ends are constrained from performing torsional movements in space.

During the course of the present work, we have found by modeling that a more frequent bending mode is for an aromatic ring to flip abruptly by about 60°. Here, a ring that was at an angle of about -30° relative to both adjoining amide groups, jumps over and settles at an angle of about +30° relative to both amides. By so doing a bend is introduced into the chain and a dramatic stress relaxation occurs at the same time. These 60° ring flips will be further discussed below.

In the calculations we have found that a very large fraction of the energy associated with overall chain bending and twisting is directly attributable to small deviations from the energy minima of the ψ and ϕ torsional angles. It is important to note here that when the chains are assumed to have the angles $\alpha = \beta = 120°$, large deviations of ψ and ϕ from their minima are required to occur in a concerted manner, for relatively small bends to appear in the stiff chain[40,41]. When $\alpha \neq \beta \neq 120°$ and $|\beta - \alpha| \sim 10°$, then small deviations from the minima of ψ and ϕ produce rather large chain bending, twisting and flexing. When such a deformation mode combines with the 60° ring flips and with the relatively rare 120° amide flips, and other possible deformations not discussed here, a remarkable level of flexibility emerges for stiff polyamide chains. Similar degrees of flexibility are expected to be present in aromatic polyester chains.

Simulations

Three sets of simulations are reported here. In the first, the amorphous builder module in the molecular modeling program PolyGraf[46] was used to calculate the effect of different valence and torsion angles on the persistence lengths of p-BA, p-PT and p-BT. In the second set of simulations, the effect of chain bending on the energy of short aromatic polyamide chain segments was studied. The importance of chain aggregation on the "rigid-rod" properties of aromatic polyamide chains is demonstrated. These simulations also give interesting clues to the modes of relaxation available to aromatic polyamide chains under stress. In a third set of simulations, short polyamide chain segments with the ends constrained to simulate segments of aromatic polyamide rigid networks[38,39,40] were subjected to bending and twisting motions while the energy and conformations of the chains were monitored. These simulations can be compared with relaxation processes observed by solid state NMR in aromatic polyamide networks[47].

Persistence Lengths

Models for chain segments of poly(benzamide)(p-BA), poly(p-phenylene terephthalamide)(p-PT), and poly(p-benzanilide terephthalamide) (p-BT) were constructed from benzamide units in various head to head and head to tail combinations using the polymer builder in PolyGraf. The benzamide building unit had fixed bond distances and valence (α and β) angles, as defined for benzanilide (BA) in Figure 1. Standard bond distances were taken from crystallographic data[48]. The starting valence and torsion angles were extracted from the work of Hummel and Flory[32] who tabulated data from the crystal structures for a number of model compounds for aromatic polyamides. They noted that the values for ϕ and ψ for most aromatic polyamides range from 30-35°.

In our first study, extended chain segments (up to 72 aromatic rings) were assembled with the torsion angles ϕ and ψ randomly assigned with equal probabilities to four states, ± 30 and

± 150° using the amorphous builder in PolyGraf. The amide group was fixed in the *trans* conformation. The ± 150° angles were included to sample the two *syn* and two *anti* conformations that relate the p-substituents of each benzene ring. 1000 to 4000 random chains each of p-BA, p-PT and p-BT were generated and the mean-square-end-to-end distance, $<r^2>_0$, was calculated for several chain segment lengths. The persistence length, **a**, was calculated for each polymer using eq.1 where L is the contour

$$<r^2>_0 = 2a[1 - (a/L)(1- e^{-L/a})] \qquad (1)$$

length of the polymer. L = R•n with R being the incremental length of a single monomer unit and n being the number of monomer units in the chain. $<r^2>_0$ was calculated for n = 42 to 72 in 6 unit increments and **a** and R were calculated by a least squares fit to $<r^2>_0$ for the six segment lengths. The calculations were repeated for several values of α and β, and for several distributions of ψ, φ and ω. The results for p-BA are summarized in Table 1. The distributions for ψ, φ and ω arise because the amorphous builder in PolyGraf can assign a range of angles about each of the chosen states, e.g. 30°±10°. The results for p-PT and p-BT were nearly identical to those for p-BA (p-PT1 and p-BT1 are included for comparison).

Our calculated persistence length for aromatic polyamides was found to be relatively insensitive to the value (compare **a** for p-BA1 and 2, p-BA6 and 7, and p-BA8 and 10 with respect to φ) and distribution (compare **a** for p-BA5 and 6 with respect to ω) of the torsion angles φ, ψ, and ω. The persistence length was also insensitive to the ordering of the aromatic amide units (similar results for p-BA, p-PT and p-BT). However, **a** was very sensitive to the values of α and β with the difference between α and β being the critical parameter. Experimental values for the persistence length of p-PT and p-BT fall in the interval 150 < **a** < 300Å[17-22] with values for p-BA being about twice the values for p-PT and p-BT[22]. In calculating p-BA9, Table 1, we changed the weighting of the torsion angle φ to favor the conformation which orients the dipoles of alternating amide groups in opposite directions by 0.3 kcal/mole, a value suggested by Hummel and Flory[32]. While this is a small interaction energy at 300°K, it is sufficient in our calculations to increase the chain persistence length for p-BA by 50%. We have not extended

Table 1

Persistence Lengths for p-BA

	α	β	ψ	ω	φ	R	a
p-BA1	117	123	30±10	0±0	30±10	6.43	860
p-BA2	117	123	30±10	0±0	10±10	6.43	995
p-BA3	117	123	30±10	0±5	30±10	6.43	940
p-BA4	117	125	30±10	0±0	30±10	6.43	580
p-BA5	115	125	30±10	0±0	30±10	6.43	380
p-BA6	115	125	30±10	0±5	30±10	6.43	385
p-BA7	115	125	30±10	0±5	10±10	6.51	359
p-BA8	115	127	30±10	0±5	30±10	6.43	258
p-BA9*	115	127	30±10	0±5	30±10	6.42	375
p-BA10	115	127	30±10	0±5	10±10	6.49	253
p-PT1	115	127	30±10	0±5	30±10	6.43	288
p-BT1	115	127	30±10	0±5	30±10	6.48	256

* weighted for dipole orientation by 0.3 kcal/mol

this weighting to p-PT or p-BT to see if a similar effect is observed there, nor can we explain why the dipole orientation should be stronger in p-BA than in p-PT or p-BT to account for the apparent increase in persistence length for p-BA relative to p-PT and p-BT. The calculations suggest that $| \beta - \alpha |$ must be equal to or greater than 10° in real aromatic polyamide chains in order to give persistence lengths consistent with experiment.

In order to view the distribution of end-to-end distances for various chain segments, we have calculated a histogram for each of six chains with end-to-end distances assigned in 10 Å increments[49]. Figure 2 is a line plot of the end-to-end distance (Å) vs frequency in relative units for the six chains, the shortest of which is 42 and the longest 72 repeat units. As the chains become longer, the end-to-end distances broaden significantly, due to the wider range of conformations that can be sampled by longer chain segments.

Chain Bending:

In our next study, we simulated the process of aromatic polyamide chain bending by building a short chain segment of p-BT with phenyl caps as shown in Figure 3. We then minimized the energy while applying a distance constraint between the terminal

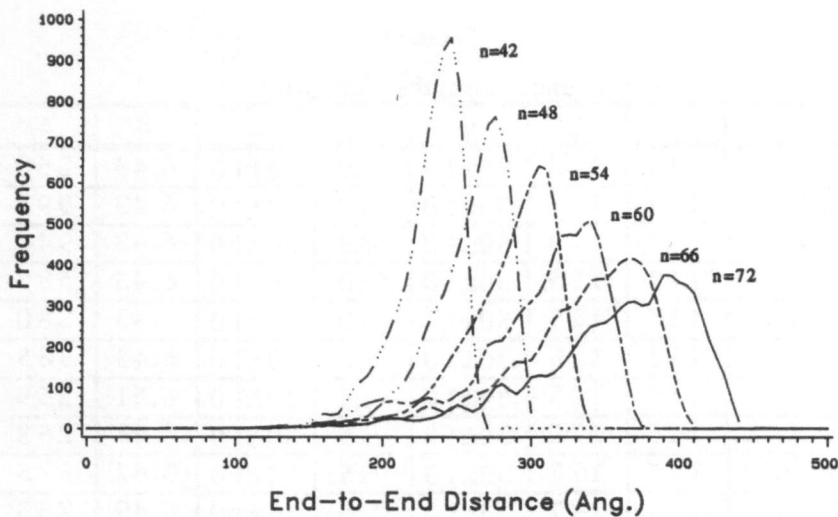

Figure 2. Aromatic Polyamide Persistence lengths for various chain lengths of p-BA8, $\alpha=115°$, $\beta=127°$.

phenyl groups at the carbons marked with asterisks. The minimized energies were calculated with molecular mechanics using a modified version[50] of the Dreiding parameter set[51] in PolyGraf. The Dreiding

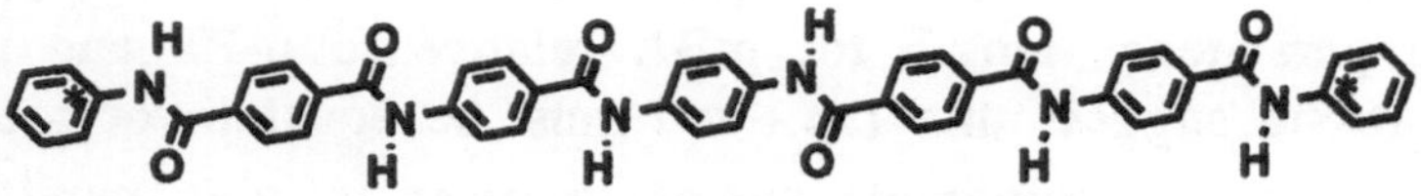

Figure 3

force field gave bond distances for a single benzanilide (BA, Figure 1) that were in reasonable agreement with the crystal structure[52], although the angles α and β were somewhat larger than observed experimentally. However, using the Dreiding parameters in PolyGraf, the torsional angles ϕ and ψ calculated for BA were each 90° compared to 31.3 and 31.6° respectively in the crystal. The corresponding torsion angles are found to be between 30° and 35° in a number of crystal structures for aromatic amide model compounds.

Hummel and Flory[22] and Tashiro et al.[29] have independently estimated the shape of the torsional potential function for ψ and ϕ in aromatic polyamides. They agree on the equilibrium values for ψ and ϕ of 30-35° but differ somewhat on the height of the energy barriers to rotation. Laupretre and Monnerie[44], using PCILO methods calculate a low rotational barrier for ψ (2 kcal/mole) and a higher barrier for ϕ (9 kcal/mole). Recently, Scheiner[53] has carried out large basis set *ab initio* calculations on BA. His calculations give minimum energies for ψ and φ at 35 ° and 10°, respectively with rotational barriers for ψ of 3.6 kcal/mol. and for φ of 2.3 kcal/mol. at 30°. We have constructed two parameter sets for our simulations (dashed lines in Figure 4) which were obtained by fitting a six term Fourier series to the torsion functions of Scheiner (Figure 4, a and b, solid lines) and Tashiro (Figure 4, c and d, solid lines).

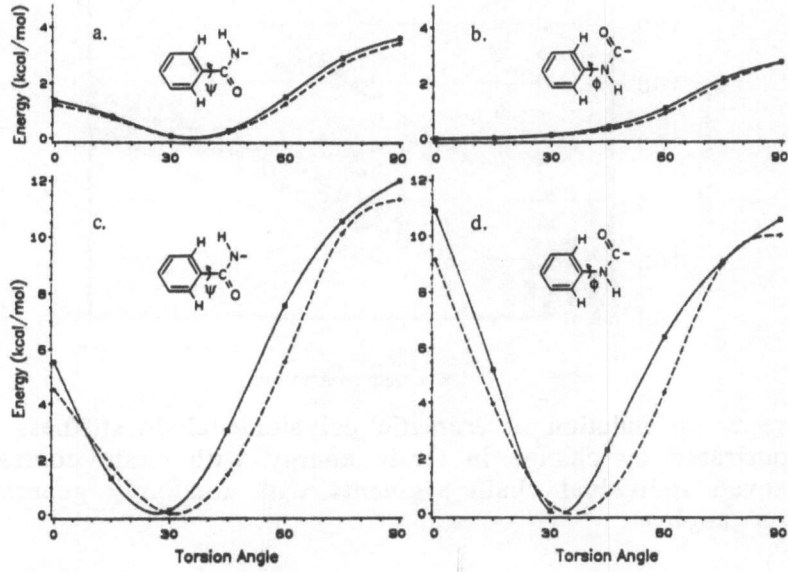

Figure 4. The solid lines in a and b are from the data of Scheiner[53]. The solid lines in c and d are from the data of Tashiro[29]. The dashed lines are our best fits to the data using BA as a model.

In order to study details of aromatic chain bending, seven chain segments of the structure depicted in Figure 3 were constructed with the values for ψ and φ randomly assigned to ±30°, or 150° and the energies of the chain segments minimized. Next, a distance constraint was imposed between the phenyl groups at the carbons marked with asterisks. This distance was reduced in 1% increments and the geometry of the structure was minimized at each step. The results of these simulations using the Scheiner parameter set are shown in Figure 5 where the total molecular mechanics energy is plotted against percent decrease of the initial end-to end distance for each chain. The results are presented for seven randomly generated chains. Six of the chains undergo small adjustments in energy in the first 5 percent of contraction with the energy changing uniformily in the remainder of the simulation. Chain segment 4 was built in a conformation where simple *syn-syn* interconversions were not available to relieve strain energy until 20 percent chain contraction had been reached. At this point the barrier

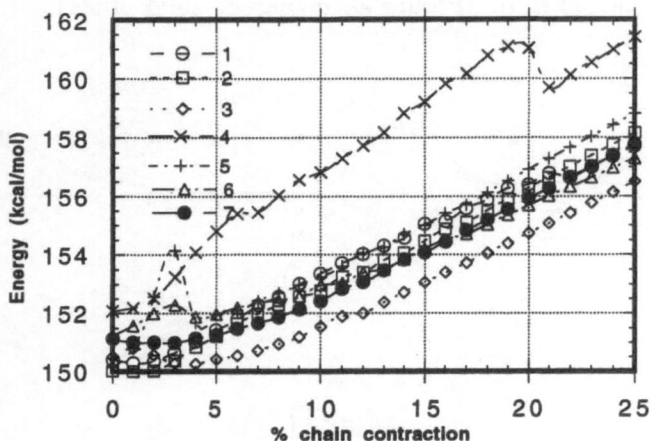

Figure 5. Simulation of aromatic polyamide chain stiffness as demonstrated by change in strain energy with chain contraction for seven individual chain segments with randomly generated torsion angles.

for a *(+)syn-(-)syn* conversion was exceeded at one amide-NH-phenyl link along the chain which allowed relaxation all along the chain to a lower energy state. When the same simulation was run using the Tashiro parameters, a steady increase in energy was observed for all seven chain segments with no *(+)syn-(-)syn* interconversions being observed. This is not surprising since the potential barriers are significantly higher for the Tashiro parameters. For both the Scheiner and Tashiro parameters most of the strain energy introduced on chain bending went into the torsion terms of the force field and for each chain the total change in energy was between 10 and 15 kcal/mole.

Using the force field and simulation methods described above, we sought to understand the effect of aggregation on the rigidity of aromatic polyamides by measuring the energy needed to bend a polyamide chain in the presence of a second, third and fourth chain segment as depicted in Figure 6. The second and third chains were added in such a manner as to form a sheet structure with hydrogen bonding between the chains. The fourth chain, shown in gray in Figure 6 was added below the plane of the sheet to make a three-dimensional structure. In the simulations, the bending constraint

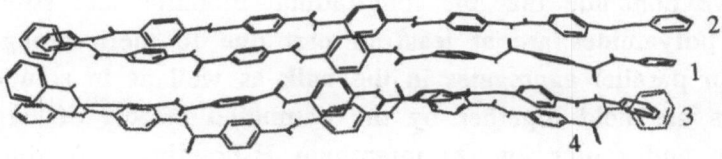

Figure 6. Constraints were applied to chain 1. In successive simulations, chains 2 and 3 were added in a manner to allow H-bonding between chains in a layer. Chain 4 (in gray) was added parallel to chain 1 but out of the plane of 2 and 3.

was applied only to chain segment 1. Figure 7 shows the effect of aggregation on the energy needed to bend an aromatic polyamide chain segment. Addition of a second chain segment increases the bending energy by less than twice that for a single chain. This is

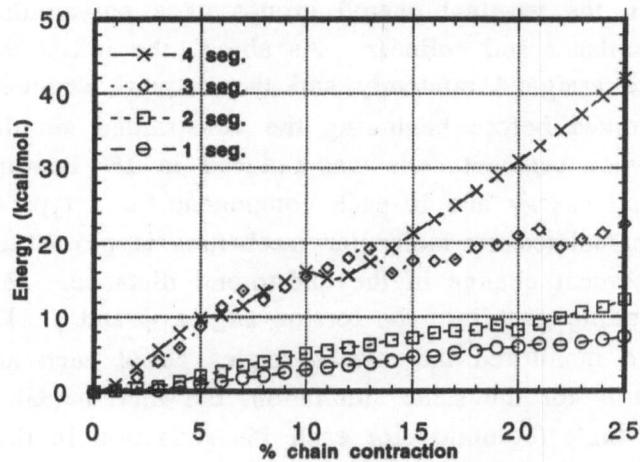

Figure 7. Simulation of aromatic polyamide chain stiffening with aggregation as discussed in text.

because the second chain does not have to yield as much as the first in response to the bending constraint. Addition of the third chain produces an initial stiffening of the polyamide sheet, but at about ten percent contraction, the sheet finds relaxation modes through *(±)syn-(∓)syn* interconversions. Addition of the fourth chain makes the aggregate three dimensional and the strain energy continues to grow to twenty-five percent contraction. The strain energy for four chains is more than five times the energy of a single chain. From these results, we conclude that the longitudinal modulus and strength of aromatic polyamides are at least in part due to their strong tendency to align in parallel aggregates in the bulk as well as in solution. Such aggregates are held together by the combined effects of large shape anisotropy and rather intense interchain H-bonding. A similar correlation, between elevated longitudinal modulus and the multi-chain aggregation of aromatic polyamides, was recently reported by Yang and Hsu[54] and Rutledge and Suter[55]. An appreciation of the intensity of the interchain H-bond can be gathered from the fact that the solubility parameter of aromatic polyamides in the absence of interchain H-bonds was determined to be 23.0 $(MPa)^{1/2}$ [56] while that in the presence of such H-bonds was calculated to be 32.0±0.8 $(MPa)^{1/2}$ [42], a 40 % difference!

Constrained Models for Networks

Since we are interested in understanding chain motion in rigid networks, our simulations were continued by placing additional constraints on the terminal phenyl groups of a polyamide segment to keep them coplanar and colinear. As above, the initial torsion angles ψ and ϕ were assigned randomly and the network segment was energy minimized before beginning the constrained simulation. The distance between segment ends was reduced in 1% increments. The change in total energy and in each component for a typical p-BT segment as calculated by molecular mechanics is plotted in Figure 8 against the percent change in the end-to-end distance. Most of the energy of bending goes into the torsion angles ψ and ϕ. During the simulation we monitored the torsion angles about each amide group. Table 2 records for the same simulation, the incremental change in each torsion angle (columns) for each 1% reduction in the constraint distance. The Scheiner parameters were used in this example. Large changes in the torsion angles represent crossing over torsional barriers. As expected with the Scheiner parameters, most of the large changes occur about the NH torsions. Inspection of the actual torsion angles show that most of the large changes are *syn/syn* conversions, passing through 0°. One can also see a cascade effect where one *syn/syn* interconversion can induce others (increment 4

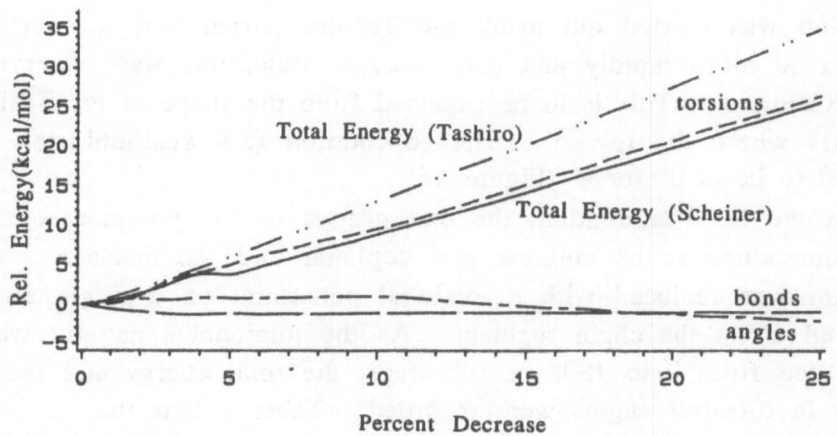

Figure 8. Change in Total Energy and component energies during bending for p-BT with the end groups constrained to be coplanar and using the Scheiner parameters. Also plotted is the change in Total Energy using Tashiro parameters.

in Table 2). A slight drop in energy can be seen in Figure 8 between steps 4 and 5 corresponding with the significant reorganization of the groups along the chain. The Tashiro parameters have higher torsion barriers and produce more resistance to bending. When the bending

Table 2

Angle Changes with Chain Bending

#	NH	CO	CO	NH	CO	NH	NH	CO	CO	NH	CO	NH
0	1	0	0	-1	-8	3	0	0	0	0	-4	12
1	5	0	1	0	0	4	0	0	0	0	-5	15
2	6	0	0	-4	4	4	0	0	0	0	-5	17
3	7	0	2	1	3	6	0	0	0	1	-4	14
4	-3	0	8	-34	9	69	2	0	-3	0	30	14
5	0	0	0	-11	2	10	0	0	-1	0	14	3
6	0	0	1	-4	1	5	0	0	0	0	10	3
7	1	0	0	-4	1	3	0	1	0	0	7	3
8	0	0	1	-3	0	4	-2	0	-1	0	8	5
9	0	0	0	-2	0	2	-2	0	-1	0	5	6
10	0	0	0	-2	0	2	-3	0	-1	1	4	7
11	0	0	0	-1	0	1	-3	0	-1	0	4	8
12	0	0	0	-1	0	1	-2	0	-1	0	3	8
13	0	0	0	-1	0	0	-1	0	0	0	3	6
14	0	0	0	-1	0	0	-1	0	0	0	2	7
15	0	0	0	-1	0	0	-1	0	0	0	2	4

simulation was carried out using the Tashiro parameters, the total energy rose more rapidly and one *syn/syn* transition was observed for a CO torsion. This is to be expected from the shape of the Tashiro potentials where the lowest barrier to rotation (5.6 kcal/mol) is predicted to be at 0° for ψ (Figure 4c).

In the third simulation, the end groups of the polymer segment were constrained to be colinear and coplanar and the distance constraint was replaced with a torsional constraint on a single .amide-ring bond along the chain segment. As the torsional constraint was incremented from 0 to 180° in 10° steps, the total energy and the changes in dihedral angles were recorded. Table 3 lists the incremental torsion angle changes observed when the highlighted

Table 3

Angle Changes with Constrained Torsion for p-BT

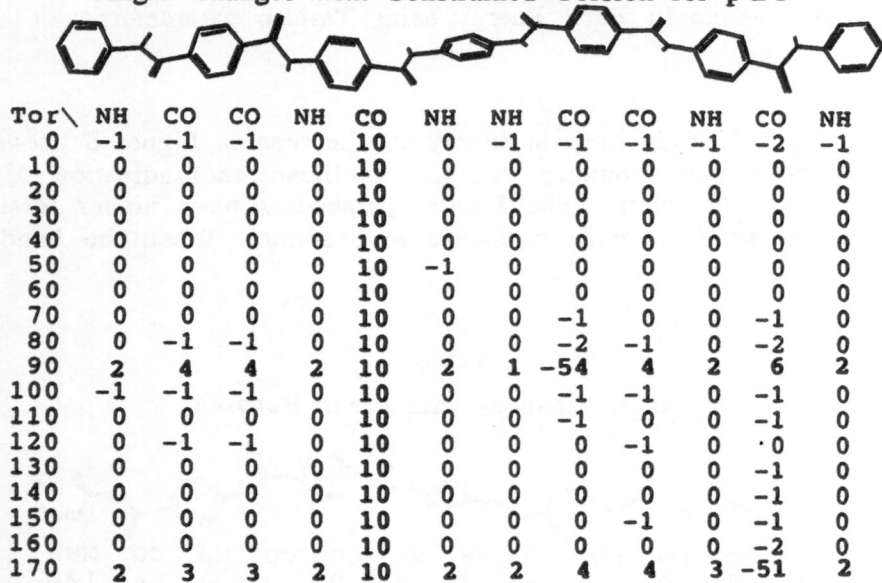

Tor\	NH	CO	CO	NH	CO	NH	NH	CO	CO	NH	CO	NH
0	-1	-1	-1	0	10	0	-1	-1	-1	-1	-2	-1
10	0	0	0	0	10	0	0	0	0	0	0	0
20	0	0	0	0	10	0	0	0	0	0	0	0
30	0	0	0	0	10	0	0	0	0	0	0	0
40	0	0	0	0	10	0	0	0	0	0	0	0
50	0	0	0	0	10	-1	0	0	0	0	0	0
60	0	0	0	0	10	0	0	0	0	0	0	0
70	0	0	0	0	10	0	0	-1	0	0	-1	0
80	0	-1	-1	0	10	0	0	-2	-1	0	-2	0
90	2	4	4	2	10	2	1	-54	4	2	6	2
100	-1	-1	-1	0	10	0	0	-1	-1	0	-1	0
110	0	0	0	0	10	0	0	-1	0	0	-1	0
120	0	-1	-1	0	10	0	0	0	-1	0	·0	0
130	0	0	0	0	10	0	0	0	0	0	-1	0
140	0	0	0	0	10	0	0	0	0	0	-1	0
150	0	0	0	0	10	0	0	0	-1	0	-1	0
160	0	0	0	0	10	0	0	0	0	0	-2	0
170	2	3	3	2	10	2	2	4	4	3	-51	2

amide-CO torsion was scanned using the Tashiro parameter set. The maximum energy of 21.5 kcal/mol was reached at ψ = 90° and led to a *syn/syn* transition further down the chain with synergistic relaxations taking place all along the chain. When the Scheiner parameters were used *syn/syn* transitions were observed for several amide-ring bonds along the chain segment.

Solid-state ^{13}C NMR

Solid-state ^{13}C NMR studies of the dynamics of aromatic polyamide networks and uncross-linked linear chains show complex relaxation pathways. At ambient temperature, the ring and carbonyl carbons of uncross-linked linear polyamides show practically identical $T_{1\rho C}$ relaxation times in the bulk and liquid crystalline solution, respectively[47]. This indicates that the carbonyl carbon is strongly coupled with the ring carbons. Infrared studies[40] show that in these systems practically all the amide groups participate in inter-chain H-bonding. In isotropic networks where intersegmental H-bonds hardly exist[40], the $T_{1\rho C}$ of the carbonyl carbon is substantially decoupled from the ring carbons[47], implying an aromatic polyamide stiff chain flexibility that does not exist when H-bonds keep them tightly together. Relaxation studies of molecularly dispersed and aggregated stiff aromatic polyamides in a matrix of flexible polyamide[57,58] indicate that the $T_{1\rho H}$ relaxation rate of the molecularly dispersed polyamides is far more rapid than that of the same polyamides in the aggregated state. This indicates that the molecularly dispersed stiff chains are substantially more flexible than when aggregated. Importantly, in all these NMR relaxation studies no indication was found at any given temperature for the existence of several discernible isomeric populations of a particular carbon atom. In light of the fact that the relaxation rates are obtained from changes in the intensity of the respective resonances, a small population of a given nucleus appearing as a shoulder to or buried under a major peak, is not expected to be detected in NMR relaxation studies such as those conducted in our laboratories[47,57-58]. We conclude that single stiff aromatic polyamide chains exhibit a substantial level of flexibility. This flexibility is greatly reduced when interchain H-bonds keep the chains closely packed in parallel arrays. NMR relaxation techniques are not yet sufficiently sensitive to differentiate between small populations of amide placements of various torsional isomers.

Conclusions

Atomistic molecular modeling methods provide new insight into the properties of aromatic polyamides. They demonstrate that single aromatic polyamide chains are relatively flexible and that their "rigid-rod" properties become more pronounced when strongly H-bonded aggregates are formed. These observations are consistent with solid-state ^{13}C NMR relaxation studies[47]. Simulations of chain bending with and without molecular constraints illustrate the importance of two modes of chain relaxation, *syn-syn* and *syn-anti* conformational changes, between the amide groups and the aromatic rings. Persistence length calculations demonstrate the importance of choosing the proper geometrical parameters in order to reproduce experimental results and illustrate the sensitivity of molecular simulations to parameterization.

While there is an extensive literature on structural properties of crystalline materials which can be used for building molecular models of polymers, much remains to be learned about simulating the dynamic properties of materials. This is illustrated by the range of energies available in the literature for torsional barriers for aromatic polyamides from theoretical calculations and the lack of appropriate experimental data for comparison. For this reason, caution should be used in drawing quantitative conclusions from molecular simulations without first giving careful consideration to molecular parameters.

Molecular modeling techniques are developing rapidly, and in the near future should provide quantitative insights into the dynamic properties of materials that can be compared directly with experiment.

References

1. H. Staudinger, *Ber.*, **53**, 1073 (1920).
2. W. Kuhn, *Ber.*, **63**, 1503 (1930)
3. E. Guth, H. Mark, *Monatsh Chem.*, **65**, 93 (1934).
4. W. Kuhn, *Kolloid Z.*, **68**, 2 (1934).
5. W. Kuhn, *Kolloid Z.*, **76**, 258 (1936)
6. P. J. Flory, "Principles of Polymer Chemistry"; Cornell University Press, Ithaca, NY, 1953; chapters X, XI, XII, and XIV.
7. H. Eyring, *Phys. Rev.*, **39**, 746 (1932).
8. P.J. Flory, *Proc. Royal Soc. London Series A*, **234**, 73 (1956).
9. P.J. Fory, Adv. *Polymer Sci.*, **59**, 1 (1984).
10. M. Doi, S. F. Edwards, "The Theory of Polymer Dynamics;" Clarendon Press, Oxford, 1986; chapters 8,9, and 10.
11. O. Kratky; G. Porod, *Rec. Trav. Chim. Pays Bas*, **68**, 1106 (1949).
12. P. J. Flory, *Proc. Royal Soc. London, Series A.*, **234**, 60 (1956).
13. H. Yamakawa, "Modern Theory of Polymer Solutions;" Harper & Row, New York, 1971; pp. 324 - 337, 358-393. G. Porod, *Monatsh Chem.*, **80**, 251 (1949).
14. A. Yu. Grosberg; A. R. Khokhlov, *Adv. Polymer Sci.*, **41**, 53 (1981).
15. A. Ciferri, in "Polymer Liquid Crystals"; Eds: A. Ciferri, W. R. Krigbaum, R. B. Meyer; Academic Press, New York, 1982; pp. 63-102.

16. D. C. Prevorsek, in "Polymer Liquid Crystals;" Eds: A. Ciferri, W. R. Krigbaum, R. B. Meyer, Academic Press, New York, 1982; pp 329-376.

17. V. N. Tsvetkov, L. N. Andreeva, *Adv. Polymer Sci.*, **39**, 95 (1981).

18. S. P. Papkov, *Adv. Polymer Sci.*, **59**, 75 (1984)

19. M. Arpin, C. Strazielle, G. Weill, H. Benoit, *Polymer*, **18**, 262 (1977).

20. M. Arpin, C. Strazielle, *Makromol. Chem.*, **177**, 581 (1976)

21. Q. Ying; B. Chu, *Makromol. Chem., Rapid Commun.*, **5**, 785 (1984).

22. K. Zero and S. M. Aharoni, *Macromolecules*, **20**, 1957 (1987).

23. S. M. Aharoni, *Macromolecules*, **20**, 2010 (1987).

24. S. J. Picken, *Liquid Crystals*, **5**, 1635 (1989).

25 S. J. Picken, *Macromolecules*, **23**, 464, (1990).

26. W. R. Krigbaum, T. Tanaka, G. Brelsford, and A. Ciferri, *Macromolecules*, **24**, 4142 (1991).

27. M. G. Northolt, J. J. Van Aartsen, *J. Polymer Sci. Poly Lett. Ed.*, **11**, 333 (1973).

28. M. G. Northolt, *Europ. Polymer J.*, **10**, 799 (1974).

29. K. Tashiro, M. Kobayashi, H. Tadokoro, *Macromolecules*, **10**, 413 (1977).

30. V. N. Tsvetkov, *Vyskomol. Soedin A.*, **19**, 2171 (1977).

31. B. Erman, P. J. Flory, and J. P Hummel, *Macromolecules*, **13**, 484 (1980).

32. J. P. Hummel, P. J. Flory, *Macromolecules*, **13**, 479 (1980).

33. S. Harkema, R. J. Gaymans, *Acta Crystallogr. B.*, **33**, 3609 (1977).

34. S. Harkema, R. J. Gaymans, G. J. Van Hummel, D. Zylberlicht, *Acta Crystallogr. B*, **35**, 506 (1979).

35. W. W. Adams, A. V. Fratini, D. R. Wiff, *Acta Crystallgr. B*, **34**, 954 (1978)

36. M. V. Schablygin, O. A. Nikitina, T. A. Belusova, G. I. Kudriavtsev, *Vysokomol. Soedin. A*, **24**, 984 (1982).

37. R. A. Gaudiana, R. A. Minns, R. Sinta, N. Weeks, H. G. Rogers, *Prog. Polym. Sci.*, **14**, 47 (1989).

38. S. M. Aharoni, S. F. Edwards, *Macromolecules*, **22**, 3361 (1989).

39. S. M. Aharoni, N. S. Murthy, K. Zero, S. F. Edwards, *Macromolecules*, **23**, 2533 (1990).

40. S. M. Aharoni, G. R. Hatfield, K. P. O'Brien, *Macromolecules*, **23**, 1330 (1990).

41. S. M. Aharoni, *Intern. J. Polym. Mater.* in press (1991).

42. G. C. Rutledge, U. W. Suter, *Macromolecules*, **24**, 1921 (1991).

43. P. Coulter, A. H. Windle, *Macromolecules*, **22**, 1129 (1989).

44. F. Laupretre, L. Monnerie, *Europ. Polymer J.*, **14**, 415 (1978).

45. J. Bicerano, H. A. Clark, *Macromolecules*, **21**, 585 (1988).

46. PolyGraf is a polymer modeling software package distributed by Polygen/MSI, Sunnyvale, CA.

47. S. A. Curran, C. P. LaClair, and S. M. Aharoni, *Macromolecules*, **24**, 5903 (1991).

48. F. Allen, O. Kennard, D. Watson, L. Bremmer, A. Orpen, and R. Taylor, *J.Chem.Soc, Perkin Trans. II*, S1 (1987).

49. We thank A. R. Khokhlov for suggesting this analysis.

50. The equilibrium distance (R_O) for the ring to nitrogen was changed to 1.40 Å and R_O for the ring to carbonyl carbon was changed to 1.495 Å.

51. S. L. Mayo, B. D. Olafson, and W. A. Goddard, III, *J. Phys. Chem.*, **94**, 8897 (1990).

52. S. Kashino, K. Ito and M. Haisa, *Bull. Chem. Soc. Japan*, **52**, 365 (1979).

53. A. Scheiner, IBM National Engineering and Scientific Support Center, Dallas, TX, personal communication.

54. X. Yang, S. L. Hsu, *Macromolecules*, **24**, 6680 (1991).

55. G. C. Rutledge, U. W. Suter, *Polymer*, **32**, 2179 (1991).

56. S. M. Aharoni, *J. Appl. Polymer Sci.*, in press (1992).

57. S. M. Aharoni, manuscript in preparation.

58. S. A. Curran, S. M. Aharoni, to be published.

NETWORKS WITH SEMI-FLEXIBLE CHAINS

B. Erman,[a] I. Bahar,[a] A. Kloczkowski,[b] J. E. Mark[b]

[a]Polymer Research Center, Bogazici University, Bebek 80815
Istanbul, Turkey

[b]Chemistry Department, University of Cincinnati, Cincinnati, Ohio
45221-0172, USA

INTRODUCTION

Classical theories of rubber elasticity[1] are based on the flexible-chain model. A flexible chain may be classified as one with a characteristic ratio of the order of unity. The elasticity of the network is primarily of intramolecular origin arising from the entropy of the individual chains. Intermolecular contributions are of secondary importance. The phantom network model in which the chains do not experience any interaction with their neighbors seems to be a good first-order approximation for real networks that consist of flexible chains. Recently, a large body of experimental work has been reported on networks made by cross-linking semi-flexible or semi-rigid chains. Stress-strain, swelling and birefringence measurements on these networks show significant deviations from the predictions of the classical network model. Among these networks are those prepared from aromatic polyamide chains[2,3] from cellulose and amylose[4-6] and from side-chain and main-chain liquid-crystalline systems.[7-11] The chains constituting these networks have characteristic ratios which are several orders of magnitude larger than those of classical flexible chains. The networks are marked with very high degree of segmental orientability under macroscopic deformation and a discontinuous stress-strain behavior indicating a phase transition under external stress. These experimental observations can not be predicted by the classical network theories. Instead, a theory recognizing the reduced flexibility of these semi-rigid chains is required.

The present paper reviews the recent theoretical models of networks with semi-flexible chains. In the following section, conformational requirements for the classification of a network chain as "semi-flexible " are discussed. Chains in this category may be described according to either (i) the worm-like chain model with a continuous curvature along its contour, or (ii) the Kuhn model chain consisting of rigid rods separated by flexible joints. Several examples of real chains which comply with either of these two models are given in this section. In the third section, existing experimental data on networks of these chains are briefly reviewed. In the fourth section, basic features of the two network models, one

Synthesis, Characterization, and Theory of Polymeric Networks and Gels
Edited by S.M. Aharoni, Plenum Press, New York, 1992

based on the worm-like chain and the other on the Kuhn chain, are discussed and their predictions are presented.

CONFIGURATIONAL FEATURES OF SEMI-FLEXIBLE CHAINS

The rigidity of a polymer chain may result from two basic sources relating to chain energetics and backbone geometry. First, the presence of large barriers to rotations about the backbone bonds may prevent the rotameric transitions of the bonds. The chain is trapped or frozen in a configuration which it has originally adopted in accordance with the distribution law of Boltzmann statistics. Such a model network of frozen chains is incompatible with the entropic picture of rubber elasticity according to which the chains should be sufficiently mobile to allow for the redistribution of isomeric states upon deformation. In order to conform with requirements of the flexible network, these chains must therefore possess sequences of higher flexibility at certain intervals along their contours. The second source of chain rigidity is associated with the geometric structure of the chain. Mainly, the presence of increased bond angles, significantly departing from the usual tetrahedral geometry, may lead to almost collinear skeletal bonds. This particular feature of the backbone geometry endows the chain with a certain degree of rigidity which persists along the direction of the first bond. Thus, the end-to-end vector of a chain whose first bond is held along the x-axis of a coordinate system may have a large x-component and small y- and z-components. In the absence of large barriers to rotameric transitions, the y- and z-components may change rapidly and randomly although the x-component remains relatively unchanged.

In Figure 1 a repeat unit of a main-chain liquid crystalline network prepared by Zentel and Reckert[8] is shown. The two phenyl groups impart a high stiffness to the repeat unit. The remaining groups provide a certain degree of flexibility. A sequence of X such units may be approximated as a chain of X stiff rods connected with flexible joints. Another example[12,13] to a semi-flexible chain and its representation as chain of rigid rods is shown in Figure 2. The examples shown in Figures 1 and 2 as well as others given by Zentel[11] conform very closely to the Kuhn chain model with rigid rods. The size of the rigid segments of these models may be described by the axial ratio $x = l/d$ of their length l to their cross-sectional diameter d. The value of x for one p-oxybenzoate unit has been calculated, for example,[14] as 3.8. Conformational calculations indicate that a series of such groups propagate approximately rectilinearly along the direction of the first unit. Thus, a sequence of X such units in a Kuhn segment results in an axial ratio of about 3.8X.

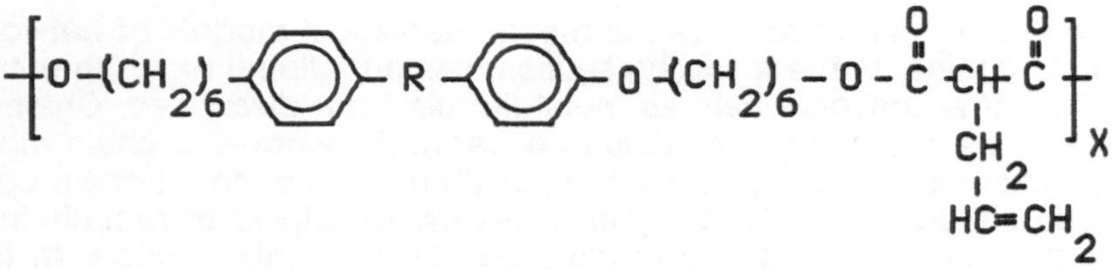

Figure 1. A repeat unit of a main-chain liquid crystalline elastomer. R may represent[8] a bond of a biphenyl, or -N=N- .

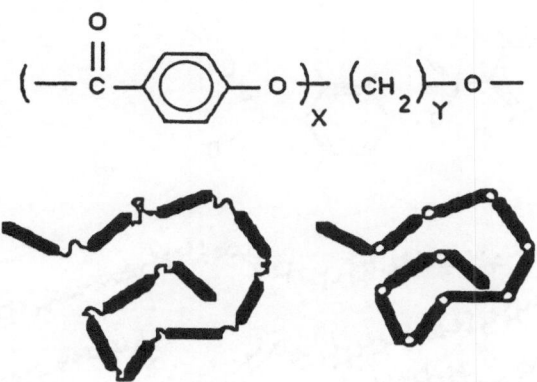

Figure 2. A sequence of stiff and flexible segments and its representation as a chain of rigid rods connected by flexible spacers. The circles in the lower right figure represent hinges that replace the flexible portions of the real chain.

Chains of p-phenylene polyamides and polyesters are other examples of semi-flexible chains. Two repeat units of a polyester are shown in Figure 3. Rotations about the O-C^{Ph} axes are subject to relatively low potential barriers,[15-17] the highest of which is in the order of 1 kcal/mol. Therefore, on a local scale, the chain enjoys a high degree of freedom arising from the rotational mobility of the phenylene rings about their axis of symmetry, directed along the chain contour. The angle between two successive O-C^{Ph} axes is 7.4°,[15] however, and this endows the chain with a very high degree of stiffness. Three random configurations of a long p-phenylene polyester chain are also shown in Figure 3. These representative configurations indicate that the end-to-end vector strongly persists along the direction of the first phenylene unit, although the chain enjoys a relatively large freedom in lateral directions. The persistence length of aromatic polyamides and polyesters, which is defined as the average projection of the end-to-end vector along the first bond of the chain, are calculated[15] to be 410 and 740 Å, respectively. Identification of the real chain with an equivalent Kuhn chain of rigid rods meets certain difficulties, as also observed from the three representative configurations of Figure 3. Indeed, the length of the equivalent segment of the freely jointed chain is calculated[15] to be 1560 and 2900 Å for the polyamides and the polyesters, respectively. Each equivalent rodlike segment contains about 240 and 450 phenyl groups for the amides and esters, respectively. These calculations indicate that these chains should conform satisfactorily to the worm-like or the Porod-Kratky chain model.

The worm-like chain is defined in terms of two parameters, L and a. The contour length L may be identified with the length r_{max} of the fully extended real chain as

$$L = r_{max} = n \, l \cos \theta/2 \tag{1}$$

Here, n and l are the number and length of bonds in the real chain, θ is the supplemental bond angle. The parameter a is representative of the persistence length of the chain and is given by

115

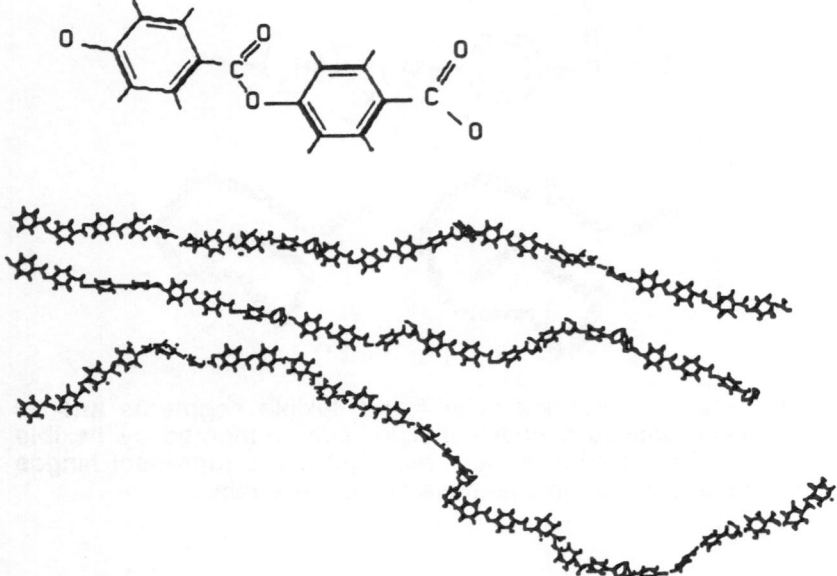

Figure 3. Two units of an aromatic polyester and three random configurations of the chain.

$$a = C_\infty \, nl^2 \, / \, 2L = (2 \cos \theta/2)^{-1} C_\infty \, l \qquad (2)$$

where C_∞ is the limiting value of characteristic ratio for the infinitely long chain. The characteristic ratio C_n of the real chain of n bonds is expressed in terms of L and a as

$$C_n = C_\infty [1 - (L / a)^{-1} (1 - \exp(-L/a))] \qquad (3)$$

Predictions of eq 3 for the p-phenylene polyester are compared with the exact solution in the upper part of Figure 4. L and a in eq 3 are calculated by substituting the previously reported values $C_\infty = 230$ and $\theta = 10.8$ into eqs 1 and 2.[15] The near coincidence of the two curves indicates that the worm-like chain is a suitable model for these molecules. Calculations performed but not reported show that agreement is perfect for the polyamide chains as well. Predictions of eq 3 are not equally satisfactory, however, for the Kuhn chain as shown in the lower part of Figure 4. Here, the horizontal line represents the characteristic ratio of the chain. The abcissa denotes the number n of equivalent rigid segments. The curve is obtained from eq 3 by taking L = nl and a = l/2, where l is the length of the equivalent rigid segment. The disagreement indicated by the curve and the horizontal line shows that a chain of rigid rods connected by flexible joints can not be satisfactorily be represented by the worm-like chain model.

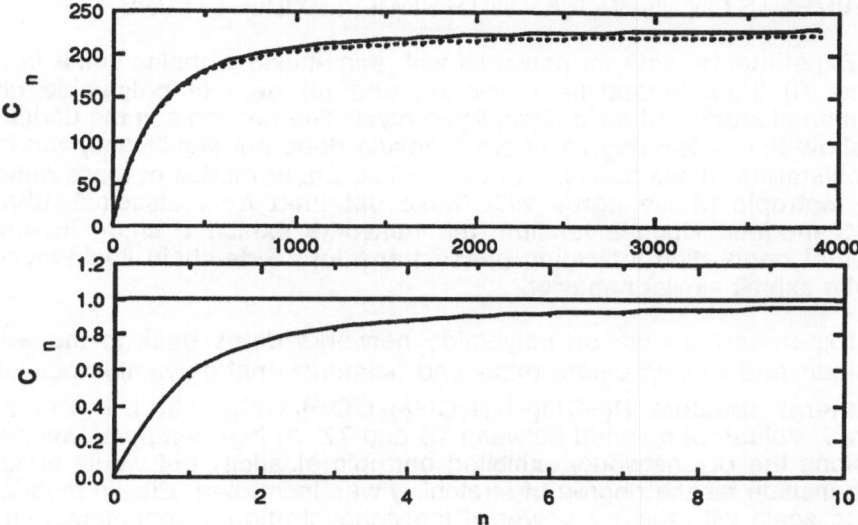

Figure 4. Upper figure: Characteristic ratio of p-phenylene polyesters as a function of number of repeat units. The solid curve results from the calculations of reference 15. The dotted curve is calculated from eq 3 for the worm-like chain. Lower figure, horizontal line: Characteristic ratio for a freely jointed chain. Curve: Prediction of eq 3 for the freely-jointed chain.

The preceding argument implies that the two models, the freely-jointed and the worm-like chains, find satisfactory applications depending on the chemical constitution of the semi-flexible chains.

A large class of semi-flexible chains other than the ones discussed above can be used in forming networks. Depending on their configurational features, they may be classified either as Kuhn or worm-like chains. Important candidates for forming networks are the cellulose derivatives,[4-6,18] and amylose.[4] The axial ratios of cellulose derivatives reported in Table I show that these molecules are relatively stiff and form liquid crystalline phases above a critical concentration v_{2c} given in the last column of Table I. The polyisocyanate family[4] is another potential candidate for networks with semi-flexible chains. These chains have[19] characteristic ratios in the order of 100 and may therefore be classified as semi-flexible .

Table I Axial Ratios and Critical Volume Fractions of Some Cellulose Derivatives

Polymer	Axial Ratio	Volume Fraction, v_{2c}
Cellulose	8.85	0.70
Cellulose trinitrate	16.50	0.43
Methylol cellulose	20.00	0.36
Hydroxypropyl cellulose	25.00	0.29

EXPERIMENTS ON NETWORKS WITH SEMI-FLEXIBLE CHAINS

Experimental data on networks with semi-flexible chains come from two sources: (i) liquid-crystalline networks, and (ii) gels of polyamide chains. Experimental studies of main-chain liquid-crystalline networks in the undeformed state show that a low degree of cross linking does not significantly modify the liquid-crystalline phase behavior of the chains. Shear moduli of these networks[8] in the isotropic phase agree with those obtained from classical rubberlike network models. Upon extension, the networks exhibit a sharp increase in segmental orientation indicating phase transition. Side-chain liquid-crystalline networks exhibit similar behavior.

Experimental work on polyamide networks dates back to the work by Schaefgen and Flory[20] where tetra- and octafunctional polyamide networks of the general structure $R\{-CO[-NH(CH_2)_5CO-]_nOH\}_b$ with b = 4 or 8 were prepared. Values of n varied between 13 and 77. At high temperatures and low extensions the dry networks exhibited entropic elasticity but would undergo a rapid transition as the degree of stretching was increased. Elastic moduli were found to scale with the 1.7 power of the concentration. Experimental study of networks with p-phenylene polyamides is new.[3] So far only networks with very short chains have been studied. The number of bonds in the experimentally-studied network chains is lower than that of a typical Kuhn segment. For this reason, these networks are relatively stiff. Sufficient flexibility may be obtained only at high dilutions and with chains whose contour lengths are much above the persistence length. In this limit, it is apparent that the chains will satisfactorily conform with the worm-like chain model.

THEORIES OF NETWORKS WITH SEMI-FLEXIBLE CHAINS

The basic features of networks with chains exhibiting nematic-like interactions were first discussed by de Gennes,[21] and a theory of nematic networks was presented by Warner et al.[22-28] The theory rests on the application of the spheroidal wave function formalism for worm-like chains which are oriented by a nematic field coupled to the local chain direction. The total free energy F is given by

$$F = F_{nem} + F_{el} + F_{ext} \tag{4}$$

where F_{nem} and F_{el} are the nematic and elastic contributions to the free energy, and F_{ext} is the elastic work done by external forces. F_{nem} is given by a Landau-de Gennes type of expansion in powers of the nematic order parameter $S = < P_2 (\cos \theta) >$ as

$$\frac{F_{nem}}{N_s k_B T} = \frac{L}{l_0} (\frac{1}{2} AS^2 - \frac{1}{3} BS^3 + \frac{1}{4} CS^4) \tag{5}$$

Here, k_B is the Boltzmann constant, T is the temperature, L is the chain arc length between cross-links, l_0 is the persistence length of the worm-like chain corresponding to the parameter a defined above, and N_s is the number density of strands. The coefficients A, B and C are functions of temperature. The elastic part of the free energy is

$$\frac{F_{el}}{N_s k_B T} = \frac{1}{2}\left(\lambda^2 \frac{l_0}{l_z} + \frac{2}{\lambda}\frac{l_0}{l_p}\right) - \frac{1}{2}\ln\left(\frac{l_0^3}{l_z l_p^2}\right) \qquad (6)$$

where λ is the extension ratio, l_z and l_p are the effective step lengths of the chain in the direction parallel and perpendicular to the ordering direction, respectively.

The elastic work per unit volume is

$$F_{ext}/V = -\sigma \ln \lambda \qquad (7)$$

where σ is the true stress defined as the force per unit deformed area. The functional relationship $S = f[S, \lambda(S)]$ between the deformation and the order parameter is obtained in the Warner theory by minimizing the free energy with respect to λ at fixed S. An explicit relationship of S to λ and σ is obtained by means of extensive numerical calculations. Predictions of the theory indicate stress-induced phase transitions, spontaneous shape changes, discontinuous stress-strain relations and non-linear stress optical laws. In Figure 5, the dependence of the orientation function or the order parameter S, on the applied stress is shown. The ordinate represents the reduced stress $\sigma^* = \sigma/N_s kT$. Curves are drawn for different values of the reduced temperature T/T_{ni}, where T_{ni} is the nematic-isotropic transition temperature. The number m of equivalent freely jointed segments in the chain is 20, provided that the length of each equivalent rodlike segment is taken to be equal to the persistence length of the chain.[27] Sharp transitions are observed at temperatures below the nematic-isotropic transition temperature, which gradually shifts to lower stresses with decreasing temperature.

The dependence of the reduced stress σ^* on deformation is shown in Figure 6 where the abcissa is the first invariant of deformation, $\lambda^2 - 1/\lambda$. The curves are obtained for a network whose chains contain 500 rodlike Kuhn

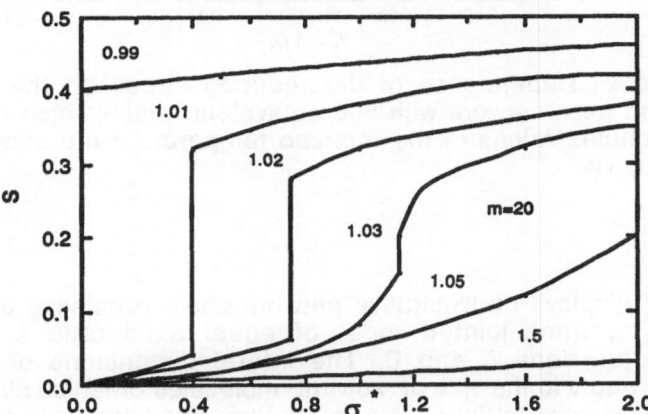

Figure 5. Dependence of the order parameter S on the applied stress for a network with 20 equivalent freely jointed rods. The reduced temperature is shown on each curve.

segments. Values of the reduced temperature are shown on each curve. At high temperatures the curves are smooth and in agreement with the statistical models of amorphous flexible rubbers. A transition is observed as the nematic-isotropic transition temperature is approached. The curves are obtained by keeping the true stress constant during the transition.

An alternative molecular model that predicts transitions in networks with semi-flexible chains based on the freely-jointed rods model was recently presented by Abramchuk and Khoklov.[29] The theory based on this model has also been developed recently to include the contributions to segmental orientation from intermolecular steric and thermotropic effects.[4,30-33] This alternate description of networks with freely-jointed rods instead of the worm-like model predicts all the unusual stress-strain and orientation features of these networks. In this model, the competition for space among neighboring chains is treated according to a lattice model. The idea of a lattice in describing the orientational behavior of the segments of flexible chains in networks was first proposed by Di Marzio.[34] His original model, as well as its modification,[35] was devoid of the concept of the length-to-width ratio, or the axial ratio, of the segments comprising the network chain embedded in a lattice. A length-to-width ratio x substantially larger than unity is required for rationalizing the unusual behavior of networks with semi-flexible chains.[30,31]

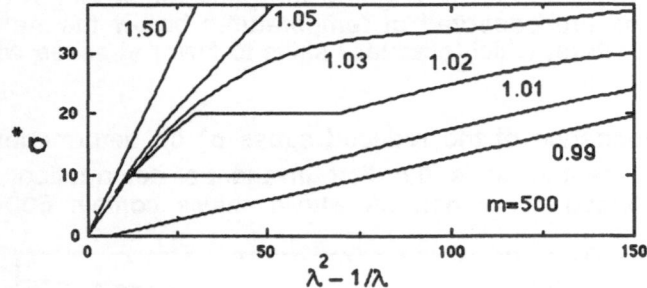

Figure 6. Dependence of the reduced stress on the strain function for a network with 500 equivalent freely-jointed rods in each chain. Values of the reduced temperature are shown on each curve.

Figure 7 displays an illustrative network chain consisting of m = 20 rigid segments (i. e., freely-jointed rods) of equal axial ratio x between two tetrafunctional junctions A and B. The lateral dimensions of the rods are assumed to be equal to the size of solvent molecules or lattice sites. The model may be readily adapted to the case of more than one lattice site occupied either by solvent molecules or the rodlike segments in lateral direction. The orientation of the segment is defined by the two Euler angles, ψ_k and ϕ_k, as shown in Figure 7 (b). The Z direction in the figure indicates the preferred direction, also referred to as the domain axis. In simple extension, this direction coincides with the direction of stretch. The subscript k indicates that the orientation of the segment lies within the kth solid angle.

Following Flory[36] and Flory-Ronca,[37,38] the accommodation of the rod in the lattice is achieved through its representation by a sequence of y_k submolecules each occupying x/y_k sites and oriented along the preferred direction as shown in Figure 7 (c). Thus, y_k is expressed in terms of ψ_k and ϕ_k as

$$y_k = x \sin \psi_k \left(\left| \cos \phi_k \right| + \left| \sin \phi_k \right| \right) \tag{8}$$

for the rod exhibiting that particular orientation. According to this definition, the value of y_k increases as the rod becomes disoriented. It is therefore referred to as the disorientation index. The junctions of the network at the extremities of each chain are assumed to displace affinely with macroscopic deformation. This assumption, which is also employed by Kuhn and Grün, is not exactly correct but simplifies the problem to a large extent. The volume of the network is assumed to be constant throughout the deformation.

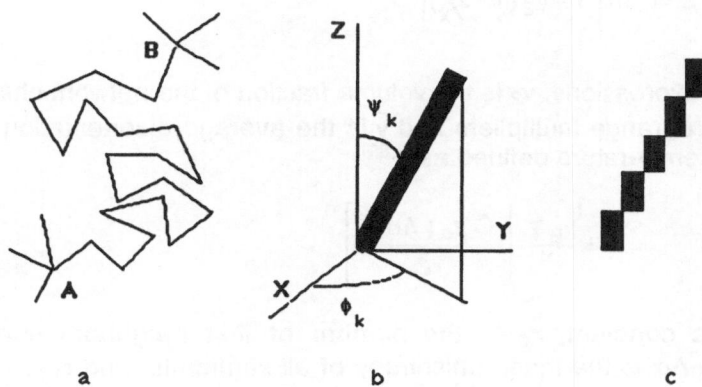

Figure 7. (a) A network chain of 20 rigid segments between the two tetrafunctional junctions A and B. (b) A rigid segment of the chain with angles ψ_k and ϕ_k representing the polar and azimuthal angles. XYZ is a laboratory-fixed system with Z showing the direction of stretch. (c) Representation of the segment by y_k submolecules, each oriented along the preferred direction.

The total configurational partition function Z_m of a mixture of n_2 polymer chains and n_1 solvent molecules in a lattice is taken as the product of combinatorial (Z_{comb}) and orientational (Z_{orient}) partition functions. The Helmholtz free energy of mixing is expressed as $\Delta A_m = - kT \left(\ln Z_{comb} + \ln Z_{orient} \right)$ The explicit form of the Helmholtz free energy is outlined in references 30 and 31. The ends of each chain are assumed to deform affinely with macroscopic strain. Such external constraints are imposed on the system by the use of Lagrange multipliers, followed by the minimization of the Helmholtz free energy. The quantity S, for example, is found from

$$S = \frac{\displaystyle\int_0^{2\pi} d\phi \int_0^\pi d\psi \; P_2(\cos \psi) \; f(\phi,\psi)}{\displaystyle\int_0^{2\pi} d\phi \int_0^\pi d\psi \; f(\phi,\psi)} \tag{9}$$

where $f(\phi,\psi)$ is the orientational distribution function which for a thermotropic system subject to uniaxial deformation assumes the form[30,31]

$$f(\phi,\psi) = \sin\psi \exp\Big\{ -a \, x \sin\psi \big[\,|\cos\phi| + |\sin\phi|\,\big] + \beta \cos\psi + $$
$$\gamma \sin\psi\,(\cos\phi + \sin\phi) + S\widetilde{T}^{-1} \, P_2(\cos\psi) \Big\} \tag{10}$$

with

$$a \equiv - \ln\left[1 - v_2\big(1 - \overline{y}/x\big)\right] \tag{11}$$

In the above expressions, v_2 is the volume fraction of the network chains, β and γ are the two Lagrange multipliers and $\overline{y}$ is the average disorientation index. $\widetilde{T}$ is the reduced temperature defined as[38,39]

$$\widetilde{T} = \frac{k_B T}{x}\left[\frac{C \, z_c \, (\Delta\alpha)^2}{r_*^6}\right]^{-1} \tag{12}$$

Here, C is a constant, z_c is the number of first neighbors surrounding a subsegment, $\Delta\alpha$ is the mean anisotropy of all segments, and r_* is the distance between subsegments for dense packing. For athermal systems, $\widetilde{T} \rightarrow \infty$.

Thermotropic interactions in this model are based on the orientational interactions of the form given by the Flory-Ronca treatment.[37,38] The reader is referred to these two papers for details of the interactions. The difference between the Warner theory and the present lattice theory for thermotropic networks lies in the treatment of the interactions. The latter adopts an orientation-dependent interaction which reflects the basic molecular characteristics of the nematogenic molecules, while the former uses the Landau expression.

The Lagrange multipliers β and γ of the lattice model are obtained from a self-consistent treatment as presented in detail in two references, 30 and 31. Their evaluation determines the orientational distribution function, and the partition function from which the Helmholtz free energy and all thermodynamic functions may be obtained. Due to the nonlinearity of the equations, numerical methods are employed in their solution.

The orientation function obtained by the use of eq 9 is presented in Figure 8 as a function of λ for various axial ratios of the segments for an athermal network with chains containing 20 freely-jointed segments. Increasing the axial ratio results in an increase in the sensitivity of the orientation function to deformation. A sharp transition to the highly oriented state takes place at a critical value $x_c = 5.83$. It is interesting to note that the Flory-Ronca theory gives $x_c = 6.42$ for the critical value for a neat liquid crystalline system of rods in which isotropic and anisotropic phases coexist. The present calculations indicate that the liquid crystalline behavior is not modified drastically in networks with low degrees of cross-linking.

Dependence of the orientation function on the axial ratios of the segments is shown in Figure 9 for three different values of the extension ratios. The values are calculated for m = 20 in an athermal network.

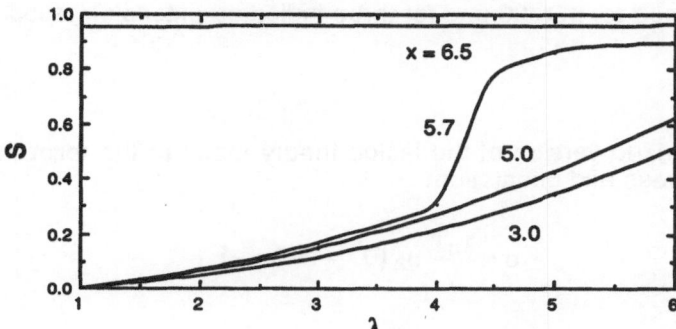

Figure 8. Relationship of the orientation function to extension ratio. Curves are obtained for m = 20 and for various axial ratios as shown in the figure.

The true stress in uniaxial deformation is obtained as:

$$\sigma = \frac{1}{V} \lambda \left(\frac{\partial \Delta A_m}{\partial \lambda} \right)_{T,V} \tag{13}$$

where the subscripts indicate that the differentiation is performed at constant temperature and volume. Calculations show, in parallel to the Warner theory, that a phase transition takes place when the force is increased above a certain critical value.

Figure 10 represents the reduced force, $[f^*] = \sigma / (\lambda^2 - 1/\lambda)$ versus the reciprocal extension ratio. A few striking features are worth mentioning. The reduced force or the modulus $[f^*]$ remains approximately constant for a very large range of deformations. In this respect the network modulus resembles that of the phantom network while in the isotropic state. The modulus exhibits a drastic decrease during phase transition followed by a very sharp increase. Calculations are performed for a dry network. Introduction of a suitable solvent into the network changes the phase transition behavior substantially.

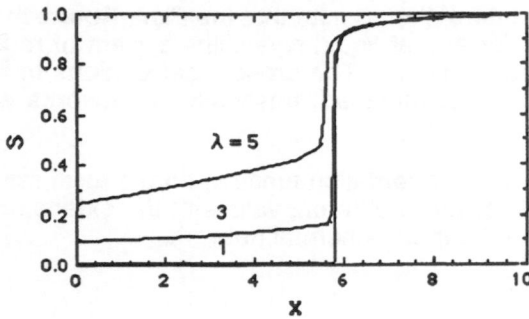

Figure 9. Dependence of the orientation function on the axial ratio of rodlike segments of the network chains. Curves are obtained for m = 20 and for three indicated extension ratios.

A linearized version of the lattice theory leads to the following expression for the true stress and orientation:

$$\sigma = \frac{k_B T}{V} n_2 (\lambda^2 - 1 / \lambda v_2) \tag{14}$$

$$S = \frac{1}{5\,mG} \left[1 + \frac{5\,x_a}{64} \frac{x\,v_2/x_a}{1 - (x\,v_2/x_a)} \right] \sigma \tag{15}$$

where G is the shear modulus given as $G = k_B T\,n_2/V$ and $x_a = 10.98$ indicates the axial ratio above which an isotropic phase can not exist in the system. Equations 14 and 15 are derived for an athermal system. A similar expression is obtained for a thermotropic system.[30] Equation 14 indicates that the linearized version of the lattice model does not predict the phase transition obtained in Figure 10 by the numerical solution of the nonlinear equations. A similar conclusion was reached previously by a phenomenological formulation.[40] Equation 15 on the other hand shows that the orientation diverges when $x v_2/x_a$ equates to unity. This ratio may be regarded as an upper bound for v_2 and x values before the network transforms into a totally ordered state. Use of the x and v_{2c} values given in Table I for cellulose derivatives gives this ratio as 0.56 for cellulose, 0.65 for cellulose trinitrate, 0.65 for methylol cellulose and 0.66 for hydroxypropyl cellulose.

CONCLUDING REMARKS

The two network theories based on the freely jointed rod and the worm-like chain models are reviewed in the present paper. Both theories predict stress-induced or strain-induced transitions in networks. The transition at constant length closely resembles the crystallization of networks where a sudden drop in the modulus is followed by a sharp increase as indicated in Figure 10. Both of the theories may be applied to athermal as well as thermotropic systems and to networks which are diluted with a suitable solvent.

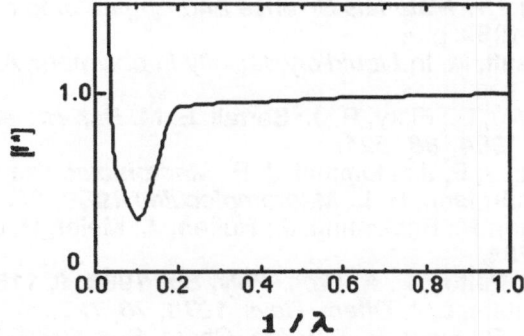

Figure 10. The relationship of the reduced force, [f*] to reciprocal extension ratio 1/λ. The ordinate values are chosen arbitrarily where [f*] is equated to unity at λ = 1.

Quantitative verification of the above theoretical approaches awaits further experimental evidence based on well-characterized networks of semi-flexible chains.

ACKNOWLEDGMENTS

It is a pleasure to acknowledge partial support by the Bogazici Research Fund through Grant 91P0029, the U. S. National Science Foundation through Grant DMR 89-18002 (Polymers Program, Division of Materials Research), and the Procter & Gamble Company through their "University Exploratory Research Program".

REFERENCES

(1)   Mark, J. E.; Erman, B. *Rubberlike Elasticity: A Molecular Primer*, Wiley Interscience: New York, 1988.
(2)   Aharoni, S. M.; Edwards, S. F. *Macromolecules* **1989**, *22*, 3361.
(3)   Aharoni, S. M.; Hatfield, G. R.; O'Brien, K. P. *Macromolecules* **1990**, *23*, 1330.
(4)   Erman, B.; Bahar, I.; Yang, Y.; Kloczkowski, A.; Mark, J. E. In *P&G*; New York, 1991;
(5)   Song, C. Q.; Litt, M. H.; Manas-Zloczower, I. In *Polymeric Materials Science and Engineering*; American Chemical Society: Washington, D. C., 1990; p 445.
(6)   Song, C. Q.; Litt, M. H.; Manas-Zloczower, I. *J. Appl. Polym. Sci.* **1991**, *42*, 2517.
(7)   Matoussi, H.; Ober, R.; Veyssie, M.; Finkelmann, H. *Europhysics Letters* **1986**, *2*,   233.
(8)   Zentel, R.; Reckert, G. *Makromol. Chemie.* **1986**, *187*, 1915.
(9)   Zentel, R.; Benalia, M. *Makromol. Chemie.* **1987**, *188*, 665.
(10)   Schatzle, J.; Kaufhold, W.; Finkelmann, H. *Makromol. Chemie.* **1989**, *190*, 3269.
(11)   Zentel, R. *Angew. Chem. Int. Ed. Engl. Adv. Mater.* **1989**, *28*, 1407.

(12) Flory, P. J. In *The Materials Science and Engineering of Rigid Rod Polymers*; 1989; p 3.
(13) Abe, A.; Ballauff, M. In *Liquid crystallinity in polymers*; A. Ciferri, Ed.; VCH: 1991.
(14) Ballauff, M.; Wu, D.; Flory, P. J.; Barrall, E. M. *Ber. Bunsenges. Phys. Chem.* **1984**, *88*, 524.
(15) Erman, B.; Flory, P. J.; Hummel, J. P. *Macromolecules* **1980**, *13*, 484.
(16) Jung, B.; Schürmann, B. L. *Macromolecules* **1989**, *22*, 477.
(17) Lautenschlager, P.; Brickmann, J.; Ruiten, J.; Meier, R. J. *Macromolecules* **1991**, *24*, 1284.
(18) Gilbert, R. D.; Patton, P. A. *Prog. Poly. Sci.* **1983**, *9*, 115.
(19) Bur, A. J.; Fetters, L. J. *Chem. Revs.* **1976**, *76*, 727.
(20) Schaefgen, J. R.; Flory, P. J. *J. Am. Chem. Soc.* **1950**, *72*, 689.
(21) de Gennes, P. G. *C. R. Acad. Sci. Ser. B* **1975**, *281*, 101.
(22) Wang, X. J.; Warner, M. *J. Phys. A: Math. Gen* **1986**, *19*, 2215.
(23) Wang, X. J.; Warner, M. *J. Phys. A: Math. Gen* **1987**, *20*, 713.
(24) Warner, M.; Gelling, K.; Vilgis, T. J. C. P. *J. Chem. Phys.* **1988**, *88*, 4408.
(25) Renz, W.; Warner, M. *Proc. R. Soc. London* **1988**, *A417*, 213.
(26) Warner, M. In *Side chain liquid crystal polymers*; C. B. Mc Ardle, Ed.; Chapman & Hall: New York, 1989; p 7.
(27) Warner, M.; Wang, X. J. *Macromolecules* **1991**, *24*, 4932.
(28) Warner, M.; Wang, X. J. *Macromolecules* **1992**, *25*, 445.
(29) Abramchuk, S. S.; Khoklov, A. R. *Dokl. Phys. Chem.* **1988**, *297*, 1069.
(30) Bahar, I.; Erman, B.; Kloczkowski, A.; Mark, J. E. *Macromolecules* **1990**, *23*, 5341.
(31) Erman, B.; Bahar, I.; Kloczkowski, A.; Mark, J. E. *Macromolecules* **1990**, *23*, 5335.
(32) Erman, B.; Bahar, I.; Kloczkowski, A.; Mark, J. E. In *Elastomeric Polymer Networks*; J. E. Mark B. Erman, Ed.; Prentice Hall: New Jersey, 1992; p 142.
(33) Kloczkowski, A.; Mark, J. E.; Erman, B.; Bahar, I. In ; I. Noda, Ed.; 1992.
(34) Di Marzio, E. A. *J. Chem. Phys.* **1962**, *36*, 1563.
(35) Tanaka, T.; Allen, G. *Macromolecules* **1977**, *10*, 426.
(36) Flory, P. J. *Proc. R. Soc., London, Ser. A.* **1954**, *234*, 73.
(37) Flory, P. J.; Ronca, G. *Mol. Cryst. Liq. Cryst.* **1979**, *54*, 289.
(38) Flory, P. J.; Ronca, G. *Mol. Cryst. Liq. Cryst.* **1979**, *54*, 311.
(39) Warner, M.; Flory, P. J. *J. Chem. Phys.* **1980**, *73*, 6327.
(40) Jarry, J. P.; Monnerie, L. *Macromolecules* **1979**, *12*, 316.

MESOPHASE POLYSILOXANE NETWORKS: MECHANICAL AND THERMO-MECHANICAL BEHAVIOR

Y. K. Godovsky

Department of Polymer Materials
Karpov Institute of Physical Chemistry
Ul.Obukha 10, Moscow 103064, Russia

INTRODUCTION

Polymer material science today is directed to the creation and study of such systems that combine various polymeric properties and functions with ordered structures. One of the most striking example of such systems is the liquid crystalline elastomers, which combine the rubber-like elasticity with the ordered structure and the mobility of liquid crystalline phases[1-3]. It is generally accepted now that attaching conventional mesogens via flexible spacer to a flexible linear polymer and subsequent crosslinking provides an approach for obtaining typical main chain and side chain liquid crystalline polymers. One of the most interesting materials in this field, both from the academic and practical points of view, is the side chain liquid crystalline elastomers[1-3], which combine the liquid crystalline order with rubber elasticity and shape stability. It is well known now that polydimethylsiloxane and some related copolymers are very suitable macromolecules for the attachment of the mesogenic side chains to their backbones. Therefore, such side chain liquid crystalline elastomers have attracted much attention during the last decade.

For the nematic liquid crystalline elastomers composed either of polymer chains of stiff rods (mesogenic groups) linked by flexible spacers, or attached as side chains, a number of unusual phenomena have been predicted theoretically[4-11] and found experimentally[2] including phase transitions, spontaneous shape changes, discontinues stress-strain relations, nonlinear stress-optical behavior as well as deviations from classical behavior of elastomers in the temperature range above the temperature of the nematic-isotropic transition (nonnematic or paranematic phase).

On the other hand, recently a new class of polymers have been developed which are able to form thermotropic mesophases without any mesogens neither in the main chain nor in the side group of the macromolecule[12-30]. In the most striking form the tendency to build up such mesophases is displayed by polymers with flexible inorganic backbones, such as polyorganosiloxanes

Synthesis, Characterization, and Theory of Polymeric Networks and Gels
Edited by S.M. Aharoni, Plenum Press, New York, 1992

and some polyphosphazenes. From the structural point of view
the mesophases in flexible polymers can reasonably be regarded
as highly disordered crystals with a high level of the molecular
mobility[29,30]. In the high enough molecular weight linear non-
crosslinked poly(diethylsiloxane) (PDES) the temperature
interval in which the mesophase can exist is normally of 280-
325 K. Although the content of the mesophase in PDES elastomers
and the temperature of isotropization is very sensitive to the
crosslinking density, nevertheless, in PDES elastomers with
typical values of the crosslinking density the existence of the
mesomorphic state leads to a very unusual mechanical, thermo-
mechanical and optical behavior.

The aim of this work is to investigate the mechanical and
thermo-mechanical behavior of a novel type of mesophase PDES
elastomers. This paper describes their mechanical, optical and
thermo-mechanical behavior, and primarily addresses the problem
of the influence of the mesophase on the stretching behavior of
the elastomers, energy and optical characteristics of the
strain-induced mesophase formation, and the phenomena of necking
and denecking resulting from the cyclic deformations.

EXPERIMENTAL

Samples

Three series of PDES samples were prepared and studied.
Two elastomers PDES samples were prepared using end-linking
reaction of hydroxil-terminated end groups and multifunctional
siloxane end-linking agent in concentration of 15 and 45 wt.%
with a catalyst[31]. The isotropization heat of the samples was
0.7 and 0.35 J/g, respectively. It corresponds to the amount of
the mesophase in the samples 24 and 12 %, respectively.

The second series of samples was prepared by "room tempera-
ture vulcanization" of high molecular weight ($\bar{M}_w = 4 \times 10^5$) linear
PDES in solution[32]. For crosslinking in solution, toluene was
used as a solvent, and the additives used for curing consisted
of various parts of partially hydrolyzed ethyl silicate and a
metallo-organic compound as a catalyst. The films cast from such
solutions were 0.5-1.0 mm thickness. The DSC characteristics of
the samples of this series are listed in Table 1. The PDES-0
sample is the initial linear PDES. Its thermodynamical characteristics are typical of high molecular weight linear PDES. The
other samples are crosslinked PDES elastomers with the degree
of crosslinking increasing from sample PDES-1 to PDES-6. It is
seen from Table 1 that the mesomorphic state in PDES is extremely sensitive to the degree of crosslinking. Similar to the
degree of crystallinity, which is used for characterizing the
crystalline state of polymers, it is worthwhile to introduce the
degree of mesomorphisity, which characterizes the amount of the
mesophase in a mesophase sample. The values listed in the last
column were estimated from the heat of isotropization according
to a simple ratio $\beta_m = \Delta H / \Delta H_{100\%}$ where $\Delta H_{100\%}$ is the heat of
isotropization of the initial non-crosslinked high molecular
weight PDES annealed at room temperature. As seen from the last
column, β_m of moderately crosslinked samples is very small. Such
a behavior is closely related to the very high value of the
critical molecular weight below which the mesophase state in
PDES does not appear at all. According to various measurements[20,25,29] the critical molecular weight ranges from 20000
to 30000. Although the ability of crystallizing also decreases

128

Table 1. DSC characteristics of PDES elastomers

Sample	T_g, K	T_m, K	ΔH_m, J/g	T_i, K	ΔH_i, J/g	Degree of mesomorphisity, β_m,
		Melting		Isotropization		
PDES-0	134	290	21	325	3.0	1
PDES-1	134	284	12.5	316	2.2	0.73
PDES-2	134	283	6.2	288	0.3	0.10
PDES-3	134	273	5.8	285	0.1	0.03
PDES-4	134	272	4.7	-	-	-
PDES-5	134	272	3.8	-	-	-
PDES-6	134	272	3.5	-	-	-

with the degree of crosslinking, nevertheless even the most crosslinked samples are, nevertheless, able to crystallize.

The third series of samples was multi-block copolymers with PDES as the soft block and either polyphenylsilsesquioxane, or polyarylate as the hard block[33]. Both these hard blocks are amorphous with very high glass transition temperatures in the vicinity of 600 K. Due to incompatibility of the components, the microphase separation occurs in these block copolymers. The microdomains of the hard blocks are the multifunctional cross-links and reinforcing fillers at a time. At an appropriate ratio of the hard and soft blocks, these block copolymers represent typical rubber-like materials. The block copolymers were obtained by means of condensation reactions of hydroxyl-terminated oligo-siloxane chains with end-functionalized polymers served as a hard block. The molecular weight of the hard blocks was 4.000-8.000 while the molecular weight of the soft PDES block was in the range of 5.000 - 20.000. The content of the PDES blocks in the block copolymers was in the range 54-82 %. The physical and mechanical behavior of both types of these block copolymers was rather similar. The main characteristics of these block copolymers are given in[33]. In the initial non-stretched state all the block copolymers studied were amorphous at room temperature because of a small length of the soft segments, which in all samples was below the critical value of the molecular weight for the mesophase formation. During stretching in all these samples the strain-induced mesophase occurred.

Methods

Stress-strain behavior of the PDES elastomers in stretching and cyclic modes was studied by means of a stretching calorimeter which also gives the possibility to measure the heat (entropic) effects resulting from the deformation and to estimate, finally, the energy effects of deformation. The construction of the calorimeter and the procedure of such measurements have been published elsewhere[34-36]. The thermo-mechanical study was made also by measuring the length of the samples as a function of temperature while the samples supported a constant load[37]. Stress-optical behavior was studied by two methods. The qualitative photoelastic measurements were carried out using an Instron 1122 testing machin equipped with a laser device[32]. The relative transmission of the sample as a function of strain simultaneously with the stress-strain curve was monitored during these experiments at room temperature. The quantitative stress-optical measurements were curried out in an apparatus based on the method of rotating polarizer (these measurements were conducted at the Institute of Organic Chemistry, Johannes Gutenberg University, Mainz, Germany). The absolute value of the bire-fringence of the sample as a function of strain may be obtained in these experiments.

MECHANICAL BEHAVIOR

Simple stretching

Figure 1 shows typical stress-strain curves for PDES listed in Table 1 obtained during the first stretching[32]. The stress-strain curve of the initially semimesomorphic sample (PDES-1 and PDES-2) resembles the stress-strain curves of crystalline polymers with a low degree of crystallinity. Initially there is a sharp rise in load with increasing extension. The modulus of elasticity corresponding to the initial part of the curve is considerably higher than that of the same film in the amorphous state. Then the slope of the curve steadily falls and the stress-softening occurs. The behavior corresponding to the next part of the curve depends on the degree of mesomorphisity of the sample and the stretching rate. In the majority of cases the samples elongate without necking. However, in some cases a well-defined necking takes place. X-ray analysis shows that, unlike the initial non-stretched state, at high extension ratios a well-defined mesophase texture occurs both for initially amorphous and mesomorphic samples. Thus, regardless of the phase state of the non-stretched samples, in the stretched samples the oriented mesophase occurs. Similar behavior also takes place for other mesophase networks studied. The energy characteristics of the mesophase formation will be described later.

A typical stress-strain curve for an isotropic amorphous samples of a moderate degree of crosslinking (PDES-3, PDES-4) consists of three well-defined parts. The initial part of the curve is identical to a stress-strain curve for a typical deformation curve of noncrystallizable elastomers. At this initial stage, the stress relaxation and hysteresis effects are very small.The second, S-shaped section appears as a result of the strain-induced mesophase formation, as it follows from X-ray, optical and thermo-mechanical data (see below). Its appearance is accompanied by a considerable stress relaxation. Finally, a sharp increase in stress occurs on the curve. In general, the shape of the stress-strain curve is determined by the relation between the deformation rate and the rate of the mesophase formation which, in turn, depends on temperature. At lower temperatures, the formation of the mesophase is a rather fast process, and the initially forming mesomorphic regions seem to

130

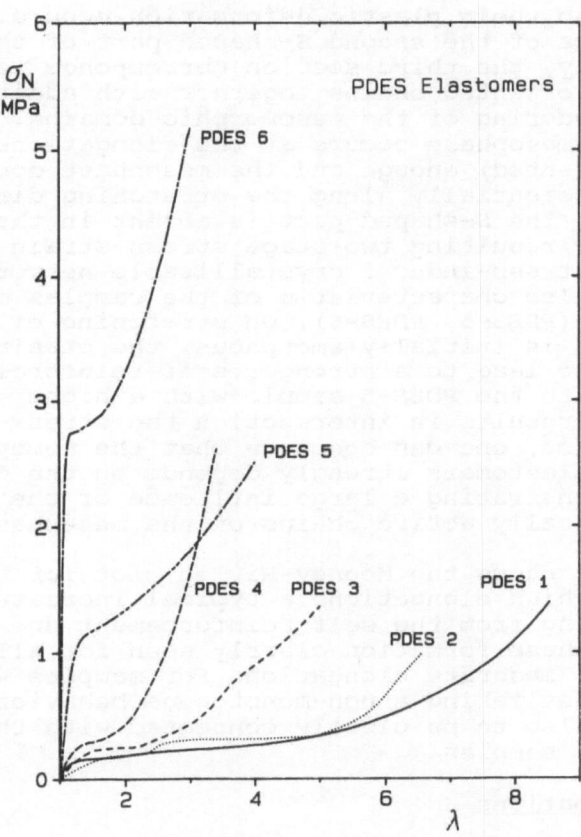

Figure 1.   Stress-strain curves for PDES elastomers.
Characteristics of the samples are listed in Table 1.

be oriented at some angle to the stretching direction. A further stretching is not only accompanied by the appearance of new mesomorphic regions but also a reorganization of these already existing regions. This process needs some additional work and leads to the three-parts stress-strain curve.

Thus, the shape of the total stress-strain curve can be explained as follows. After the initial equilibrium stretching, the mesophase occurs. Due to the growth of the oriented meso-morphic domains chains are straightening which results in level-ling off the stress. On further elongation along with additional mesophase formation a further orientation of the mesomorphic chains through their plastic deformation occurs.This results in the appearance of the second S-shaped part of the stress-strain curve. Finally, the third section corresponds to further stret-ching of the oriented chains together with additional reorgani-zation and ordering of the mesomorphic domains. At higher temp-eratures the mesophase occurs at the elongations when chains are extended (oriented) enough and the mesophase domains are oriented preferentially along the stretching direction. At such temperatures, the S-shaped part is absent in the stress-strain curve and the resulting two-stage stress-strain curve looks like that of the stress-induced crystallizable networks. Similar behavior is also characteristic of the samples of high degree of crosslinking (PDES-5, PDES-6). On stretching of the PDES-4 sample, which is initially amorphous, the strain-induced meso-phase seems to lead to a stronger self-reinforcing effects (in comparison with the PDES-5 sample with a higher crosslinking density). It results in intersection the stress-strain curve of the PDES. Hence, one can conclude that the mesophase behavior of the PDES elastomers strongly depends on the degree of cross-linking, demonstrating a large influence of the molecular weight of the elastically active chains on the mesophase formation.

Figure 2 shows the Mooney-Rivlin plot for various PDES networks. At high elongations a typical increase in reduced force resulting from the self-reinforcement due to the strain-induced mesophase formation clearly seen for all samples. How-ever, even at moderate elongations for samples with a moderate degree of crosslinking a non-monotonous behavior takes place, which seems also to be closely connected with the mesophase nature of the samples.

Cyclic deformations

An important feature of filled elastomers and rubber-like block copolymers is stress softening whereby they exhibits lower tensile properties at extension less than previously applied[34] As a result of this effect, a hysteresis loop on the stress-strain curve is observed. This effect is irreversible. The meso-phase PDES elastomers also exhibit stress softening behavior.

A typical behavior during the loading-unloading cycles is shown in Figure 3. The hysteresis effects during the first loading-unload cycle are very large for all these elastomers. Only for PDES-1, which has a high degree of mesomorphisity at room temperature, a larger part of the deformation during the first cycle is irreversible. In this respect it resembles the cyclic behavior of segmented polyurethanes and polyesters with a high content of the rigid block[34] and even that of crystalline polymers[34]. For all other PDES samples the deformation during

the first cycle is practically reversible which is typical of
the filled rubbers and block copolymers with a low or moderate
(less than approximately 50 %) content of the rigid block.

A very unusual feature of the PDES elastomers behavior
during the cyclic deformation is the denecking phenomenon. It
consists in the following. When a sample, stretched to a rather
large extension ratio (either with the formation of a neck or
without it) comes back to the initial non-stretched state,
always the denecking occurs: a neck is formed in the contracting
sample, which moves along the sample engulfing the thinner part
of the sample until the whole sample transforms to the initial
state. This phenomenon in polymers was discovered for the first
time in our studies[29-33,38,39]. Hence, the important feature of
the cyclic deformation of PDES elastomers is the strain-induced
mesophase formation during stretching follow by the immediate
isotropization during contraction. With increasing temperature
the hysteresis loops resulting from the cyclic deformations
decrease considerably due to the decreasing of the amount of
the strain-induced mesophase. At high enough temperatures, when
the strain-induced mesophase does not occur at all the cyclic
deformations of PDES elastomers are similar to conventional
elastomers.

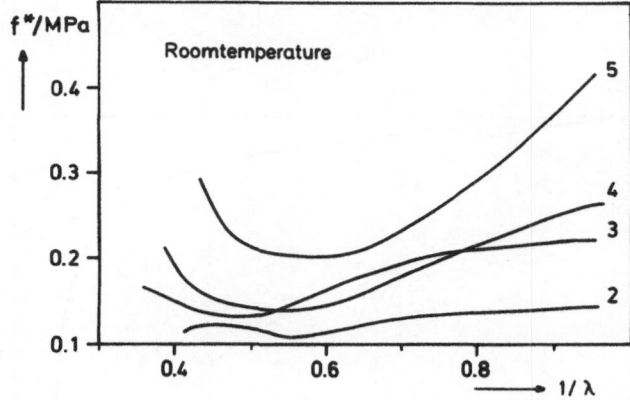

Figure 2. The Mooney-Rivlin representation of stress-strain
isotherms for various PDES samples. The numbers
near the curves correspond to the samples in Table 1.

OPTICAL BEHAVIOR

The stress-optical properties of conventional elastomers
are closely related to their mechanical behavior. It has been
established that the liquid crystalline elastomers, due to the
coupling of anisotropic order of the mesogenic gropus to
network elasticity, show unusual photoelastic properties[2].
However, the stress-optical behavior of liquid crystalline

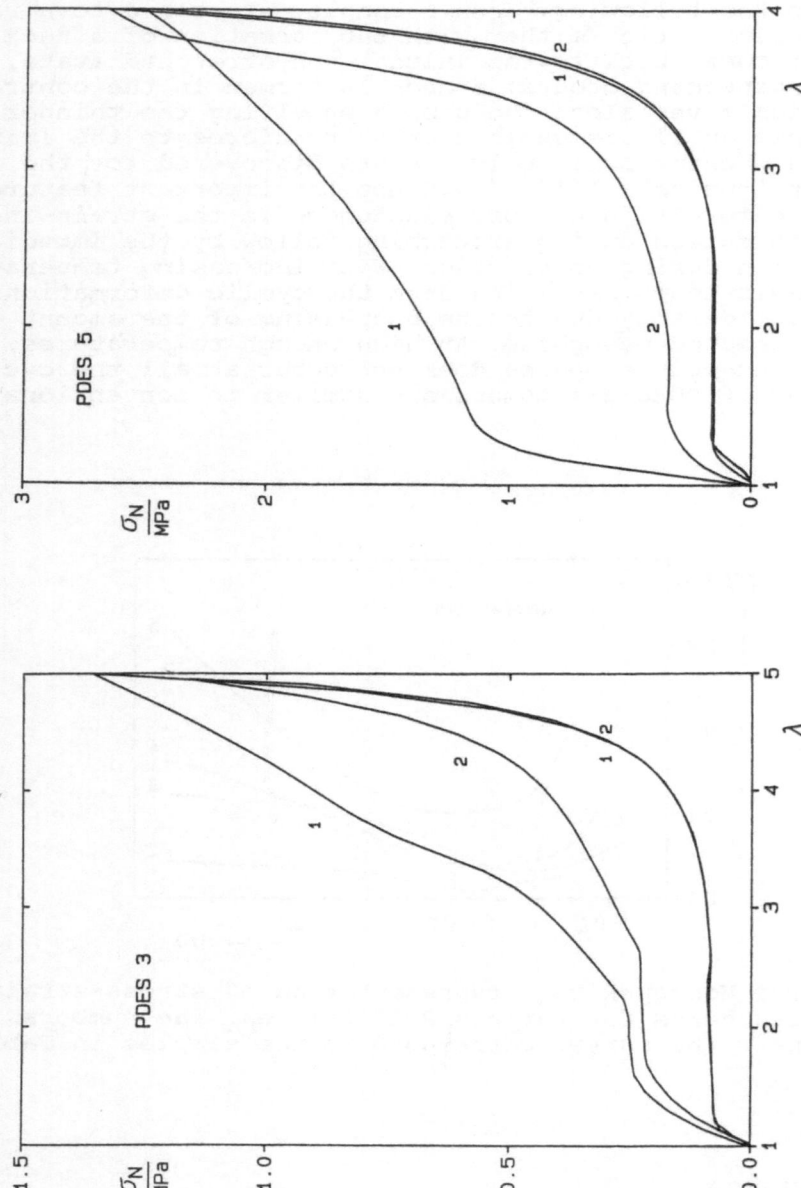

Figure 3. The cyclic deformation of PDES-3 and PDES-5 elastomers.
1 - 1-st cycle; 2 - 2-nd cycle.

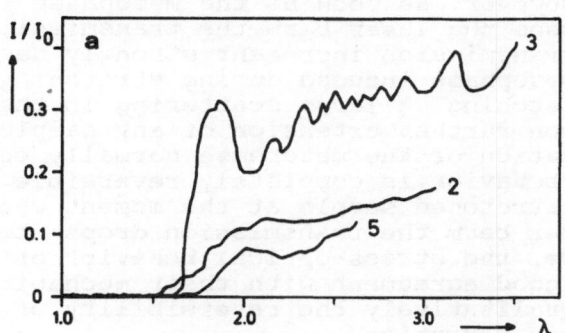

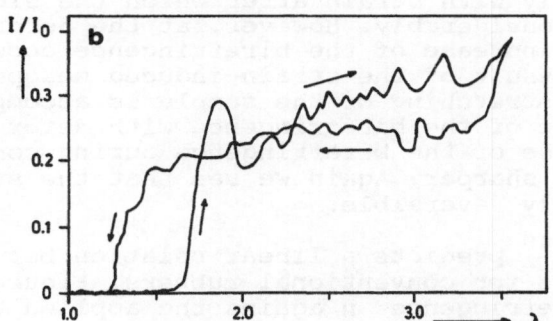

Figure 4. a) Relative transmission as a function of deformation
at room temperature. 2 - PDES-2; 3 - PDES-3;
5 - PDES-5;
b) Relative transmission as a function of deformation
in the cyclic mode at room temperature for PDES-3.

elastomers in the liquid crystalline state is determined mainly by the optical properties of the liquid crystalline phase. A similar behavior is also characteristic of the PDES mesophase elastomers.

Although upon deformation of conventional PDMS and PDES elastomers their birefringence is rather low, the strain-induced mesophase formation is accompanied by a considerable increase of that due to the appearance of the anisotropic mesophase domains[32]. Typical experimental results shown in Figure 4 demonstrate this very clearly. At the first stage of the stretching of the amorphous samples there is no change in transmission. However, as soon as the mesophase induced upon stretching crosses the laser beam the transmission increase jumpwise. The transmission increment strongly depends on the amount of the mesophase induced during stretching, as well as the rate of stretching. A large scattering in the relative transmission upon further extension of any sample shows that a local reorganization of the mesophase normally occurs. The stress-optical behavior is completely reversible. During contraction of the stretched sample at the moment when the neck crosses the laser beam the transmission drops stepwise to the zero line. Hence, the stress-optical behavior of the PDES networks is in good agreement with their mechanical behavior, demonstrating qualitatively the reversibility of the strain-induced mesophase formation.

Figure 5 shows a plot of the birefringence Δn against strain for PDES-3. At the initial stage the birefringence increase linearly with strain after which the slope of the line decrease considerably. However, at the deformation $\lambda = 1.7-1.8$ a stepwise increase of the birefringence occurs. This increase is a result of the strain-induced mesophase formation. The subsequent stretching of the sample is accompanied with only a small increase of the birefringence with deformation. The stepwise decrease of the birefringence during contraction is pronounced even sharper. Again we see that the stress-optical behavior is fully reversible.

The theory[40] predicts a linear relation between birefringence and stress for conventional rubbers. Figure 6 shows a plot of the birefringence n agains the applied true stress (true stress is defined as force per unit actual, strained cross-section on assumption of constant volume deformation). Even at the very initial stresses a very strong deviation from this linearity occurs. The positive sign of the birefringence $\Delta n = n(z) - n(x)$, where z corresponds to the axis of stress, and x to the direction normal to z and to the direction of the laser beam shows that segments become ordered even before the stepwise increase due to the stress-induced mesophase formation. During contraction the birefringence drops discontinuously at the moment when the neck is crossing the laser beam. A more detail analysis of these dependencies will be given in our further publications.

THERMOMECHANICAL BEHAVIOR

All the results discussed above point to a large contribution of the mesophase to the mechanical behavior of PDES elastomers. An additional valuable information concerning the

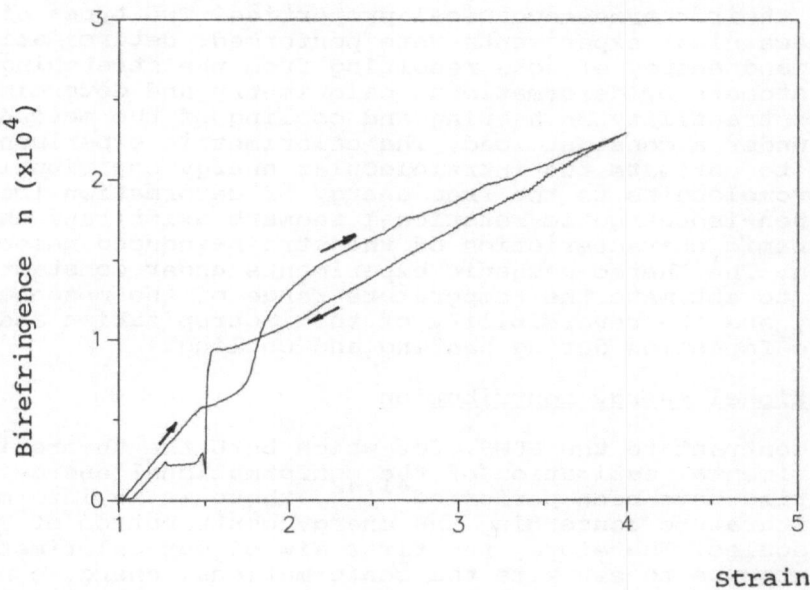

Figure 5. Birefringence Δn as a function of strain for PDES-3 at room temperature.

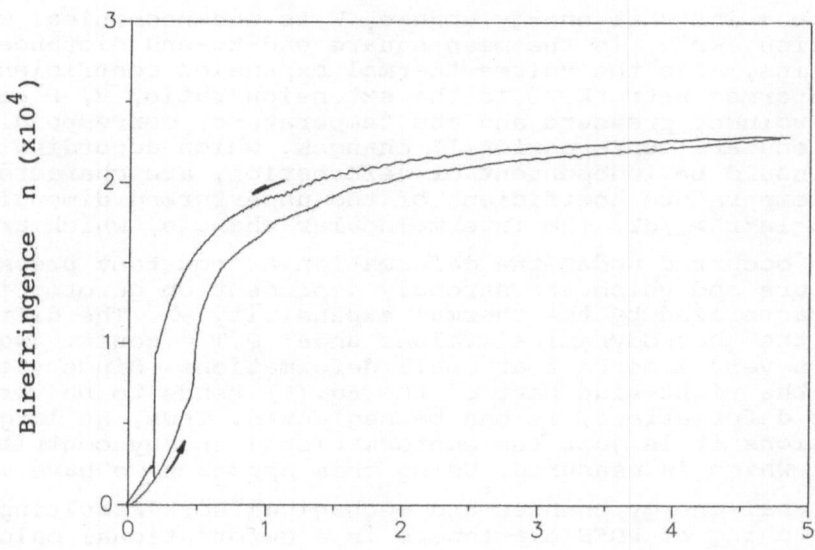

Figure 6. Birefringence Δn as a function true stress for PDES-3 at room temperature. The true stress is defined as force per unit actual, strained cross-section.

deformational behavior of PDES elastomers can be obtained by studying their thermo-mechanical properties. Two types of thermo-mechanical experiments were performed: determination of the heat and energy effects resulting from the stretching of the PDES elastomers by deformational calorimetry and determination of the contractility on heating and cooling of the mesophase samples under a constant load. The calorimetric experiments allow us to estimate the intramolecular energy contribution of PDES macromolecules to the free energy of deformation (because in PDES non-isoenergetic rotational isomers exist) and the thermodynamic characteristics of the strain-induced mesophase formation. The thermo-mechanic experiments under constant load allow us to estimate the temperature range of the mesophase stability and the reversibility of the isotropization and the mesophase formation during heating and cooling.

Conformational energy contribution

In contrast to the PDMS, for which both the theoretical and experimental estimation of the conformational energy contribution have been performed[41,42], there is no information in the literature concerning the energy contribution of PDES macromolecules. Therefore, the first aim of our calorimetrical experiments was to estimate the conformational energy contribution of PDES macromolecules. According to the thermodynamical theory of rubber-like elasticity the energy contribution to the free energy of deformation in stretching calorimetric experiments can be described as follows

$$(\Delta U/W)_{V,T} = T \frac{d \ln<r^2>_o}{dT} = (\Delta U/W)_{P,T} - \frac{2 \alpha T}{\lambda^2 + \lambda - 2} \qquad (1)$$

ΔU is the internal energy change, W is the mechanical work of deformation, $<r^2>_o$ is the mean square end-to-end distance of the free chains, α is the volume thermal expansion coefficient of the undeformed network, λ is the extension ratio, V, P and T are the volume, pressure and abs.temperature, correspondingly. Intramolecular (conformational) changes, which according to the theory should be independent of deformation, are characterized by the temperature coefficient of the unperturbed dimensions of chains d $\ln<r^2>_o$/dT. The intermolecular changes, which are normally occurred under the deformation at constant pressure and temperature and which are strongly dependent on deformation, are characterized by the thermal expansivity α. The difference between the thermodynamical values under P,T = const. and V,T = const. is very important at small deformations. Since the last term in the right-side part of the Eq.(1) tends to be very small at large deformations, it can be neglected. Thus, at large deformations it is just the conformational energy contribution $(\Delta U/W)_{V,T}$ which is measured. Using this approach we have measured the internal energy changes and mechanical work resulting from the stretching of PDES elastomers in a deformational calorimeter.

Our first measurements were carried out at the slightly cross-linked films[29-30,38]. To exclude the contribution of the mesophase, the experiments were performed with the amorphous samples at room temperature at small and moderate deformation ratios as well as at the temperatures 100-120°C which are considerably higher than the isotropization temperature of the non-stretched samples. Both these series of measurements gave an extremely high negative value of the energy contribution equal to -1.2. It is unusually high value of the energy contribution in comparison to other macromolecules studied corresponding to a very high negative temperature coefficient of unperturbed chain dimensions d $\ln<r^2>_o/dT = -4\times10^{-3} \deg^{-1}$. Recently[32,33,39], this value was reconsidered because it has been established that a well-defined amount of the strain-induced mesophase appeared at elongations at temperature as high as 130°C (see below). To exclude the influence of the mesophase on the intramolecular energy contribution the measurements were carried out at high enough temperatures at which even the small amounts the strain-induced mesophase seem to be excluded totally. However, even more important from this point of view are the measurements[33] performed on the rubber-like PDES block copolymers with the values of the molecular weight of the PDES blocks considerably lower than the critical molecular weight of the mesophase formation in PDES. In this case even the strain-induced mesophase cannot appear at all. Similar measurements were performed also on the PDES elastomers of a high crosslinking density when the mesophase is also absent even at the highest degree of extension which can be reached for these elastomers without breakage.

It has been established that the behavior of all the samples free of the mesophase follow the theoretical prediction. The most important conclusion drawn from these measurements is that that the conformational energy contribution for PDES macromolecules is negative one and independent of the deforma-

tion ratio similar to other macromolecules. The quantitative energetic characteristics of PDES and PDMS macromolecules are listed in Table 2. It is noteworthy that the sign of the energy contribution in PDES is negative unlike PDMS which is not able to exist in a mesophase state. The negative value of the energy contribution in PDES seems to mean that the extended conformations are energetically more favorable. It is also worth mentioning that in PE[34,42] and some linear polyphosphazenes[43], which are also capable of forming the mesophase under an elevated pressure (PE) or normal pressure (polyphosphazenes), the conformational energy contribution is also negative. It is interesting to prove whether the negative sign of the intramolecular energy contribution is a characteristic of the ability of the flexible macromolecules to exist in a mesophase state.

Energy characteristics of strain-induced mesophase formation

There are two negative contributions to the internal energy change upon stretching of PDES elastomers. According to the negative conformational energy contribution of PDES macromolecules stretching of all samples (after stress softening) have to be accompanied by a decrease of the internal energy. Moreover, the strain-induced mesophase formation should be accompanied by a considerably large decrease of internal energy. The

Table 2. Energy characteristics of PDMS and PDES macromolecules

Polymer	Energy contribution	$d \ln<r^2>_o/dT \times 10^3$
PDMS	0.27 (+ 0.05)	0.86
PDES	-0.25 (+ 0.05)	-0.80

internal energy changes during stretching of various types of PDES elastomers were measured by means of deformation calorimetry.

Figures 7-8 show typical curves of the mechanical work, heat evolved and internal energy changes at two different temperatures as a function of deformation. For all samples studied the internal energy decreases both at the room and high temperature. There are either minima or the tendency to reach a minimum on the curves. Such a behavior seems to indicate that in these minima the larger part of the strain-induced mesophase is formed and on further stretching the main contribution to the decrease of the internal energy should arise from the conformational contribution.

Using the data on the internal energy changes presented in Figure 8 it is possible now to estimate the amount of the mesophase (degree of mesomorphisity $ß_m$) induced during stretching of the initially amorphous samples as a function of deformation The results of the estimation are shown in Figure 9. It is seen that at the room temperature the mesophase starts to induce at the very initial deformations. For lightly cross-inked PDES-2 sample $ß_m$ can reach rather high values at the room temperature. However, may be the most surprising results have been obtained at higher temperature. It is seen that the mesophase can be induced in both samples at the temperatures considerably higher than the isotropization temperature of the mesophase without the action of external stresses. It means that due to the stress induced mesophase formation the self-reinforcement of such samples can occur at a very high temperatures.

Reversible contractability under load

Now a very interesting question arise: up to what temperatures the mesophase induced by stretching at room temperature is stable? To answer this question the thermomechanical experiments with samples loaded by a constant load were carried out. As it seen from Figure 10 where typical results for PDES-2 are shown, the shrinkage of samples is very small in the initial temperature range. With rising temperature it increases, however, progressively and finally the contraction occurs. Dependence of the temperature of the contraction on the extension ratio is extremely large. The temperature at which the contraction of the sample was finished was chosen as the isotropization temperature, T_i, under a load. The contraction of the mesophase samples was quite reversible because during cooling of the loaded isotropic samples the oriented mesophase occurred again

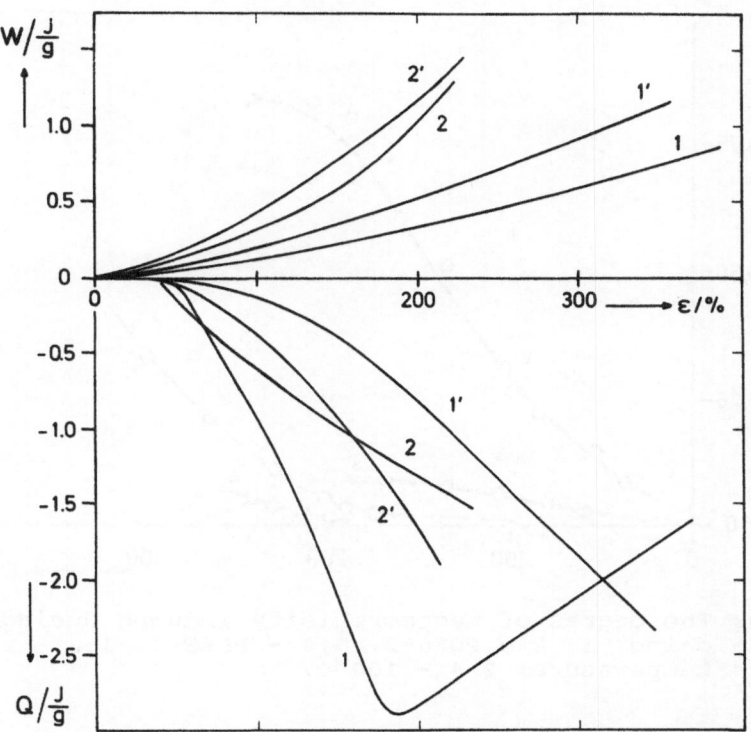

g. 7. Mechanical work W and heat Q on stretching PDES-2 and
PDES-5 from the unstrained state to ε . 1,1'-PDES -2;
2,2'- PDES-5. 1,2 - room temperature; 1'2' - 100°C.

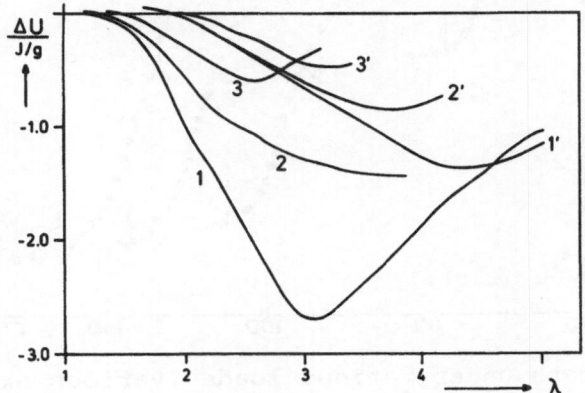

g. 8. Internal energy changes as a function of strain for
PDES elastomers. 1,1'- PDES-2; 2,2' - PDES-3; 3,3' -
PDES-5. 1, 2, 3 - 20°C; 1',2',3' - 100°C.

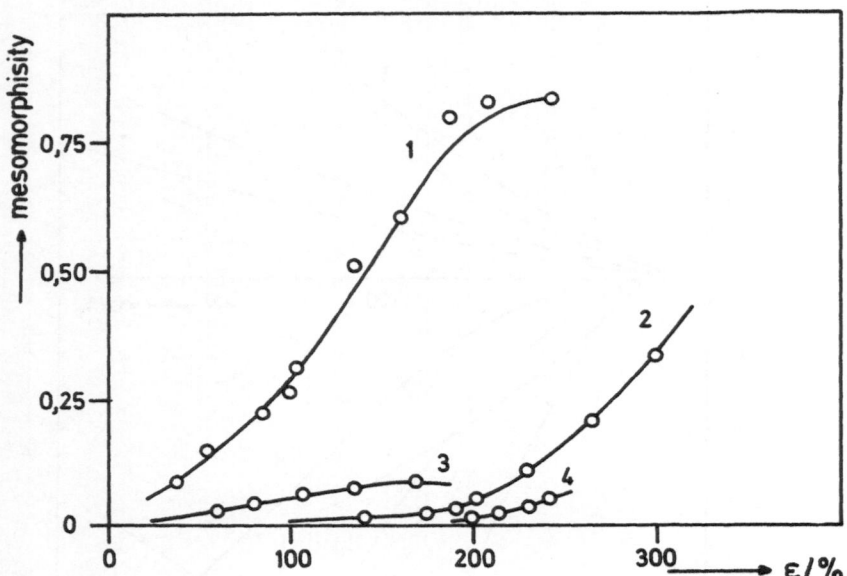

Figure 9. The degree of mesomorphisity induced during stretching. 1, 2 - PDES-2. 3,4 - PDES-5. 1,3 - room
temperature; 2,4 - 100°C.

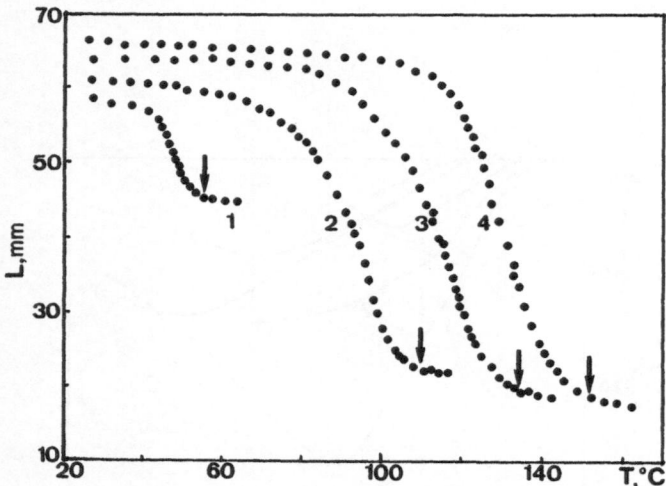

Figure 10. Length under various loads (various extension
ratios) as a function of temperature for PDES-2.
The respective values of the the extension ratios
are: 1 - 1.92; 2 - 3-10; 3 - 3.95; 4 - 4.53.
Arrows indicate the isotropization temperature T_i.

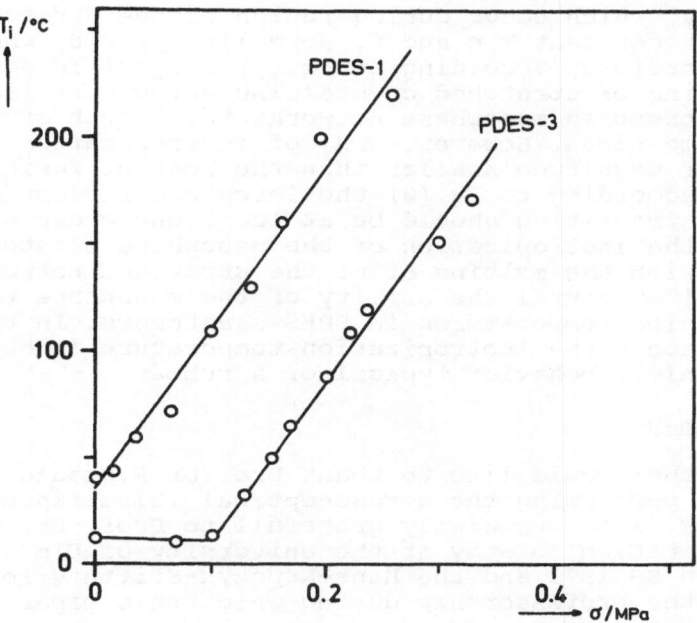

Figure 11. Isotropization temperature, T_i, vs.extension ratio
for PDES-1 and PDES-3.

spontaneously[32],[37]. It is remarkable that the degree of under-
cooling at which the spontaneous elongation normally occurs in
all runs is of about $\Delta T=20-25\,^{\circ}C$ which is typical for crystallization of fast crystallizing flexible polymers.

Figure 11 shows the dependence of T_i on the extension
ratio for PDES-1 and PDES-3. Before measurements PDES-1 was
drawn plastically approximately 5 times after which the sample
was loaded with a load. PDES-3 was practically amorphous at
room temperature, therefore, a load was applied directly to the
initial sample. It is seen that for both these samples the
dependence is linear in the range of the extension ratios
studied. The slope of the dependences is practically the same
for both samples and is very large which seems to be the direct
consequence of a low value of the heat of isotropization.

The application of the Clausius-Clapeyron equation for the
influence of pressure on the temperature of the first order
phase transition T_{tr} (melting or isotropization) to the uniaxialy
stretched rubber-like materials which undergo the first order
phase transition under a load (force) gives

$$dT_{tr}/df = -(T\Delta L/\Delta H) \tag{2}$$

where ΔL and ΔH are the changes in length and enthalpy, respectively, which occur during fusion of the stretched elastomer at constant T,p and f. Normally, $\Delta L < 0$, whereas $\Delta H > 0$. Therefore, according to Eq.(2) dT_{tr}/df is positive. During melting of stretched crystalline network or isotropization of stretched mesophase networks the length of the samples decrease some times. However, ΔH_i of isotropization is at least one order of magnitude smaller than the heat of fusion, ΔH_m. Therefore, according to Eq.(2) the force coefficient of the temperature transition should be at least one order of magnitude larger for the isotropization of the mesophase elastomers in comparison with the melting of of the stretched networks. It explains quite natural the ability of the mesophase to be induced at high temperatures in PDES elastomers. In the temperature range above the isotropization temperature is observed the thermomechanical behavior typical of a rubber.

Acknowledgements

The author would like to thank Prof.Dr.R.Stadler (Mainz) for his help in performing the stress-optical (birefringence) measurements. I am especially gratefull to Prof. Dr. W. Pechhold for his invitation to stay at the University of Ulm as visiting professor in SS 1992 and the Hans-Kupczyk-Stiftung for financial support of the professorship during which this paper was completed.

REFERENCES

1. G.W. Gray, in: Side Chain Liquid Crystal Polymers, C.B. McArdle, ed., Blackie & Son, Glasgow (1989).
2. W. Gleim, H. Finkelmann, in: Side Chain Liquid Crystal Polymers, C.B.McArdle, ed., Blackie & Son, Glasgow (1989).
3. R. Zentel, Angew. Chem.Adv.Mater. 101 1437 (1989).
4. P.-G. de Gennes, C.R.Acad.Sci.Ser. B281 101 (1975).
5. P.-G. de Gennes, in: Polymer Liquid Crystals, A. Ciferry, W.R. Krigbaum, R.B. Meyer, eds., Academic Press, New York, (1982).
6. V.V. Rusakov, in: Structural and Mechanical Properties of Composite Materials, Sci.Papers of Ural Sci.Center of Akad. Nauk SSSR, Sverdlovsk (1984).
7. S.S. Abramchuk, A.R. Khokhlov, Dokl.Akad.Nauk SSSR, 297 385 (1987).
8. S.S. Abramchuk, I.A. Nyrkova, A.R. Khokhlov, Polymer Science USSR, 31 1936 (1989).
9. M. Warner, K.P. Gelling, T.A. Vilgis, J.Chem.Phys.88 4008 (1988).
10. B. Deloche, E.T. Samulski, Macromolecules, 21 3107 (1988).
11. M. Warner, X.J. Wang, Macromolecules, 24 4932 (1991).
12. C.L. Beatty, J.M. Pochan, M.F. Froix, D.D. Hinman, Macromolecules 8 547 (1975).
13. M.F. Froix, C.L. Beatty, J.M. Pochan, D.D. Hinman, J.Polym. Sci.,Polym.Phys.Ed. 13 1269 (1975).
14. C.L. Beatty, F.E. Karacz, J.Polym.Sci.,Polym.Phys.Ed. 13 971 (1975).
15. J.M. Pochan, C.L. Beatty, D.D. Hinman, F.E. Karasz, J.Polym. Sci.,Polym.Phys.Ed. 13 977 (1975).
16. J.M. Pochan, D.F.Hinman, M.F. Froix, Macromolecules 9 611 (1976).
17. V.S. Papkov, Yu.K. Godovsky, V.M. Litvinov, V.S. Svistunov, A.A. Zhdanov, 4-th International Conference on Liquid Crystals, Tbilisi (1981).

- V.S. Papkov, Yu.K. Godovsky, V.S. Svistunov, V.M. Litvinov, A.A. Zhdanov, J.Polym.Sci.,Polym.Chem.Ed. <u>22</u> 3617 (1984).
- D.Ya. Tsvankin, V.S. Papkov, V.P. Zhukov, Yu.K. Godovsky, V.S. Svistunov, A.A.Zhdanov, J.Polym.Sci.,Polym.Chem.Ed. <u>23</u> 1043 (1985).
- Yu.K. Godovsky, V.S. Papkov, Makromol.Chem.,Macromol.Symp. <u>4</u> 71 (1986).
- V.S. Papkov, V.S. Svistunov, Yu.K. Godovsky, A.A. Zhdanov, J.Polym.Sci.,Polym.Phys.Ed. <u>25</u> 1859 (1987).
- J. Friedrich, J.F. Rabolt, Macromolecules <u>20</u> 1975 (1987).
- G. Kögler, A. Hasenhindl, M. Möller, Macromolecules <u>22</u> 4190 (1989).
- G. Kögler, K. Loufakis, M. Möller, Polymer <u>31</u> 1538 (1990).
- E.W. Fischer, K. März, N. Willenbacher, M. Ballauff, M. Stamm, 23rd Europhysics Conference on Macromolecular Physics, Stockholm, Vol.15B, L3 (1991); see also N.Willen-bacher, Dissertation, MPI, Mainz, (1990).
- W. Pechhold, P. Schwarzenberger, in: Frontiers in High Pressure Research, Plenum Publ.Corp.,N.Y. (1992)
- Yu.K. Godovsky, N.N. Makarova, V.S. Papkov, N.N. Kuzmin Makromol.Chem.,Rapid Commun. <u>6</u> 443 (1985).
- Yu.K. Godovsky, I.I. Mamaeva, N.N. Makarova, V.S. Papkov, N.N. Kuzmin, Makromol.Chem.,Rapid Commun. <u>6</u> 797 (1985).
- Yu.K. Godovsky, V.S. Papkov, Adv.Polym.Sci. <u>88</u> 129 (1989).
- V.S. Papkov, Yu.P. Kvachev, Progr.Colloid Polym.Sci. <u>80</u> 221 (1989).
- Yu.K. Godovsky, I.A. Volegova, L.A. Valetskaya, A.V. Rebrov, L.A. Novitskaya, S.I. Rotenburg, Polymer Science USSR <u>30</u> 329 (1988).
- Yu.K. Godovsky, Angew.Makromol.Chem., in press
- Yu.K. Godovsky, I.A. Volegova, A.V. Rebrov, L.A. Novitskaya, S.B. Dolgoplosk, I.G. Kolokol'tzeva, Polymer Science USSR <u>32</u> 726 (1990).
- Yu.K. Godovsky,"Thermophysical Propereties of Polymers", Springer-Verlag, Berlin, Heidelberg, New York, (1992).
- Yu.K. Godovsky, Polymer <u>22</u> 75 (1981).
- Yu.K. Godovsky, Progr.Colloid Polym.Sci. <u>75</u> 70 (1987).
- Yu.K. Godovsky, L.A. Valetskaya, Polymer Bull. <u>27</u> 221 (1991)
- V.S. Papkov, Yu.K. Godovsky, V.S. Svistunov, A.A. Zhdanov, Polymer Science USSR <u>31</u> 1729 (1989).
- Yu.K. Godovsky, L.A. Valetskaya, V.S. Papkov, Makromol. Chem., Macromol.Symp. <u>48/49</u> 433 (1991).
- L. Treloar,"The Physics of Rubber Elasticity", Clarendon Press, Oxford (1975).
- P.J. Flory,"Statistical Mechanics of Chain Molecules", Wiley-Interscience, New York (1969).
- J.E. Mark, B. Erman,"Rubberlike Elasticity: A Molecular Primer", Wiley-Interscience, New York (1988).
- E. Saiz, J.Polym.Sci.,Polym.Phys.Ed. <u>25</u> 1565 (1987).

ORIENTED LIQUID CRYSTALLINE NETWORK POLYMERS

B.A.Rozenberg, L.L.Gur'eva

Institute of Chemical Physics
Russia Academy of Sciences
Chernogolovka, Moscow region 142432, Russia

INTRODUCTION

Liquid crystalline network polymers (LCNP's) were discovered about 20 years ago [1-3], simultaneously with liquid crystalline polymers of linear type (LCP's). Nevertheless, we can state now that LCP's have been studied considerably better than LCNP's. The results of these investigations are documented in numerous books and reviews [3-10]. Formation of highly oriented, macroscopically uniform LCP's under influence of external mechanical, electric or magnetic fields and fixation of chain orientation at cooling of polymers below glass transition temperature allows one to obtain films, fibers or molded plastics with high degree of anisotropy of physical characteristics and outstanding mechanical properties, thermal and chemical stability. LCP's also exhibit interesting optical properties revealed under the action of electric or magnetic fields. These properties are perspective for the development of new type of materials for recording, displaying and storing information [8,11]. So, if the usefulness and opportunities of LCP's application now is understood quite well, the revealing of LCNP's potentials are only in a progress.

Advances in investigations of the LCP's chemistry, structure and properties and especially in LCNP's application stimulated extension of LC concept to the network polymers. It was shown [11] that ordered anisotropic comb-like elastomers can be used as optical switches, wave guides and materials with non-linear optical properties.

Recently groups of researchers [11-26] concentrated on synthesis and characterization of a new type of monomers (oligomers) with rigid, rod-like central mesogenic group, capped from both ends by the reactive functional groups, and LCNP's on their basis. This new class of thermosets (LCT's) has to be capable to undergo orientation in electric or magnetic fields and to retain the orientation of mesogenic groups in the resulting cured polymer. The same approach can be used for fabrication of oriented LC thermoplastics from corresponding monomers. It is worth noting that orientation in the electromagnetic field is the only way for ordering network chains in the case of LCT's used as polymer matrices for composites. The opportunity to arrange the molecular reinforcement in the interlaminar direction for the fiber reinforced composites, seems to be very important. Because of low molecular weight, LCT's as starting substances for LCNP's can be oriented in the electromagnetic field much easier than high molecular LCP's [3,8].

Synthesis, Characterization, and Theory of Polymeric Networks and Gels
Edited by S.M. Aharoni, Plenum Press, New York, 1992

Moreover, using LCT's as matrices for composites it would be possible to achieve the following advantages:
- to orient the molecules of the polymer matrix in any necessary direction;
- to decrease the coefficient of thermal expansion in the direction of the orientation;
- to increase the matrix thermal resistance;
- to avoid largely the rheological problems and provide good wetting and impregnation of fibers by the oligomer;
- to decrease the matrix shrinkage during cure and
- as a result to simplify the processing of composites and to improve their mechanical properties.

In this paper synthesis and characterization of new diepoxy type LCT and LCNP's on its basis are presented. Some common problems of LCTs' synthesis and phase transitions are also in the focus of our discussion.

EXPERIMENTAL SECTION

Synthesis of monomer

Synthesis of the diepoxy LC monomer - diglycidyl ester of terephtaloil-bis-(4- hydroxybenzoic acid) (DGET) was carried out according to the following scheme [16,17]:

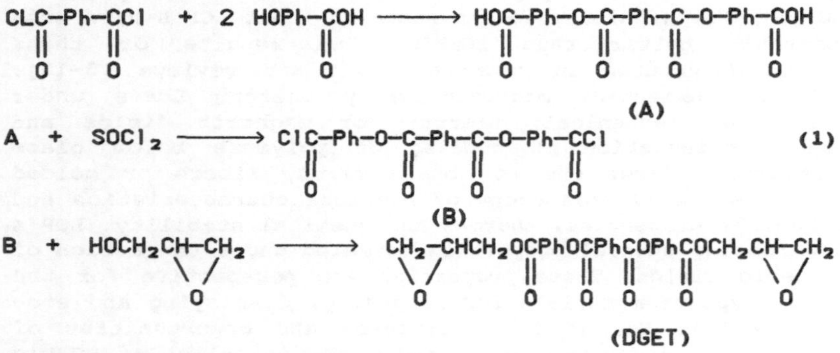

$$CLC-Ph-CCl \; + \; 2 \; HOPh-COH \longrightarrow HOC-Ph-O-C-Ph-C-O-Ph-COH$$

(A)

$$A \; + \; SOCl_2 \longrightarrow ClC-Ph-O-C-Ph-C-O-Ph-CCl \qquad (1)$$

(B)

$$B \; + \; HOCH_2CH-CH_2 \longrightarrow CH_2-CHCH_2OCPhOCPhCOPhCOCH_2CH-CH_2$$

(DGET)

Substances A and B were synthesized according to the description given in ref. [27]. The target product (DGET) was synthesized as follows:

Solution of freshly distilled 5.3 g of glycidyl alcohol in 20 ml of THF added by drops during 1 h to the mixture of 11 g of A and 8 ml of triethylamine (TEA) in 700 ml of freshly purified THF or 1,4-dioxane under intensive mixing in the dry argon flow at $23°C$. The reaction mixture was stirred in the argon flow during 24h. The precipitate of chlorohydrate of TEA was filtered. The reaction product was evolved from the reaction mixture by precipitation in 1.5 l of methanol, hexane or distilled water with the following filtration, washing with the corresponding precipitating agent and drying. White crystal powder of the DGET obtained was reprecipitated again. Yield is 10.2 g (75%).

The purification of starting reagents was fulfilled in the following manner. The glycidyl alcohol was distilled in vacuum (266 Pa) at $70°C$ and its purity was controlled by chemical analysis on the content of epoxy group and by gas chromatography. TEA was refluxed over the dry KOH and then distilled at $87°C$. THF was purified by refluxing over the dry KOH and then distilled under the fresh portion of KOH and finally under the $LiMgH_3$.

Elemental and chemical analysis, gel-permeation chromatography (GPC), IR-spectroscopy, polarized light microscopy, differential scanning calorimetry (DSC), high-resolution proton and ^{13}C NMR and X-ray analysis were used for sample characterization.

Techniques

Molecular weight was determined with a HPLC Millichrom-1 equipped with column with Silasorb SPH-600 treated with silane as stationary phase. 1,4-dioxane was used as eluent; the rate of elution was 200 microliter/min. The measurements were made by using UV detector.

IR-spectra were recorded with a IR-spectrometer Specord M-80.

High resolution proton and ^{13}C NMR spectra were recorded with BS-567A instrument operating at 80 and 20 MHz.

Thermal analysis was performed with a differential scanning calorimeter Mettler, calibrated in the usual manner, and with the derivatograph Q-1500 D.

Investigations of cure kinetics were performed by using an isothermal calorimeter DAC-1-1A.

Textures of the materials were studied with a polarized light microscope Boetius and X-ray diffractometer DRON-UM-2 using Cu K_α radiation both equipped with a hot stage and temperature programmer. Order parameters of the samples oriented in the magnetic field with H=1.45T were determined from X-ray diffractogram as in ref. [28].

Characterization of monomer

IR (KBr): ν/cm^{-1} = 1605, 1730 and 910. They correspond to the valence vibrations of benzoic ester and epoxy groups.

^{1}H NMR (CDCl$_3$): δ (in ppm from tetramethylsilane) and I (in Hz) = 8.27 (s, 4H, aromatic); 8.11 (d, 4H, aromatic, I=8.9); 7.29 (d, 4H, aromatic, I=8.9); 4.63 (dd, 2H, aliphatic, I=12.2 and 3.1); 4.11 (dd, 2H, aliphatic, I=12.2 and 6.1); 3.27 (m, 2H, epoxy, I=2.7; 3.1; 4.1; 6.1); 2.84 (dd, 2H, epoxy, I=4.8 and 4.12); 2.67 (dd, 2H, epoxy, I=4.8 and 2.7).

^{13}C NMR (CDCl$_3$): δ (in ppm from tetrametyhylsilane) and I (in Hz)= 44.5 (^{1}C, t); 549,3 (^{2}C, d); 65.5 (^{3}C, t); 121.5 (^{6}C,$^{6'}$C, dd, ^{1}I=161.7, ^{2}I=5.3); 127.5 (^{5}C, t, ^{5}I=8.5); 130.3 (^{11}C,$^{11'}$C, d, I=167.2); 131.3 (^{7}C, $^{7'}$C, dd, ^{1}I=173.6, ^{2}I=7.3); 133.6 (^{10}C,t, ^{2}I=3.9); 154.4 (^{8}C, s); 163.4 (^{4}C, s); 165.2 (^{9}C, s). ^{13}C NMR spectrum corresponds to the structure:

These data indicate that the synthesized substance corresponds to the molecular structure of DGET (scheme 1). However, the epoxy group content of DGET, as determined by chemical analysis, is 80-95 % of the calculated value. GPC analysis also shows definite disagreement with the expected structure of DGET. Chromatograms contain two peaks. One of them is DGET; the molecular weight of the second compound is twice as large as the DGET one. The side product content depends on the conditions of the DGET synthesis and drastically increases with the reaction temperature (Fig 1).

Nevertheless, the elemental analysis of DGET is rather good. Thus, for a sample with 20% of the side product it is found that C=64.86% and H=4.28% (calculated for the DGET structure C=64.80% and H=4.18).

These experimental results indicate clearly that the side product at DGET synthesis is its dimer, i.e., dimerization of DGET takes place under the reaction conditions. As expected, the comparison of the epoxy group contents in the synthesized samples of DGET revealed a good agreement between direct experimental measurements of these values and those calculated from GPC measurements of the dimer fraction in the samples.

According to the polarized light microscopy data DGET transforms to a fluid, at 158°C displaying an optical anisotropy, which at $230-250^{\circ}$C converts to a semi-transparent solid, insoluble in organic solvents.

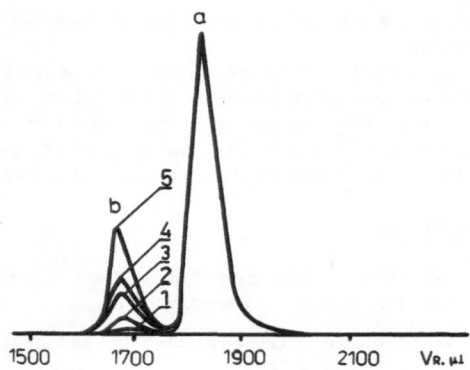

Figure 1. Dependence of the reaction product composition at DGET synthesis on the reaction temperature, $^{\circ}$C: 1-19; 2-21; 3-23; 4-25; 5-30.

Just at these temperatures we observe on DSC thermogram the endothermic peak at 158°C with the enthalpy $\Delta H = 31$ kJ/mol and wide exothermic peak in the region of $220 - 260^{\circ}$C and $-\Delta H = 160-190$ kJ/ mol depending on the DGET's purity and the way of extraction [17]. from the reaction system. It was found that the DSC curves may occasionally contain, in addition to phase transition at 158°C, the endothermic peak at 120°C with $\Delta H = 20$ kJ/mol (Fig.2). An occurrence of this phase transition is connected with the nature of the precipitating agent used and the preheating of the sample; it displays at the first temperature scanning and disappears during the second one when the sample had been heated to $T>120^{\circ}$C, but it can be appeared again after reprecipitation from the THF to methanol. One can conclude from this results that this phase transition is responsible for the metastable crystal - stable crystal transition. IR-spectroscopy and X-ray analysis of DGET samples with different prehistory gave an additional direct confirmation to this conclusion [17].

The system of sharp reflexes covering wide ($15-20^{\circ}$) and small (3.1°) angles is observed on the X-ray scattering patterns for DGET at room temperature (Fig. 3, curve 1). When the temperature was raised to 158°C, the small-angle reflex disappeared (curves 2 and 4) while the sharp wide-angle reflexes widened into the diffuse halo. (curve 3). The preservation of week traces of crystal structure may be connected with the presence in DGET small amount of its dimer.

The above data (endothermic maximum on the thermograms, the presence of optical anisotropy after a fluidity appears, the diffuse halo at wide-angle X-ray scattering and the absence of small-angle scattering maximum after DGET crystals melted) indicates conclusively that DGET is thermotropic LC monomer with the nematic-type mesophase. The temperature of the DGET crystal-LC transition is 158°C . The absence of longitudinal packing of DGET molecules above the melt point supports that the DGET crystal structure transforms into the nematic phase upon melting.

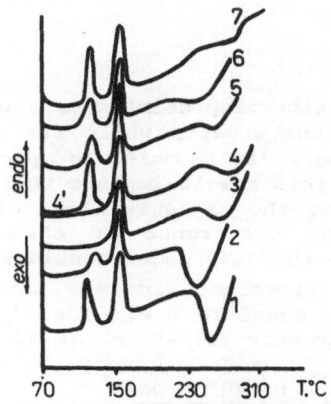

Figure 2. DSC traces of DGET samples prepared by using different precipitating agents or subjected to heating. Heating rate is 16 K/min. Precipitating agent: 1 - methanol; 2 - hexane; 3 - water; 4 - after preliminary scanning of the sample 1; 5 - sample 4 reprecipitated from THF to methanol; 6 - sample 3 reprecipitated from THF to methanol; 7 - sample 1 many times reprecipitated from THF to methanol.

An additional confirmation for the nematic structure of DGET mesophase is provided by data of DGET behavior in the magnetic field (Fig.4). DGET orientation was carried out by 5 min exposition in the magnetic field (H=1.45 T) at 160°C with the subsequent cooling to room temperature with dT/dt = 4 K/min. The occurrence of wide and small X-ray scattering reflexes on the X-ray patterns for DGET samples exposed in the magnetic field is the obvious evidence for the correctness of the conclusion on the nematic character of the DGET mesophase. Moreover, these results show that DGET orientation in the nematic phase is not destroyed at the crystallization owing to partial polymerization of DGET in the mesomorphic state.

The analysis of the texture-pattern of oriented DGET samples shows

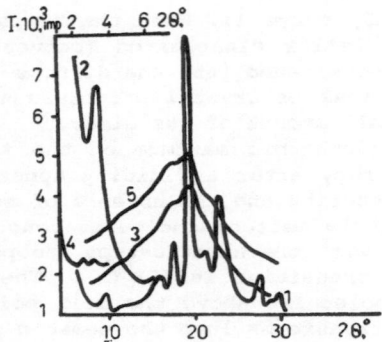

Figure. 3. DGET X-ray scattering pattern at wide (1, 3, 5) and small (2, 4) angles at 20°C(1, 2), at 158°C(3, 4) and at 200°C(5).

that the wide-angle reflection responsible for the side packing of mole-cules is concentrated on the equator while the small-angle reflection responsible for the longitudinal molecule packing is concentrated strictly on the meridian. This result implies that the main axis of DGET molecules is oriented along the magnetic field direction. This is ano-ther confirmation for the occurrence of the orientationally well-regulated nematic phase in the case under consideration.

The degree of order determined from the azimuthal distribution of X-ray scattering intensity equal to 0.81, i.e. local deviations of the DGET molecules from the director are as low as 20°.

Characterization of networks based on DGET

Network polymers obtained by DGET self cure in the LC temperature range possess anisotropy (Fig. 3, curves 3 and 5) and preserve the LC state characteristic for the monomer while for DGET cured above isotro-pization point no anisotropy was observed. Anisotropic polymers were synthesized in the LC state of DGET by the action of aromatic diamines

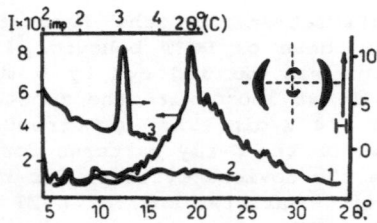

Fig. 4. Texture-pattern of the DGET sample oriented in the magnetic field (H=1.45 T) and scanning curves along the equator on the wide angles (1), along the meridian on the wide (2) and small angles (3).

4,4′-diaminodiphenylsulphone) or ternary amine as catalyst. In the last case LCNP can be modified by addition of the bisphenol A. As expected the cure reaction in solution (DGET + isomethyltetrahydrophtalic anhydide) results in the isotropic polymer network.

The network polymer obtained by DGET self cure in the LC state has $T_g = 233 °C$ and $M_c = 530$ (the data of TMA analysis) that is in the good agreement with the calculated molecular weight of the interknot chain for the molecular structure characteristic to the anionic polymerization of epoxies.

Table 1. Mesogens of minimal size for LCT's.

1. —⟨◯⟩—C≡C—⟨◯⟩—   6. —⟨◯⟩—⟨◯⟩—

2. —⟨◯⟩—CH=C(R)—⟨◯⟩—   7. —⟨◯⟩—C(=O)—O—⟨◯⟩—
   R = H, CH₃

3. —⟨◯⟩—N=N—⟨◯⟩—   8. —⟨◯⟩—C(=O)—NH—⟨◯⟩—

4. —⟨◯⟩—N=N(→O)—⟨◯⟩—   9. —⟨◯⟩—CH₂—CH₂—⟨◯⟩—

5. —⟨◯⟩—CH=N—⟨◯⟩—   10. —⟨◯⟩—CH₂—O—⟨◯⟩—

The long-range ordering of DGET in the magnetic field does not restrict its polymerization. The densely-crosslinked polymer formed preserves the monomer orientation. The uniaxially ordered polymer is stiff semi transparent solid no longer shows liquid-crystalline transition and remain oriented up to the temperatures of its thermal decomposition >320 °C).

SYNTHESIS OF LCT's

The principles of LCT's synthesis combine well-known ones of the reactive functional groups construction for conventional thermosets and the design of rigid, rod-like mesogenic groups. Sometimes spacer as element of LCT is desirable or even it is essential from the viewpoint of synthesis and especially processing of LCT's (low solubility and high melting temperature and, as consequence, high cure temperature).

One can find mesogens of minimal size that can be used for LCT's synthesis [11-26, 29,30] in Table 1.

All of mesogenes listed here are rigid, rod-like on nature except of two last [29], which are flexible but rod-like (anti conformer) owing to the conformational isomerism with kinked-like (gauche) conformer. These new types of flexible rod-like mesogens were not used so far for the LCT's synthesis. Nevertheless, it would be interesting to use them, since the conformational equilibria have to be shifted to the extended rod-like anti conformer in the magnetic field and will be fixed in the resulting network after cure.

The choice of the terminal functional reactive groups for the LCT's synthesis is quite limited because of strong requirements to their thermostability and reactivity and, as mentioned above, to the cure mechanism. Terminal functional reactive groups, which are suitable for the synthesis of LCT's [11-26,29,30], are listed in the Table 2.

Table 2. Terminal functional reactive groups for LCT's

One of the most suitable terminal functional groups for LCT's is the epoxy one. The main weak point of this group is its limited thermostability.

All combinations of terminal groups and mesogens are not suitable for the synthesis of LCT's because in many cases the reactivity of the functional terminal groups is very high in the particular liquid crystalline temperature range of LCT's. Another problem one runs into when choosing terminal groups for LCT, is the elimination of low molecular compounds during cure of LCT's. So, for example, the curing process of LCT's with imide of nadic acid as terminal functional groups of LCT accompanied by the decomposition of this group due to the proceeding of the side reverse Diels-Alder reaction with elimination of gaseous cyclopentadiene at high cure temperature [20,21].

From the viewpoint of mechanical properties the preferable mesomorphic structure of oriented LCNP's is nematic one. The preferable arrangement of mesogenic groups in the backbone (interknot) chains follows from the same considerations. It is possible if the LCT cure proceeds via polycondensation reaction. It is quite evident that any low-molecular compounds have not to be evolved at the reaction of these functional terminal groups with the corresponding crosslinking agents. LCT's have not to contain in their structure functional groups in the

154

main chain capable to side reaction under the conditions of high temperature of LCT's cure. Finally, it is rather desirable to have of LCT's with a wide temperature range of LC state.

Some of LCT's described in literature and their phase transition temperatures are given in Tables 3 and 4.

PHASE TRANSITIONS OF LCT's

LCT's phase transitions combine features of low molecular liquid crystals and LCP's. Because of reactive character of LCT's terminal groups even the very first phase transitions at heating, that are polymorphic transformations of crystals (metastable crystal to stable one) [17], are accompanied by LCT's cure. When the molecular mobility of LCT is occurred after monomer crystal melt and mesomorphic phase formation, the process of cure proceeds with sufficiently high rate that is increased in the isotropic phase (see next section). This is one of the characteristic peculiarities of the LCT's behavior during their phase transformations at heating. It means that LCT's phase transitions used to be accompanied by changes of the system composition. When the LCT cure rate is sufficiently high at heating, LCT is solidified in the mesomorphic state and phase structure of curing LCT is freezed in the resulting network.

Because of continuous changes of LCT composition during heating, all phase transition points are changed, i.e., phase diagrams obtained, are essentially nonequilibrium [21]. The typical view of such diagram, obtained for DGET, is presented in the Fig.5.

Transition point from crystal to nematic fluid is decreased at the beginning of the LCT cure, but it is increased after attaining of definite conversion (α=0.4-0.5). Such melting temperature – conversion dependence is quite similar to the well-known melt temperature – composition of mixture with eutectic point. Isotropization point usually is increased with conversion growth at the LCT transformation to the LCNP. Therefore, the temperature range of LC state is increased with conversion. This phenomenon can be used in practical purposes for the decreasing of the LCT cure rate in the LC state after definite conversion by decreasing the temperature below the melt temperature of the starting LCT.

The second important peculiarity connected with the reactivity of LCT's and formation of the multicomponent polydispersive oligomeric system during heating consists in the broadening of the range of all phase transitions. It is not the phase transition point as for individual low-molecular liquid crystals but it is indeed a range which width depends on the conversion and, as a consequence, on the polydispersivity of oligomers. Since the isotropization point of oligomer is increased with the increasing of conversion up to the definite value, the LC state temperature range of the curing oligomeric system is divided on two parts. One of them after crystal melting (between curves 2 and 3) is pure mesomorphic state (usually nematic type) and another is biphasic (N + I, between curves 3 and 4), which temperature range is sufficiently erratic. The temperature range of pure nematic state is decreased during cure, while the temperature range of the biphasic region is widen. At definite conversion the curing system is solidified and phase transition measurements by polarized light microscopy are become impossible.

Finally, the third peculiarity, connected with the reactivity of LCT's, consists in impossibility to observe all phase transitions are characteristic for the investigating system. The isotropization point of LCT's often can not be observed (see Table 4) because of great reactivity of LCT's and usually rather narrow mesomorphic temperature region. DGET described in given article is one of fortunate exceptions. Namely sufficiently wide mesomorphic temperature range and comparatively low

Table 3. LC thermosets

N	Chemical structure	References
1.	$CH_2\text{--}CHCH_2O\text{--}\langle\bigcirc\rangle\text{--}C\text{--}O\text{--}\langle\bigcirc\rangle\text{--}OCH_2CH\text{--}CH_2$ (with epoxide O bridges and C=O)	25,26
2.	$CH_2\text{--}CHCH_2OCPhOCPhCOPhCOCH_2CH\text{--}CH_2$ (epoxide O's, four C=O)	16,17
3.	$CH_2\text{--}CHCH_2OPhCH=CPhOCH_2CH\text{--}CH_2$, with CH_3 branch	11
4.	$CH_2\text{--}CHCH_2O\text{--}[PhCH=CPhO(CH_2)_nOO]_j\text{--}PhCH=CPhOCH_2CH\text{--}CH_2$, CH_3 branches; $R=1\text{--}10$	11
5.	$CH_2=CHCO(CH_2)_6OPhCOPh(R)OCPhO(CH_2)_6OCCH=CH_2$ (C=O groups); $R=\text{--}H,\ \text{--}CH_3$	15
6.	$CH_2=CHCOPhCH=NPhN=CHPhOCCH=CH_2$ (two C=O)	24
7.	$CH_2=CHCOPhCH=N\text{--}N=CHPhOCCH=CH_2$ (two C=O)	2,24
8.	$CH_2=CHCOPhC\equiv CPhOCCH=CH_2$ (two C=O)	2
9.	$CH_2=CHCOPhN=CHPhCH=NPhOCCH=CH_2$ (two C=O)	24
10.	$HO(CH_2)_nOCPhOCPhCOPhOCO(CH_2)_nOH$ (four C=O), $n=4\text{--}12$	31
11.	$HOPhCH=CPhOH$, with CH_3 branch	11,24

12. HO—[PhCH=CPhO(CH$_2$)$_n$O]$_j$—PhCH=CPhOH   11,24
         | |
         CH$_3$ CH$_3$

13.   30

14.   25

15. OCN—⟨⟩—C—O—⟨⟩—NCO   25
             ‖
             O

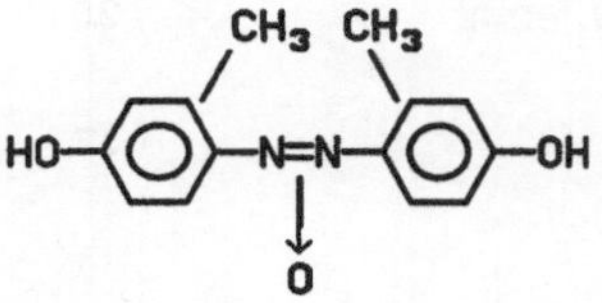

16. H$_2$N—⟨⟩—C—O—⟨⟩—NH$_2$ 25
             ‖
             O

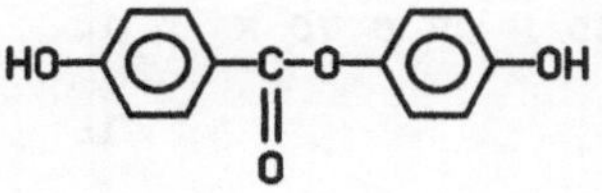

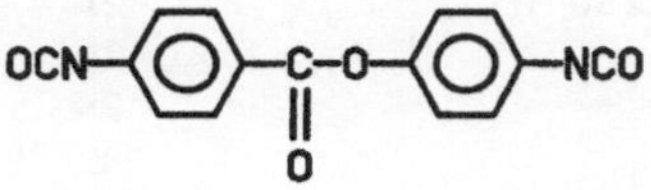

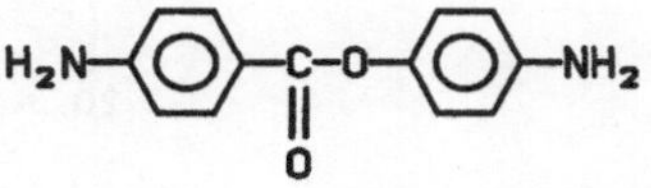

17.   19,20

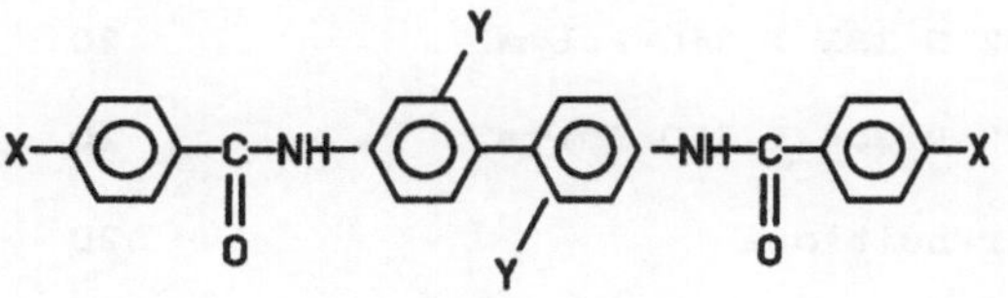

X= , , ; Y= —CH$_3$, —CF$_3$

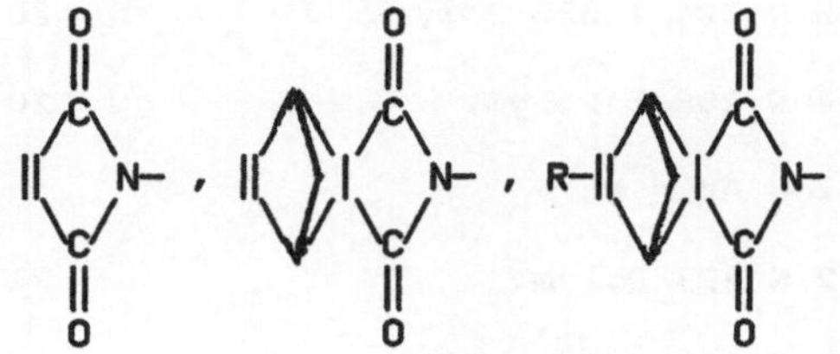

18. X—⟨⟩—C—O—Ar—C—⟨⟩—X   19,21
          ‖   ‖
          O   O

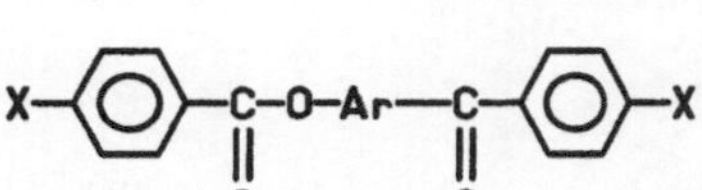

X= , , ;

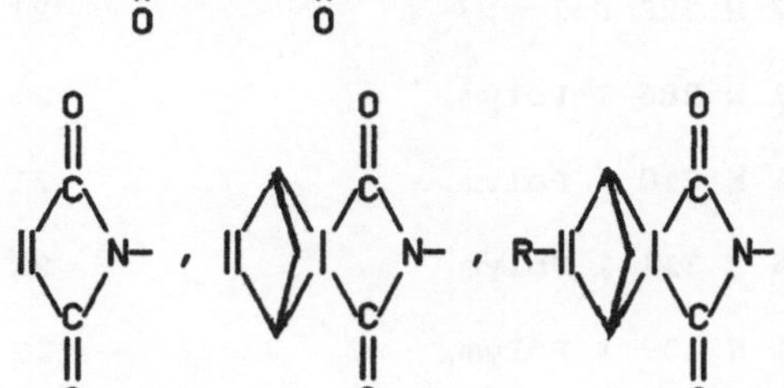

Ar= , ,

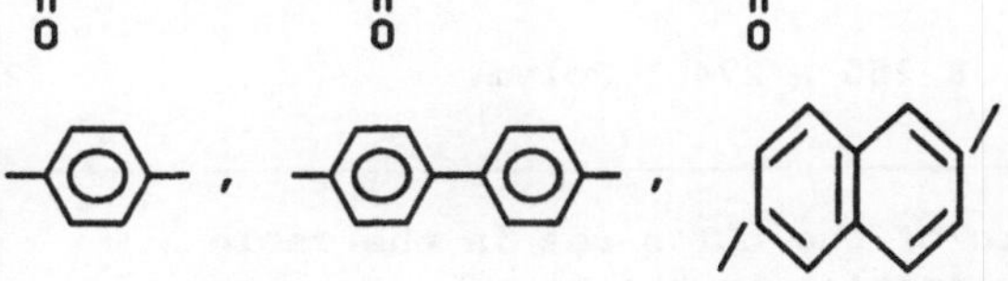

Table 4. Phase transitions of LCT's

N[*]		Phase transition, °C	References
1.		K 118 I I 93 N 80 K	25
2.		K 158 N 250 I	17
3.		K 62 N 109 I	11
5.	R = H	K 108 N 155 I I 155 N 88 S 70 K	14
	R = CH$_3$	K 86 N 116	15
6..		K 180 S Polym. (S 250 N)[**]	24
7.		K 138 N Polym.	24
15.		K 117 N 173	25
17a.	Y = CH$_3$	K 342 N 345 Polym.	20
17b.	Y = CH$_3$	K 292 N 333 I 350 Polym.	20
17c.	Y = CH$_3$	K 290 N 304 I 350 Polym.	20
17a.	Y = CF$_3$	no transitions	20
17b.	Y = CF$_3$	K 265 N 324 I 350 Polym.	20
17c.	Y = CF$_3$	K 284 N 288 I Polym.	20
		Below Ar = Ph	
18a.	R = H	K 282 N 293 Polym.	21
18a.	R = CH$_3$	K 245 N 280 Polym	21
18a.	R = Cl	K 215 N 270 Polym.	21
18b.	R = H	K 307 N 328 Polym.	21
18b.	R = CH$_3$	K 271 N 286 I Polym.	21
18b.	R = Cl	K 271 N 310 I Polym.	21
18c.	R = H	K 288 N 311 I Polym.	21
18c.	R = CH$_3$	K 211 N 259 I Polym.	21
18c.	R = Cl	K 255 N 274 I Polym.	21

* The structure of the LCT's see in the Table 3.
** See the explanation in the text.

reactivity of DGET allows us to measure experimentally its phase transition points during cure.

Here it is worth also noting an interesting phenomenon that takes place during phase transition measurements by using polarized light microscopy. It was noted that measurements of isotropization point depend on the character of preliminary treatment of the glassy plate surface. Treatment of the latter by polydimethylsiloxane allows us to determine the isotropization point from DGET samples that have been cured on the untreated glassy surface at temperatures considerably lower. This effect has perfect reproducibility. The most probable reason of the "surface effect" may consist in sharp increasing of the polymerization rate due to the cocatalysis of anionic polymerization of epoxies by hydroxyl groups of glass surface.

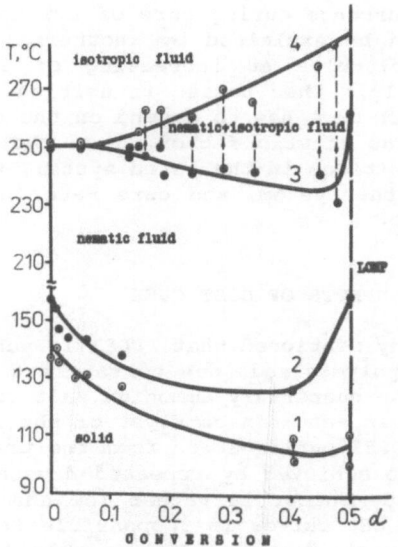

Figure 5. Phase transition points of DGET during cure obtained by polarized light microscopy. Heating rate equal to 40 K/min. DGET was heated and exposed at 164°C up to the corresponding conversion then cooled and heated again. Crystal-nematic transition: 1 - at cooling after heating and exposition at 164°C to the corresponding conversion; 2 - at second heating; Nematic-isotrop transition at heating: 3 - the beginning; 4 - the end.

All LCT's published in the literature except one (see N 6 in Tables 3 and 4) display the crystal to nematic phase transition. LCT N6 melts with smectic phase formation that is preserved in the resulting network. However, this sample placed on the heated till 250°C plate of polarized optical microscope exhibit nematic texture [24], while the smectic to nematic transition in the case under consideration can not be observed by DSC even at a heating rate of 80 K/min since the system loose the molecular mobility due to the LCT cure.

One of the interesting and important problem in the field of LCT's phase transitions is the structural organization of resulting networks. LCNP's usually inherit the molecular organization that are characteristic for the starting LCT's [2, 18, 24]. However, in some cases it was shown that cure produces a more ordered polymer [24]. Thus, high temperature cure of the acrylic LCT in the nematic state results in the LCNP displayed a smectic texture [24]. Very similar results were obtained at polymerization of monofunctional acrylic monomers in the nematic and even in the isotropic state [24]. In both cases the resulting polymers are characterized by smectic texture. This phenomenon according to Blumstein et al. [24] is connected with the ability of interacting side chain groups to pack into an orderly, layered arrangement without disturbing the regular disposition of the polymeric backbone. It is worth noting also that LCT's cure in the isotropic state or in the solution usually give LCNP's without any anisotropy.

These results look on the first sight as contradictory ones. Nevertheless, we consider that it is typical situation for all reactive liquid crystalline systems including LCT's.

An anisotropy occurrence during cure of LCT in the isotropic state or in the solution can be explained by isothermal phase transition of rigid rod-like LCT molecules at increasing of its molecular weight during cure. Naturally, the phase transition from isotropic to mesomorphic state at LCT cure has to depend on the cure temperature and the solvent quality. The kinetic factor, i.e. the relationship between the rate of phase transitions in the cured system, that is determined by molecular mobility of the system, and cure rate in the isotropic phase has to be important too.

KINETIC AND MECHANISM ASPECTS OF DGET CURE

It has been already mentioned that DGET is subjected to self cure upon heating. DGET is polymerized due to the fact that it contains uncontrolled traces of the quaternary ammonium salt formed during the DGET synthesis. Such salts can act as a catalyst of the anionic polymerization of epoxies [32]. DGET purification from the traces of the quaternary ammonium salt can be achieved by repeated crystallization. The reactivity of such purified monomer is rather low and the high-temperature exothermic peak on the DSC curves is responsible for DGET cure is shifted to higher temperatures. Purified DGET sample is convenient for determination of the DGET isotropization point ($T_i = 250^{\circ}$C). According to expectation this value of isotropization point is highest.

The principal problem of the polymerization kinetics consists in the influence of monomer phase structure on its reactivity. The answer can be given if we would measure an activation energy of the reaction in a wide temperature range covering all phase structures of thermotropic LC monomer.

Kinetic investigation by isothermal calorimetry shows that the cure reaction proceeds at $T \geq 120^{\circ}$C. Typical view of DGET polymerization kinetic curves and temperature dependence of the initial reaction rate in Arrhenius plot is given in the Figures 6 and 7.

As it was expected, the temperature dependence of the apparent rate constant of DGET cure does not obey to the Arrhenius law in all investigated temperature range. However, as it was shown in the Fig.7, some parts of this curve can be described as linear ones. These parts correspond to existence of different phases: crystalline (120-158°C) nematic and nematic + isotropic (158-250°C), depending on the heating rate. Activation energies calculated for these temperature regions are equal 92

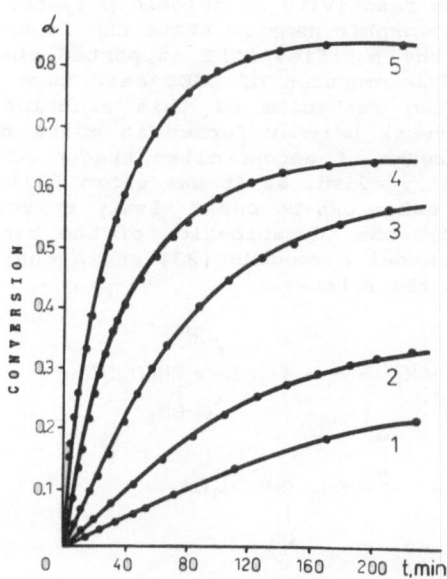

Figure 6. Typical view of DGET polymerization kinetic curves at different temperatures. T, °C: 1-130; 2-140; 3-158; 4-164; 5-184.

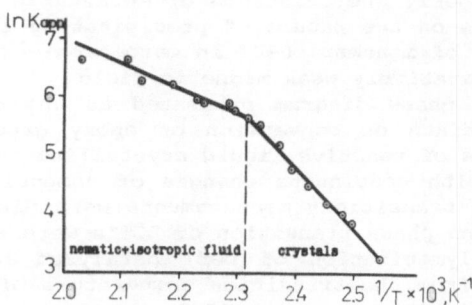

Figure 7. Temperature dependence of the apparent rate constant of DGET cure in Arrhenius plot. K_{app} is given in kal/g.equiv·min.

and 46 kJ/mol correspondingly. It is practically impossible to make correct evaluation of activation energy of DGET cure in isotropic phase because of very high cure rate. These measurements were done by using DSC with heating rate 20 K/min. Measured value of activation energy of DGET cure in isotropic state equal 40 kJ/mol. These results clearly show that LCT epoxy group reactivity at anionic polymerization does not differ very much in mesomorphic nematic state and isotropic one.

The work with the purified DGET supported the view that at high temperatures very slow reaction of DGET self cure can proceed without any catalysts but the mechanism of this reaction and the molecular structure of the network polymer formed is quite different from those obtained in the presence of uncontrolled traces of the catalyst or if the latter is specially added. As it was shown [33], the diglycidyl esters of dicarbonic acids can be cured slowly at sufficiently high temperatures (180-250°C). The investigation of the kinetics and mechanism of such reaction on model compounds [33] shows that it proceeds as insertion according to the scheme:

$$
-\overset{\overset{\displaystyle O}{\|}}{C}-OR \;+\; Ar-OCH_2CH\overset{}{\underset{O}{\diagdown\!\diagup}}CH_2 \longrightarrow -\overset{}{C}\underset{\diagdown O-\overset{|}{C}H_2}{\overset{\diagup OR}{-O-}}CHCH_2OAr \longrightarrow
$$

$$
\longrightarrow \quad -\overset{\overset{\displaystyle O}{\|}}{C}-OCH_2-\underset{\underset{\displaystyle OR}{|}}{CH}-CH_2OAr \qquad\qquad (2)
$$

Cyclic orthoether is formed as intermediate in this reaction.

CONCLUSIONS

Liquid-crystalline diepoxy monomer with a wide nematic temperature range was synthesized. The synthesis of DGET involves dimerization under the reaction conditions used. The quantity of this side product can be controlled by reaction temperature decrease. Crystalline polymorphism is characteristic for DGET. The existence of metastable crystalline phase considerably depends on the nature of precipitating agent used and temperature prehistory of monomer. DGET in mesomorphic state can be highly ordered in the comparatively weak magnetic field.

Nonequilibrium phase diagram presented as the dependence of DGET phase transition points on conversion of epoxy groups were obtained. Some common problems of reactive liquid crystalline systems' phase transitions connected with continuous changes of composition and molecular weight during phase transitions measurements were discussed. Literature data on synthesis and phase transition of LCT's were summarized.

An anionic polymerization of DGET catalyzed by tertiary amines proceeds beginning from the transition temperature of crystalline metastable phase. Slow high-temperature DGET cure can proceed also via the mechanism of insertion reaction of glycidyl group to ester group.

Polymerization and polycondensation reactions of DGET in the LC state result in the anisotropic polymer with high conversion of functional groups. There are no phase transitions in densely-crosslinked anisotropic network up to the destruction temperature (>320°C).
No anisotropy was observed for DGET cured in isotropic state or in solution.

Densely-crosslinked polymer networks with a wide temperature range of existence of uniaxial ordering can be obtained by cure of oriented in the magnetic field monomer.

162

FERENCES

P. G. de Gennes, Possibilities of Liquid Crystalline Network Polymers, Phys. Letters A28:725 (1969).

L. Strzelecki, L. Liebert L., Synthesis Some New Mesomorphic Monomers. Polymerization of p-acryloiloxybenzilidene p-carboxyaniline, Bull. Soc. Chim. France, (1973), 605.

P. G. de Gennes, "The Physics of Liquid Crystals", Clarendon Press, Oxford (1974).

S. P. Papkov and V. G. Kulichihin, "Liquid Crystalline State of Polymers", (in Russian), Chemistry, Moscow (1977).

V. G. Kulichihin, Orientation Phenomena in Polymeric Liquid Crystals, in: "Orientation Phenomena in Polymer Solutions and Melts" (in Russian), Chemistry, Moscow (1980).

Yu. B. Amerik and B. A. Krenzel, "The Chemistry of Liquid Crystals and Mesomorphic Polymers", (in Russian), Science, Moscow (1981).

N. A. Plate and V. P. Shibaev, "Comb-like Polymers and Liquid Crystals, Plenum Press, New York-London (1987).

"Liquid Crystalline Polymers", (in Russian), N. A.Plate, ed., Chemistry, Moscow (1988).

V. V. Tsukruk and V. V. Shilov, "The Structure of Polymer Liquid Crystals, (in Russian), Scientific Thought, Kiev (1990).

"Polymeric Liquid Crystals", A. Blumstein, ed., Plenum Press, New York (1985).

G. Barclay, C. K. Ober, K. Papathomas and D.Wang, Liquid Crystalline Epoxy Networks, Proc. ACS Div. Polymeric Materials: Science and Engineering 63:356 (1990).

G. G. Barclay, S. G. McNamee and C. K. Ober, Mechanical and Magnetic Orientation of Liquid Crystalline Epoxy Networks, Proc. ACS Div. Polymeric Materials: Science and Engineering 63:387 (1990).

C. K. Ober, G. G. Barclay, K. L. Papathomas and D. W. Wang, Curing and Alignment of Liquid Crystalline Epoxy Networks, Mat. Res. Symp. Proc. 203:265 (1991)

D. J. Broer, J. Boven, G. N. Mol, G. Challa, In-situ Photopolymerization of Oriented Liquid Crystalline Acrylates, 3.Oriented Polymer Networks from Mesogenic Diacrylate, Makromol. Chem., 190:2255 (1989).

D. J. Broer, R. A. M. Hikmet and G. Challa, In-situ Photopolymerization of Oriented Liquid Crystalline Acrylates, 4. Influence of a Lateral Methyl Substituent on Monomer and Oriented Network Properties of a Mesogenic Diacrylate, Makromol. Chem., 190:3201 (1989).

I. I. Serebryakova, L. L. Gur'eva, V. V. Tsukruk, V. V. Shilov, V. P. Tarasov, L. N. Erofeev and B. A. Rozenberg, Diglycidyl ester of terephtaloil-bis-(4- hydroxybenzoic acid) as monomer for synthesis of thermostable polymers, USSR Pat. 1541209 (1988).

L. L. Gur'eva, G. P. Belov, G. N. Boyko, P. P. Kushch and B. A. Rozenberg B.A., Synthesis and Phase Transitions of Liquid Crystalline Diepoxy Monomer, Vysokomol.Soed., in press.

V. V. Tsukruk, L. L. Gur'eva, V. P. Tarasov, V. V. Shilov, L. N. Erofeev and B. A. Rozenberg, Liquid Crystalline Diepoxide as monomer for synthesis of Anisotropic Network Polymers, Vysokomol. Soed., 33B:168 (1991).

A. E. Hoyt, B. C. Benicewicz and S. J. Huang, Rigid-rod Molecules as Liquid Crystal Thermosets, J. Polym. Prepr. Am.Chem. Soc. Div. Polym. Chem., 30:536 (1989).

A. E. Hoyt and B. C. Benicewicz, Rigid-rod Molecules as Liquid Crystalline Thermosets. I. Rigid-rod Amides, J. Polym. Sci. Polym. Chem. Ed.,28:3403 (1990).

A. E. Hoyt and B. C. Benicewicz, Rigid-rod Molecules as Liquid Crystalline Thermosets. II. Rigid-rod Esters, J. Polym. Sci. Polym. Chem. Ed.. 28:3417 (1990).

22. K. Peter and M. Ratzsch, Crosslinkable Liquid Crystalline Main-chain Polymers, Makromol. Chem., 191:1021 (1990).
23. S. M. Aharoni, Gelled Networks Prepared from Rigid Fractal Polymers, Macromolecules, 24:235 (1991).
24. S. B .Clough, A. Blumstein and E. C. Hsu, Structure and Thermal Expansion of Some Polymers with Mesomorphic Ordering, Macromolecules, 9:123 (1976).
25. R. Dhein, H. P. Muller, H. M. Meier and R. Gipp, Process for the Preparation of Polymeric Networks Having Superstructures, Corresponding Polymeric Networks and the Use thereof, US Pat. 4762901 (1988).
26. S. Kirchmeyer, A. Karbach, H. P. Muller, H. M. Meier and R. Dhein, Ordered Epoxy Networks on the Basis of 4-glycidylphenyli-4-glycidylbenzoat and 4-aminophenyl-4-aminobenzoat, Angew. Makromol. Chem., 185/186:33 (1991).
27. A. Yu. Bilibin, A. V. Tenkovtsev, O. N. Piraner and S. S. Skorochodov, Synthesis of the High Molecular Polyesters on the Basis of Mesogenic Monomer for Polycondensation, Vysokomol. Soed., 26A:2570 (1984).
28. Yu.S.Lipatov, V.V.Shilov, N.E.Kruglyak and Yu.P.Gomza, "Methods of X-ray Analysis for Study of Polymeric Systems", Scientific Thought, Kiev (1982).
29. G. Ungar, J. L. Feijoo, V. Percec and B. Yourd, Liquid Crystalline Polyethers Based on Conformational Isomerism, Macromolecules 24:1168 (1991).
30. A. Blumstein, S. Vilasagar, S. Ponrathnam, S. B. Clough, G. Maret, R. B. Blumstein, Nematic and Cholesteric Thermotropic Polyesters with Azoxybenzene Mesogenic Units and Flexible Spacers in the Main Chain, J. Polym. Sci., Polym. Phys. Ed., 20:877 (1982).
31. A. F. Dimian, F. N. Jones, Model crosslincable oligoesterdiols as binders for coatings, Polym. Mater. Sci. and Eng. Proc. ACS Div. Polym. Mater. Sci. and Eng., 56:640 (1987).
32 .B. A. Rozenberg, Kinetic, Thermodynamics and Mechanism of Reactions of Epoxy Oligomers with Amines, Adv. Pol. Sci., 75:113 (1986).
33. S. A. Kosyhina, L. L. Gur'eva, P. P. Kusch, B. A. Rozenberg, Reaction between epoxy and ester groops as the reason of diglycidyl ester of terephtaloil-bis-(4- hydroxybenzoic acid) self cure, Vysokomol. Soed., in press.

CROSSLINK PRODUCTS, MECHANISM, AND NETWORK STRUCTURE OF BENZOCYCLOBUTENE TERMINATED BISPHENOL A POLYCARBONATES

Maurice J. Marks

Texas Polymer Center
Dow Chemical U.S.A
Freeport, TX 77541

Introduction

Studies on crosslinked benzocyclobutene terminated bisphenol A polycarbonates (BCB PC's) (1) have provided insight into both their crosslinking chemistry and network structure.

1

Characterization of the BCB reaction products formed upon crosslinking reveal important information about the structure and functionality of the crosslink unit and on the mechanism of BCB homopolymerization. Also, studies on the network structure of BCB PC's illustrate the limitations of conventional network models based on the theory of rubbery elasticity as applied to rigid networks. In this study crosslinked BCB PC's were hydrolyzed to produce mixtures of bisphenol A (BA) and poly-BCB phenolic products. These mixtures were separated using reverse phase liquid chromatography and the components were identified by mass spectroscopy (LC/MS). The selectivity of the poly-BCB products observed suggest a BCB homopolymerization mechanism involving a series of pericyclic reactions. In addition, the crosslink density (M_c) of BCB PC networks having varying initial molecular weights was determined by both equilibrium swelling measurements and by dynamic mechanical analysis (DMA) of the rubbery plateau.

Although considerable research on BCB polymers has been conducted in the last several years,[1] the products formed by BCB homopolymerization have not been previously identified. Earlier studies have established that the first step in thermally activated BCB reactions is an electrocyclic ring opening rearrangement to an o-quinodimethane[1,2] (2, Fig. 1). In the presence of Diels-Alder acceptors 2 readily undergoes 4 + 2 cycloaddition reactions.[2,3] In the absence of such olefins, 2 forms the thermally labile spiro-dimer 3.[4] From this point the fate of the BCB self-reaction products has been unclear. A homolytic cleavage of 3 to a diradical was proposed

Synthesis, Characterization, and Theory of Polymeric Networks and Gels
Edited by S.M. Aharoni, Plenum Press, New York, 1992

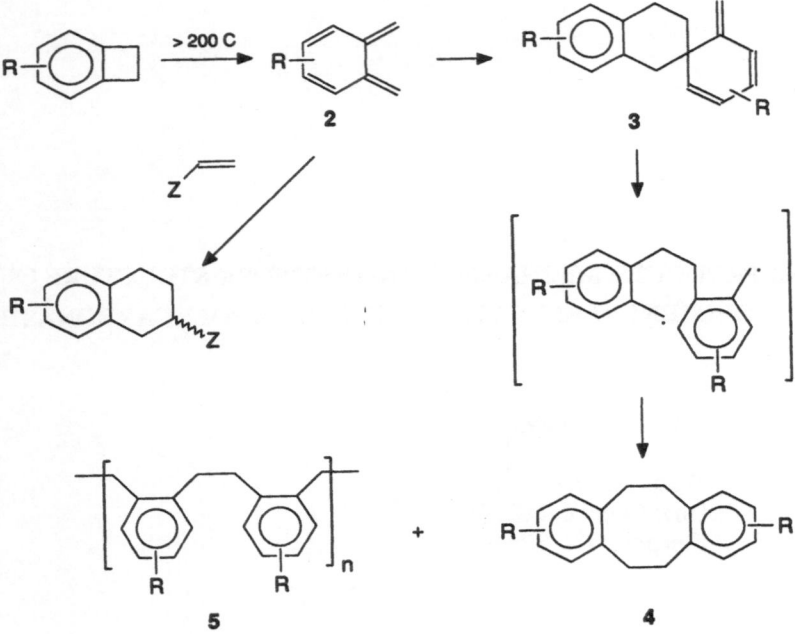

Figure 1. Previously Known or Proposed BCB Reactions

to yield dibenzocycloocta-1,5-dienes **4** and poly(o-xylene)s **5** by radical recombinations.[5,6] NMR data suggested methyl group termination of poly(o-xylene).[6]

The network structure of some BCB polymers has recently been investigated. Rheological gel point studies on two low molecular weight bis-BCB's concluded that either tetrafunctional or a mixture of tri- and tetrafunctional network junctions were formed.[1] Poly(o-xylene) was presumed to be the predominant BCB homopolymerization product.

BCB Polymerization Products and Mechanism

LC analysis of a BCB PC hydrolysate (Fig. 2) shows the appearance of 16 compounds which arise from BCB reactions, the remaining are due to BA (Fig. 3). Peak #1 in this LC is BCB-OH as shown by comparison to an authentic sample. The mass spectra of peaks #2 and #3 show M+ at m/e 240 (Table 1). Both dihydroxy-dimethylstilbene (**7**, 6 isomers possible) and dihydroxy-dibenzocylooctadiene (**4**, two isomers possible) are consistent with the observed fragmentation patterns of these two peaks. The mass spectra of the hydrocarbon analogs of **4** and **7** show almost identical fragmentation patterns.[7,8] Since stilbene compounds are known to completely scramble before fragmentation,[9] the MS data does not distinguish between the two possibilities for either peak. Another isomer which would also have a similar mass spectrum[10] is **3**, but, since the hydrocarbon analog is known to be thermally unstable,[10] this dimer would not survive the high temperature crosslinking reaction conditions.

Peaks #4, #11, #20, and #21 in the LC of BCB PC hydrolysate are BA isomers and coproducts, each of which are present in the monomer before polymerization.

166

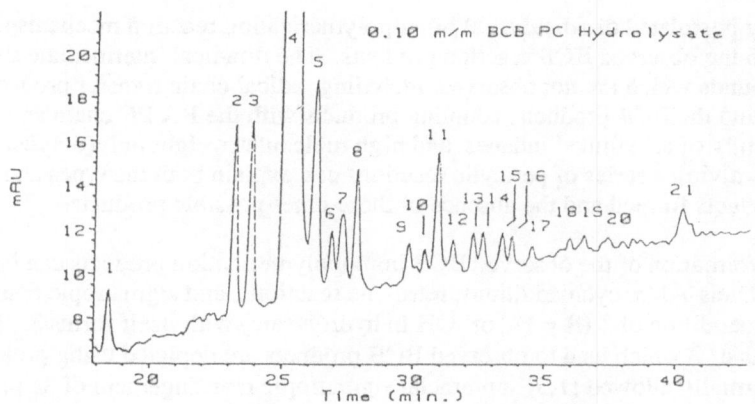

Figure 2. LC of BCB PC Hydrolysate

Table 1. Mass Spectra Data from BCB PC Hydrolysate

LC Peak #	Observed m/e (% Rel. Abundance)
2, 3	240 (81), 225 (100), 213 (20), 211 (24), 210 (26), 132 (22), 121 (32) 120 (55), 119 (22)
5 - 8	360 (32), 240 (10), 239 (19), 225 (25), 200 (19), 172 (8), 160 (12), 159 (19), 158 (14), 146 (17), 144 (13), 121 (20), 120 (100)
9, 10, 12 - 16	360 (37), 358 (7), 356 (7), 253 (13), 252 (57), 239 (13), 121 (27), 120 (100)
17 - 19	480, 476, 465, 372, 120 (from SIC)

The MS of peak #5 shows a M+ at m/e 360 and a fragmentation pattern consistent with dehydro-tris(hydroxy-xylene) 11 (16 possible isomers). MS of peaks #6 - #8 show similar results. Fragmentation of this compound through the central phenyl ring gives characteristic ions at m/e 200 and 159. The MS of peaks #9, #10, and #12 - #16 also show a M+ at m/e 360, but the fragmentation pattern indicates these compounds are isomers of 2,3-bis(hydroxy-tolyl)-hydroxy-1,2,3,4-tetrahydronaphthalene 8. Hydrogen losses lead to aromatization of the unsaturated ring, the intensities of which are similar to those found in the MS of tetrahydronaphthalene.[11]

Mass spectra of peaks #17, #18, and #19 could not be obtained due to their low concentration, but selected ion chromatographs (SIC) of m/e 480, 476, 465, and 372 ions tentatively indicate that these last peaks have M+ of m/e 480 and are isomers of tetrahydronaphthalene 12. The apparent fragmentation observed is consistent with hydrogen loses to aromatize the tetralin ring and loss of an hydroxy-tolyl fragment, both characteristic of the parent structure. Further characterization of each poly-BCB product is underway.

The LC/MS analysis of the poly-BCB products show that the network junction functionality ranges from one, corresponding to unreacted BCB-OH, to four. Since the response factors for these products are not known, the relative concentration of each component cannot be rigorously quantified. Assuming each component has the same response factor, the average functionality of the poly-BCB products observed is about 2.5.

The previously postulated diradical BCB homopolymerization reaction mechanism is not consistent with the observed BCB reaction products. The diradical intermediate should produce several compounds which are not observed, including radical chain transfer products with the BA PC chain and the BCB products, coupling products with the BA PC chain and/or the BCB products, a family of substituted indanes, and high molecular weight poly(o-xylene). A mechanism involving a series of pericyclic reactions can explain both the types and range of poly-BCB products formed and the absence of these other possible products.

The selective formation of the observed BCB homopolymerization products can be explained by a series of Diels-Alder cycloadditions, retro-ene reactions, and sigmatropic rearrangements (Fig. 4). Cycloaddition of 2 (R = PC or -OH in hydrolysate) with itself forms 3. Two types of rearrangements of 3 which lead to observed BCB products are depicted using projections 3a and 3b. A thermally allowed [1,3] suprafacial sigmatropic rearrangement of 3a proceeds with inversion of configuration of benzocyclohexane C-1 to form 4. The geometry of 3a is appropriate for this rearrangement, with models indicating a C-1 to C-3 distance of about 2.8 A.

An analogous [1,3] rearrangement occurs in substituted bicyclo-[1.1.2]-hex-2-enes at 150° - 200°C.[12]

In 3b the exocyclic methylene group of the pseudo-axial cyclohexadiene points directly at the axial hydrogen on the opposite benzylic carbon. Models of 3b show both a reasonable geometry and appropriate through-space C-H distance (about 2.0 A) for an antarafacial [1,7] retro-ene reaction to 6. While the great majority of retro-ene reactions are 6 electron, [1,5] rearrangements,[13] the 8 electron [1,7] retro-ene rearrangement of certain homoallyl ethers is known.[14] Subsequent aromatization of the second ring by a suprafacial [1,5] H sigmatropic rearrangement of 6 gives the observed 7. Tetrahydronaphthalene 8 forms by the Diels-Alder cycloaddition of o-quinodimethane 2 and stilbene 7.

Further cycloaddition of 3 with 2 yields 9, which can exist in two conformers 9a and 9b. In 9b C-9 is about 2.1 A from the opposite benzylic axial hydrogen, well within range of the depicted suprafacial [1,9] retro-ene rearrangement to 10. No previous examples of [1,9] retro-ene reactions appear in the literature, but [1,9] sigmatropic rearrangements are known.[15] Such "long range" types of pericyclic reactions appear to require both the appropriate functionality and geometrical contraints in order to occur. From 10 to 11 requires only facile [1,5] and [1,3] H sigmatropic shifts. The [1,5] H rearrangement was discussed above, and [1,3] H shifts in toluene isomers are well known.[16]

The highest molecular weight poly-BCB product observed arises from a final Diels-Alder cycloaddition of 2 to 11 to yield 12. No higher molecular weight poly-BCB product is formed since neither 9 nor 12 contain reactive double bonds to further undergo cycloaddition with 2.

BCB PC Network Structure

Understanding the structure of network polymers, ranging from lightly crosslinked elastomers to rigid thermosets, is a key aspect in the design and utilization of these useful materials. BCB PC's are new "ideal" network polymers in that the linear precursors can be prepared with known molecular weight and crosslinking occurs only at the chain end to yield known products. In comparison to the much studied crosslinked polydimethylsiloxane (PDMS),[17] BCB PC's are rigid, high T_g networks. In this study the crosslink density (M_c) of BCB PC networks was determined by both equilibrium swelling measurements and by dynamic mechanical analysis (DMA) of the rubbery plateau.

LC Peak #	Structure	MW

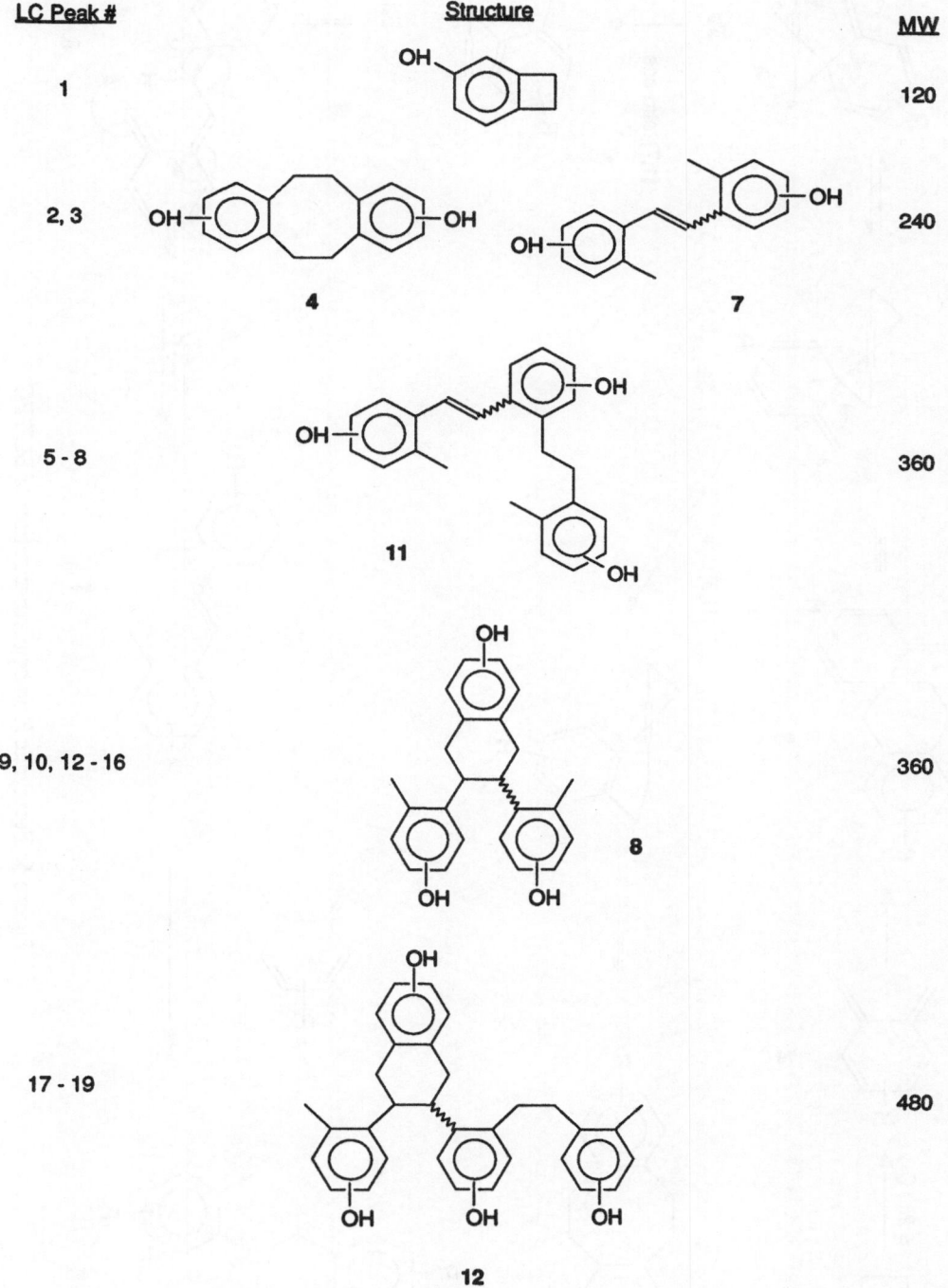

Figure 3. Structures of Poly-BCB's in BCB PC Hydrolysate

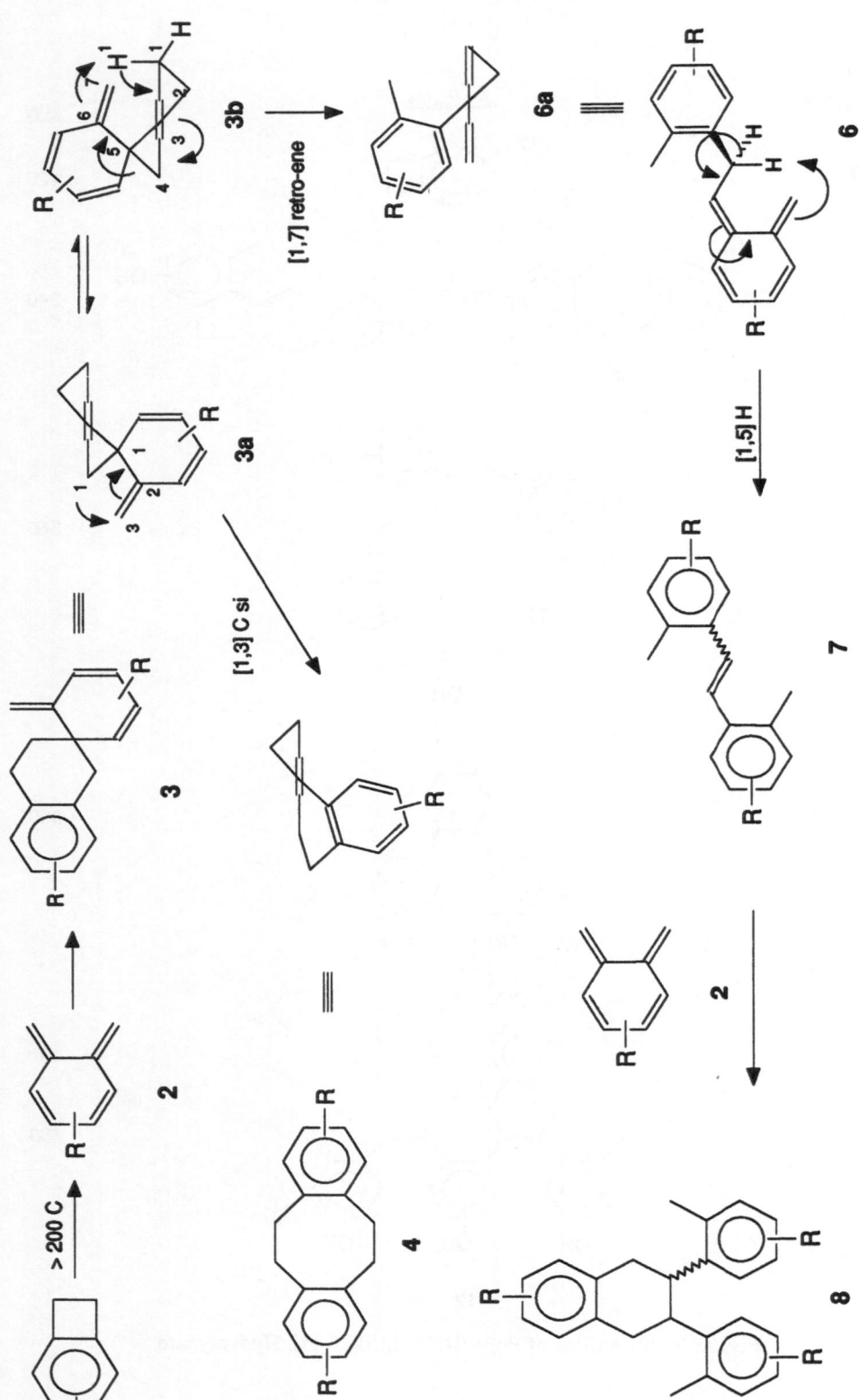

Figure 4. Pericyclic Homopolymerization of BCB PC

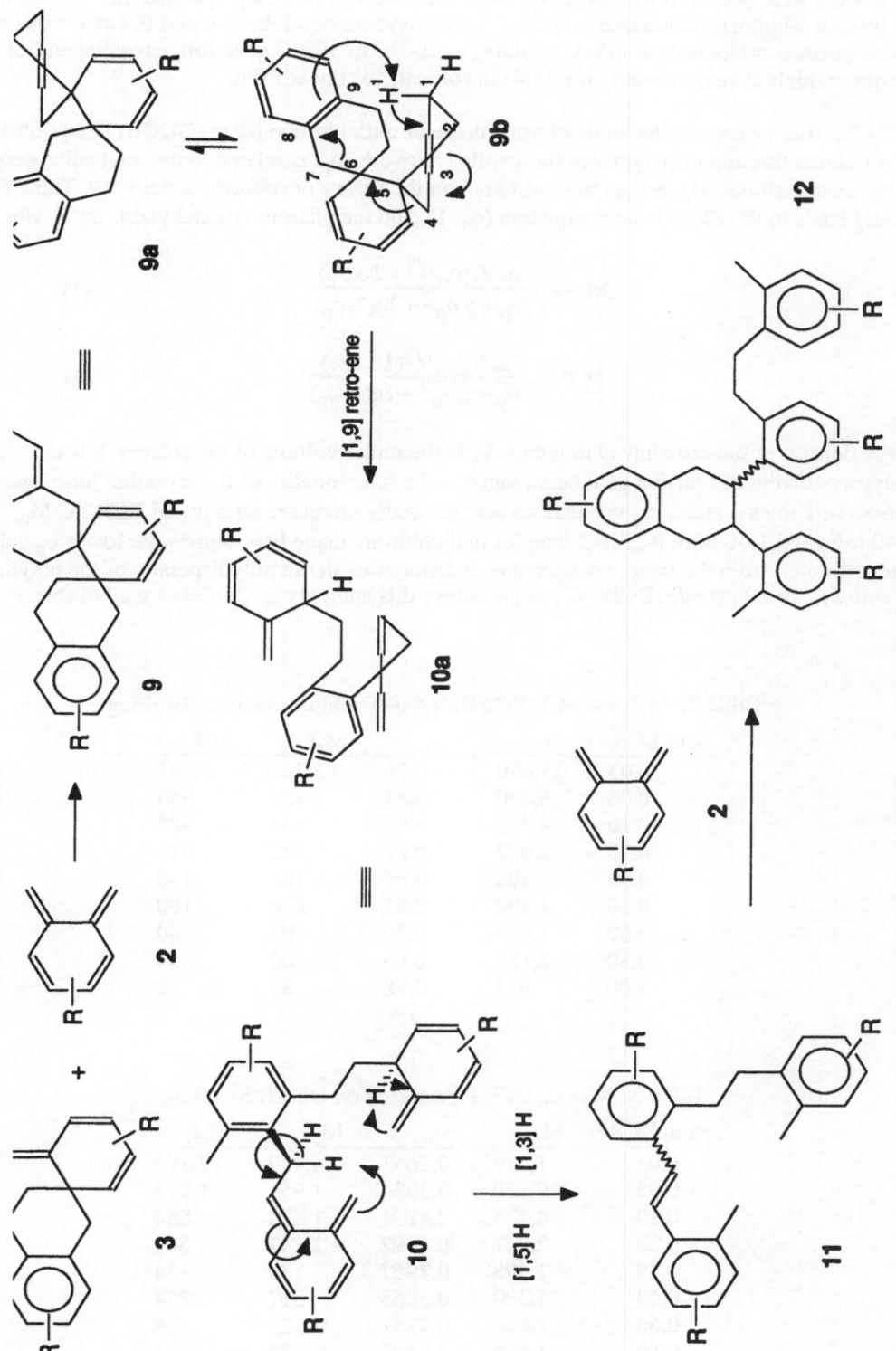

Figure 4 (cont.). Pericyclic Homopolymerization of BCB PC

BCB PC's were prepared with M_n from about 1,000 to 12,000 and polydispersities of about 2.5 by interfacial phosgenation of mixtures of 3-hydroxybenzocyclobutene and BA at various mole ratios (termed mole/mole or m/m). Molding at 200°C to 300°C gave fully crosslinked films or plaques which were completely insoluble in solvents for linear PC's.

BCB PC films were swollen in dichloromethane or o-dichlorobenzene (ODCB) to equilibrium. The volume fraction of polymer in the swollen network, υ_p, is related to the molecular weight between crosslinks, M_c, using relations based on the theory of rubber elasticity.[18-22] The affine model leads to the Flory-Rehner equation (eq. 1), and the phantom model yields eq. 2, where ρ_p

$$M_c{}^a = - \frac{\rho_p V_s (\upsilon_p{}^{1/3} - 2\upsilon_p/\phi)}{\upsilon_p + \chi\upsilon_p{}^2 + \ln(1-\upsilon_p)} \tag{1}$$

$$M_c{}^p = - \frac{\rho_p V_s \upsilon_p{}^{1/3}(1 - 2/\phi)}{\upsilon_p + \chi\upsilon_p{}^2 + \ln(1-\upsilon_p)} \tag{2}$$

is the density of the crosslinked polymer, V_s is the molar volume of the solvent, χ is the polymer-solvent interaction parameter, and ϕ is the functionality of the crosslink junction. As shown in Tables 2 and 3, υ_p in either solvent generally increases with initial BCB PC M_n. The 0.40 m/m and 1.00 m/m BCB PC samples in dichloromethane have somewhat lower υ_p values than expected from the trend, possible due to differences in the polydispersity of the polymers (Table 2). The 0.40 m/m BCB PC sample shows this anomaly in ODCB as well (Table 3). In

Table 2. M_c's for BCB PC's Based on Dichloromethane Swelling

m/m BCB	M_n	υ_p	$M_c{}^a$	$M_c{}^p$
0.03	11,470	0.26	3,027	1,897
0.06	6,870	0.34	1,450	958
0.10	4,425	0.48	535	385
0.20	2,027	0.51	425	313
0.30	1,408	0.65	185	148
0.40	1,390	0.61	230	180
0.60	1,046	0.76	93	80
0.80	1,155	0.83	60	54
1.00	957	0.78	83	72

Table 3. M_c's of BCB PC's Based on ODCB Swelling

m/m BCB	M_n	υ_p	$M_c{}^a$	$M_c{}^p$
0.03	11,470	0.2651	4,439	2,796
0.06	6,870	0.3634	1,957	1,313
0.10	4,425	0.4291	1,222	854
0.20	2,027	0.4392	1,142	803
0.30	1,408	0.7967	130	114
0.40	1,390	0.6465	317	254
0.60	1,046	0.9741	29	29
0.80	1,155	0.9799	27	26
1.00	957	0.9971	15	15

either solvent the calculated M_c's are well below the initial M_n of the starting BCB PC, even assuming the maximum possible BCB network functionality of 4 (as described in the previous section). M_c's calculated for the highest crosslink density BCB PC's are unreasonably low, being less than the repeat unit M_n.

Of the many assumptions and uncertainties in network swelling models based on the theory of rubber elasticity is the probable variation of χ with crosslink density. A strong dependence of χ on υ_p was seen in PDMS networks, where χ varied by a factor of 7 with a doubling of υ_p.[23] To further examine the network structure of BCB PC's and eliminate the effect of χ, M_c's were also calculated from rubbery moduli data as measured by DMA.

The affine network model relates the equilibrium modulus G_e, which is usually estimated by the storage modulus G', to M_c (eq. 3), where ϕ is a constant dependent on the functionality of

$$M_c = \phi \rho RT/G' \qquad (3)$$

the network junctions (equal to 1.0 for tetrafunctional functions and equal to 0.53 for trifunctional junctions), ρ is the density of the polymer, R is the gas constant, and T is the absolute temperature.[19] G' was shown to be a good estimate of G_e by DMA frequency sweeps above T_g. To calculate M_c, G' values of the BCB PC's at 40° above T_g were normalized based on the glassy modulus of BA PC at -158°C to remove baseline differences in the separate DMA measurements (Table 4, where G' is in units of dyne/cm^2). At 40° above T_g the modulus of each BCB PC network is in a linear region of the elastic plateau (Fig. 5). Densities of the materials at the elevated temperatures were estimated from the coefficient of linear thermal expansion (CLTE). The CLTE of 0.40 m/m BCB PC was found to be 3.9×10^{-5} in./in., somewhat less than the 4.0×10^{-5} in./in. of linear BA PC. Assuming equal CLTE's for each of the BCB PC 's, the density of these materials is 1.16 g/cm^3 in the temperature range from 210°C to 235°C.

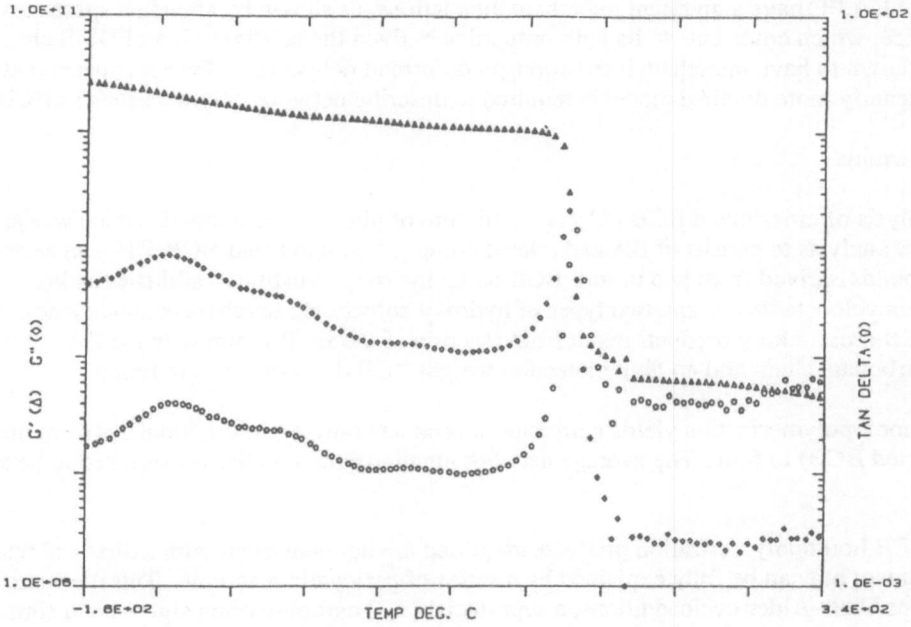

Figure 5. Dynamic Mechanical Analysis of 0.03 m/m BCB PC

The M_c's of BCB PC's calculated from eq. 3 are all at least one order of magnitude lower than the initial M_n's of the starting linear polymers (Table 4). As dictated by the affine network model, the tetrafunctional network ($\phi=1$) has a greater M_c than the corresponding trifunctional one ($\phi=0.53$). As it is known that poly-BCB junctions can be at most tetrafunctional and since there is no evidence of junction aggregation, the difference in the calculated M_c's and the initial M_n's cannot be due to a very large network functionality.

Table 4. M_c of BCB PC's from G

m/m BCB	M_n initial	Norm.G'@ $T_g + 40^\circ C$	M_c $\phi=1$	M_c $\phi=.53$
.03	11,470	6.881E+07	679	360
.06	6,870	1.480E+08	315	167
.20	2,510	1.828E+08	261	138
.30	1,895	2.494E+08	193	102
.40	1,390	2.658E+08	184	98

The M_c results from the DMA experiments show that the low M_c's calculated from the swelling measurements cannot be due to variations in χ with crosslink density. Rather, these results indicate that network models based on the theory of rubber elasticity are inapplicable to rigid network polymers such as BCB PC's. Several aspects of these types of networks are not taken into account by such models, including dangling chain ends, chain entanglements, polydispersity, non-Gaussian distributions of chain lengths, and interchain interactions. In end linked networks such as BCB PC's, the fraction of dangling chain ends is related to the degree of end group conversion rather than initial M_n as formulated for randomly crosslinked elastomers.[19] Chain entanglements have been recently shown to contribute 75% of the elastic modulus of crosslinked 1,2-polybutadiene.[24] Studies on bimodal PDMS networks illustrated the difficulty in applying the established network theories to polydisperse systems, especially with regard to apportioning the effective strain between the long and short chains.[25] Rigid polymers such as BA PC have significant interchain interactions, as shown by abundant spectroscopic evidence, which contribute to its bulk properties.[26] Even the relatively inert PDMS chains have been shown to have interchain interactions in deformed networks.[27] Thus it appears that a significantly more detailed model is required to describe network polymers such as BCB PC's.

Conclusions

Hydrolysis of crosslinked BCB PC gave a mixture of phenolic compounds which was shown by LC/MS analysis to consist of BA and related compounds, unreacted BCB-OH, and several compounds derived from two to four BCB units: hydroxy-substituted stilbenes and/or dibenzocyclooctadienes, and two types of hydroxy-substituted tetrahydronaphthalenes. None of the BCB crosslinking products result from reaction of the BCB group with the BA polycarbonate chain and no high molecular weight BCB derivatives were found.

BCB homopolymerization yields a mixture of products ranging in functionality from one (for unreacted BCB) to four. The average network junction functionality is estimated to be about 2.5.

The BCB homopolymerization products identified are not consistent with a diradical reaction mechanism but can be fully explained by a series of pericyclic reactions. This mechanism involves Diels-Alder cycloadditions, a suprafacial [1,3] carbon-carbon sigma bond sigmatropic rearrangement, an antarafacial [1,7] retro-ene reaction, a suprafacial [1,9] retro-ene reaction, suprafacial [1,5] hydrogen sigmatropic rearrangements, and a [1,3] hydrogen shift. Each of the sigmatropic rearrangements and retro-ene reactions are thermally allowed processes. This mechanism fully accounts for the types and range of products observed in crosslinked BCB PC's and explains why additional products are not formed.

Crosslink densities of BCB PC's calculated from both equilibrium swelling measurements and elastic storage moduli using the affine or phantom network models gave M_c's much less than the initial M_n's of the precursor linear BCB PC, results inconsistent with the known endlinking nature of the crosslinking reaction. The similarity of the results obtained from the equilibrium swelling study and the DMA experiments show that the probable variation in χ with v_p does not alone account for the discrepancies observed between M_c and M_n. In both instances the network models are derived from the theory of rubber elasticity and do not adequately take into account factors such as chain entanglements and rigidity, polydispersity and non-Gaussian distributions of chain lengths, and interchain interactions. A more sophisticated network model which takes these factors into account needs to be developed to describe these types of polymer networks.

Experimental

Chemicals. BCB PC's were prepared by interfacial phosgenation of BA and BCB-OH using dichloromethane/water, aqueous sodium hydroxide, and triethylamine catalyst. Linear BCB PC's thus obtained were crosslinked by molding at 200°C to 300°C. Dichloromethane was used as received from Dow Chemical Co. production facilities. ODCB was obtained from Aldrich Chem. Co. and was distilled prior to use.

Hydrolysis of Crosslinked BCB PC. A mixture of 1 g crosslinked BCB PC, 9 g THF, and 5 g 20 wt.% KOH in methanol was shaken for 18 hr. at room temperature. A 0.5 ml aliquot of the supernatent solution was diluted with 10 ml THF, acidified with 2 drops of conc. HCl, and then filtered for LC analysis.

LC/MS Analysis of BCB PC Hydrolysate. LC analysis was done on a Hewlett-Packard Model 1090 system with UV detection at 278 nm and fitted with a Scientific Glass Engineering glass-lined column (150 x 4mm) containing Spherisorb ODS2 (3 micron). A mobile phase gradient of 20% THF/80% water at 0 min., 30% THF at 10 min., 60% THF at 30 min., 50% THF at 40 min., and 100% THF at 50 min. was used with a flow rate of 0.5 ml/min. MS analysis was conducted through a moving polyimide belt interface connected to a Finnigan 4500 quadrapole mass spectrometer. The vaporizer and cleanup heaters were set at 270°C. Electron ionization mass spectra were obtained in a mass range of 115-600 amu scanned at 2 second intervals with the electron multiplier set at 1600V.

BCB PC Swelling. BCB PC films having 0.01" - 0.02" thickness were compression molded at 200°C to 300°C. Approximately 1 mm x 2 mm sections of the films were placed in covered petri dishes containing the solvent. Swelling measurements were taken until equilibrium was achieved, which took up to 24 hours for the highest crosslink density samples. Swelling of films in dichloromethane was measured by determining their increase in surface area while immersed in the solvent by a ruler to the nearest 0.05 mm. Swelling was assumed to be isotropic and the volume was calculated by taking the 3/2 root of the area. ODCB swollen films were removed from the solvent, patted dry to remove solvent on the film surface, and weighed on an analytical balance.

The various material constants used in the relations to calculate M_c are as follows: ρ_p (PC) = 1.20; ρ_s = 1.325 (CH_2Cl_2), 1.30 (ODCB); V_s = 64.1 (CH_2Cl_2), 112.56 (ODCB); χ (PC) =

0.411 (CH_2Cl_2), 0.390 (ODCB). χ was determined by measuring the phase distribution of PC/polystyrene/ODCB.[28]

<u>DMA Analysis of BCB PC's.</u> DMA temperature sweep experiments were performed under N_2 on 2" x 1/2" x 1/16" bars using a Rheometrics 7700 Dynamic Spectrometer in its torsional rectangular mode. Spectra were obtained from -160°C to about 400°C using an oscillatory frequency of 1 Hz and a strain of 0.05%. Measurements were taken at 5° increments after 0.5 min. equilibration. DMA isothermal frequency sweep experiments were performed under N_2 on 1/2" x 1/16" disks using a Rheometrics 605 Mechanical Spectrometer at 225°C and an oscillatory frequency range of 0.1 to 100 rad./sec. and a strain of 0.05%

Acknowledgements

The author gratefully acknowledges the assistance of the Texas Analytical and Engineering Sciences Laboratory, Dow Chemical U.S.A., for their several contributions to this research. Financial support provided by the Dow Chemical Co. is gratefully acknowledged.

References

1. K. Bruza, C. Carriere, R. Kirchoff, N. Rondan, R. Sammler, <u>J. Macromol. Science - Chemistry</u>, 1991, A28, **11-12**, 1079.

2. J. Charlton and M. Alauddin, <u>Tetrahedron</u> 1987, **43**, 2873.

3. W. Oppolzer, <u>Synthesis</u> 1978, 793.

4. L. Errede, <u>J. Amer. Chem. Soc.</u> 1961, **83**, 949.

5. L. Tan, <u>Amer. Chem. Soc. Poly. Preprints</u> 1985, **26**, 176.

6. L. Tan and F. Arnold, <u>J. Poly. Sci. A. Poly. Chem. Ed.</u> 1988, **26**, 1819.

7. M. Mintas, et al, <u>Org. Mass Spectrom.</u> 1977, **12**, 544.

8. Sample from K. Bruza, MS by D. Patrick, both of Central Research, Dow Chemical Co.

9. J. Bowie and P. White, <u>Org. Mass Spectrom.</u> 1972, **6**, 135.

10. Y. Ito, M. Nakatsuka, and T. Saegusa, <u>J. Amer. Chem. Soc.</u> 1982, **104**, 7609.

11. Mass Spectral Data, American Petroleum Institute Research Project 44, Texas A&M University, serial #201-m, 10-31-68.

12. W. Roth and A. Friedrich, <u>Tetrahedron Letters</u> 1969, 2607.

13. G. Desimoni, G. Tacconi, A Barco, and G. Polline, "Natural Products Synthesis Through Pericyclic Reactions," ACS Monograph 180, Washington, 1983.

14. A. Viola, et al, <u>J. Chem. Soc. Chem. Comm.</u> 1974, 842.

15. S. Sugiyama, A. Mori, and H. Takeshita, <u>Chem. Lett.</u> 1987, 1247.

W. Bailey and R. Baylouny, J. Org. Chem. 1962, **27**, 3476.

J. Mark, Acc. Chem. Res. 1985, **18**, 202.

J. Mark and B. Erman, "Rubberlike Elasticity, A Molecular Primer," Wiley, New York, 8.

P. Flory, "Principles of Polymer Chemistry," Cornell Univ. Press, Ithaca, N.Y., 1953.

P. Flory, "Networks," in Encyclopedia of Polymer Science and Engineering, vol. 10, ey, N.Y., 1986, pp. 95 - 112.

J. Queslel and J. Mark, "Rubber Elasticity and Characterization of Networks," in nprehensive Polymer Science, vol. 2, Pergamon, N.Y., 1989, pp. 271 - 309.

B. Erman, in Crosslinking and Scission in Polymers, O. Guven, ed., Kluwer, Boston, 1990, 153 - 169.

N. Neuburger and B. Eichinger, Macromolecules 1988, **21**, 3060.

T. Twardowski and O. Kramer, Macromolecules 1991, **24**, 5769.

S. Clarson, V. Galiatsatos, and J. Mark, Macromolecules 1990, **23**, 1504.

V. McBierty, "NMR Spectroscopy of Polymers in the Solid State," in Comprehensive mer Science, vol. 1, G. Allen, ed., Pergamon, NY, 1989, p. 417.

P. Solta and B. Deloche, Makromol. Chem. Makromol. Symp. 1991, **45**, 177.

E. Gurnee, Central Research, Dow Chemical Co.

RIGID HYPERCROSSLINKED POLYSTYRENE NETWORKS
WITH UNEXPECTED MOBILITY

Vadim A. Davankov and Maria P. Tsyurupa

Institute of Organo-Element Compounds
Russian Academy of Sciences
Moscow, 117813, Russia

INTRODUCTION

Due to manifold application possibilities in important
modern technologies, three-dimensional polymeric networks at-
tract attention of many researcher groups. Simultaneously,
many attempts have been made to describe theoretically the behavior of polymeric networks. Generally accepted modern theories predict that networks acquire a noticeable flexibility
only in the case that the distance between two neighbor junction points exceeds the length of chain segments. Therefore,
slightly or moderately crosslinked polymers, only, should be
in position to swell with thermodynamically good solvents.
With the degree of crosslinking rising, the high elasticity
properties of a network should gradually disappear, the glass
transition temperature being shifted rapidly toward the temperature region of thermal decomposition of the material. These
fundamental properties are well documented with examination
results of many types of polymeric networks, in particular,
that of crosslinked polystyrene. Indeed, conventional styrene-
divinylbenzene (DVB) copolymers are known[1,2] to lose their
ability to swell with solvents or acquire high elastic state
on heating at a DVB content exceeding 15 %.

However, we were able to prepare crosslinked styrene copolymers which do not follow the above requirements of the
theory and should therefore be considered as a new type of polymer network.

SYNTHESIS OF HYPERCROSSLINKED POLYSTYRENE

Whereas conventional polystyrene networks are prepared by
copolymerization procedure of styrene with DVB (or other di-
enes) with or without a diluent, we used solutions of high-
molecular-weight atactic polystyrene as the starting material
for crosslinking. As crosslinking agents, several bifunctional
compounds were employed: p-xylylene dichloride (XDC), 4,4'-
bis-(chloromethyl)-diphenyl (CMDP), monochlorodimethyl ether

Synthesis, Characterization, and Theory of Polymeric Networks and Gels
Edited by S.M. Aharoni, Plenum Press, New York, 1992

(MCDE), dimethyl formal (DMF), 1,4-bis-(p-chloromethylphenyl)-butane (CMPB), tris-(chloromethyl)-mesitylene (CMM). In
the presence of a Friedel-Crafts catalyst, they all react with
two (or three for CMM) phenyl groups of polystyrene chains and
form bridges that appear two phenyl groups longer than the
crosslinking molecule itself. The most typical reagents are
MCDE and XDC which form bridges of the following structures:

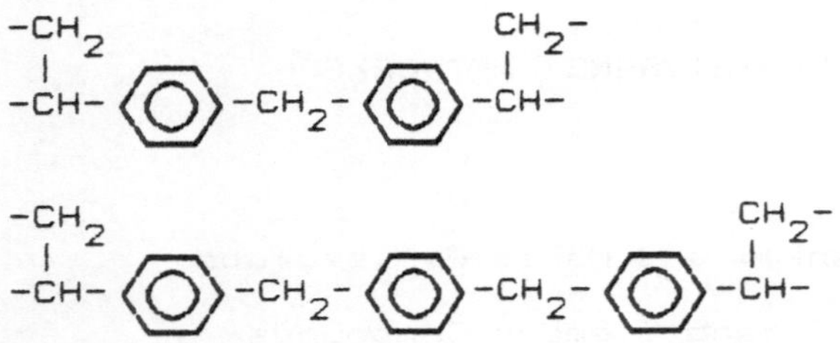

It is important to emphasize that the bridges are conformationally rigid, with the only exception of CMPB which contains a flexible tetramethylene fragment.

Polystyrene, an industrial product of circa 300,000 Dalton molecular weight and a broad molecular weight distribution, was dissolved in dichloroethane, nitrobenzene, tetrachloroethane (all of which being thermodynamically good solvents), or in cyclohexane. In the majority of cases, a 10 %
(w/v) solution in dichloroethane was used. A desired amount of
a bifunctional crosslinking agent was added to the solution.
Before adding stannic chloride (which served as the catalyst),
the solution was cooled, in order to have enough time for obtaining homogeneous distribution of the catalyst in the viscous solution. The latter quickly converts into a transparent
gel which volume gradually decreases thus releasing a certain
amount of free solvent (macro syneresis).

Besides the dissolved linear polystyrene, slightly crosslinked copolymers of styrene with 0.3-2.0 % DVB, taken in a
state of strong swelling with a good solvent, have been used
as the initial polymer for a subsequent intensive crosslinking
with MCDE. Whereas the first system results in obtaining
blocks of polymers which have to be disintegrated into irre-
gularly shaped particles (before washing them from the catalyst and drying), the second material is used in the form of
spherical beads. Spherical particles are preferred for many
practical applications of the final product.

By conducting the reaction at 60 to 80 $^{\circ}$C for 5 to 10 hours and in the presence of 0.5 to 1.0 mol catalyst, it is possible to achieve a complete conversion of the crosslinking
agent. The final polymer does not contain any pending chloromethyl groups, and no free crosslinking agent is left in the
reaction media, which can be easily checked by means of
thin-layer or gas chromatography.

The complete conversion permits calculating the crosslinking degree of the network formed from the molar ratio of the
reacting components. The degree of crosslinking is the ratio
of cross bridges number to the total number of bridges and unsubstituted phenyl groups:

180

$$X = \frac{M}{M + (1-fM)} \cdot 100\ \%.$$

where M is the number of moles of the cross agent taken for 1 mol of styrene repeating units and f is the functionality of the agent. Thus, if 0.5 mol of a bifunctional agent binds to 1 mol of polystyrene, the formal degree of crosslinking of the network appears to be 100 %, implying that practically all phenyl groups of the initial polystyrene should be involved into formation of cross bridges.

IR-spectra of products crosslinked to 100 % appear to be identical to those of copolymers of 40 % styrene and 60 % p-DVB. Absorption band at 820 cm^{-1} is characteristic of p-di-substituted benzene rings. Generally, intensity of this band increases with the degree of crosslinking rising. A band at 765 cm^{-1} which is usually ascribed to non-planar deformational vibrations of mono-substituted phenyl groups (a very strong band) is weakly represented in the spectra of highly crosslin-ked materials. Indeed, a certain amount of phenyls should be expected to remain in the product of reacting 1 mol polysty-rene with 0.5 mol of crosslinking agent, implying that the latter partially interacts with aromatic rings already invol-ved into formation of a cross bridge. Bridges of more complex structure should result from such reactions, but their amount is rather low since no bands could be observed at 820-765, 730-625, and 865-820 cm^{-1} which are characteristic of tri-substituted and o- and m-disubstituted benzene, respectively. Obviously, probability of these types of substitution is lo-wer than the above postulated p-substitution. By the way, the number of unmodified styrene units can be further reduced by involving more than 0.5 mol of crosslinking agent into reac-tion, which should help to attain real 100 %-crosslinking but would simultaneously increase the portion of more complex junction points in the network.

In the case that more then 25 mol-% of the above mentio-ned bifunctional crosslinking agents were introduced into re-action with polystyrene to produce networks with a cross-linking degree of more than 33 %, final products display pecu-liar properties and, therefore, are referred to as hypercross-linked.

SWELLING OF HYPERCROSSLINKED POLYSTYRENE

Generally accepted modern theories predict the swelling of a polymeric network to depend on the thermodynamic quality of the solvent, the degree of crosslinking and functionality of junction points. However, only few examples are known for a good correlation between the theory and properties of real networks[3]. Especially often are discrepancies with polymers prepared in the presence of a diluent, which forced Dušek and Prins[4] to involve a "memory term" into consideration.

Studying properties of hypercrosslinked polystyrene reve-aled the behavior of networks to be more complicated and de-pending on much more parameters.

1. Influence of the Dilution of the Reaction Mixture

Fig. 1 represents swelling capacity of products of cross-linking linear polystyrene in dichloroethane by means of MCDE and XDC as a function of concentration of the initial polymer solution. At any given crosslinking degree, the swelling rises dramatically on diluting the system that is subjected to structurization. A similar dependence is also valid for cross-linking products of styrene-DVB beads which are taken in a swollen with dichloroethane state (Fig. 2). The higher the DVB content and, accordingly, smaller the solvent amount within the initial copolymer, the lower the swelling ability of final products of any given degree of crosslinking.

The increased swelling capacity of networks prepared in diluted systems is well known. It is usually explained by a partial formation of intramolecular links or loops which are thought to be inefficient in reducing the swelling properties. On crosslinking long polymeric chains, however, this side re-action should play a noticeable role in extremely diluted so-lutions, only. The polymer concentrations in our initial sys-tems amounted up to 18 % for linear polystyrene and 26 and 47 % for copolymers with 1 and 4 % DVB (curves 2 and 5 in Fig. 2), respectively. At concentrations that high, the polymeric coils should completely penetrate each other, so that the pro-bability of an intramolecular reaction should be negligible. Moreover, for a final network with a crosslinking degree of 43 and 100 % (curves 3 and 2 in Fig. 1), there is no possibility and no sense to distinguish between intramolecular and inter-molecular bridges, et all.

A polymeric network should be rather considered as a sys-tem of mutually condensed and interpenetrating cycles. Each cycle comprises portions of polymeric chains as well as cross-

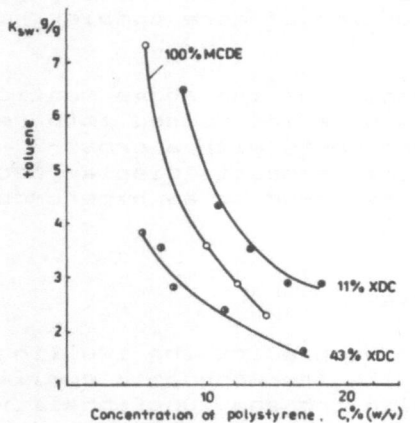

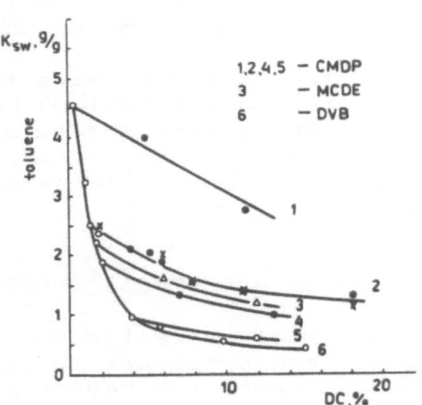

Fig. 1. Swelling in toluene of products of crosslinking linear polystyrene solutions in dichloroethane as a func-tion of concentration. Cross agents, XDC or MCDE. Degree of crosslinking, 11 %, 43 %, 100 %.

Fig. 2. Swelling in toluene of styrene-DVB copolymers crosslinked additionally in dichloroethane with CMDP or MCDE as a function of total crosslinking degree.
6 - Initial styrene-DVB co-polymers.

links between them. For densely crosslinked networks, it does
not matter whether these portions belong to one single chain
or different chains. Very important, however, is that each
cycle can embrace chains belonging to neighbour cycles. This
unavoidable interpenetration of the network meshes is the factor that strongly depends on the concentration of polymeric
material in solution during the closure of cycles. To our
opinion, the decisive role of the above topological factor,
namely, interpenetration of the network meshes, in reducing
the swelling ability of networks is not generally recognized
and considered, thus far.

This topological factor largely accounts for the fact
that, at equivalent crosslinking degrees, the swelling ability
of products of crosslinking linear polystyrene in solution or
swollen styrene copolymers appears to be much higher than that
of conventional styrene-DVB copolymers (see Fig. 2). The latter are prepared in the absence of any diluent and, therefore,
meshes of their network are densely filled with other polymeric chains and meshes. Contrary to this, the interpenetration
of meshes in the loose hypercrosslinked network (that is prepared in the presence of an about 10-fold excess of a solvent)
must be lower considerably.

2. Influence of Internal Strains in the Network

Suppose two batches of copolymer beads are subjected to
additional crosslinking: one copolymer contains 1 % DVB and is
swollen with dichloroethane to its ultimate extent, the other
copolymer contains 0.3 % DVB but is provided with a limited
amount of the solvent, only, in such a manner that the initial
polymer/solvent ratios in the two systems considered are
equal. One could expect the swelling properties of final
crosslinked products in the both sets of experiments to be
identical. Fig. 3 (curves 1 and 2) shows, however, that materials obtained in the second series swell stronger.

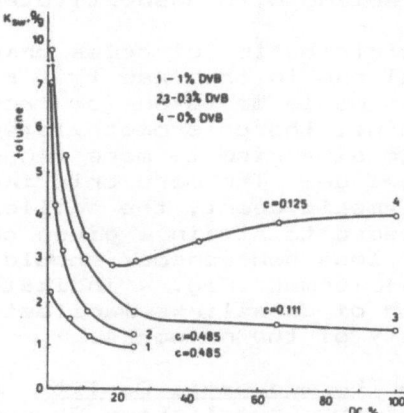

Fig. 3. Swelling in toluene of styrene-DVB copolymers (1-3)
         crosslinked additionally in dichloroethane and li-
         near polystyrene (4) crosslinked in dichloroethane
         as a function of total crosslinking degree. Cross-
         linking agent, MCDE. Content of DVB, 1.0 % (1).
         0.3 % (2, 3). Concentration of the polymer in the
         initial systems, g/ml, is given in the Figure.

In the case that more dichloroethane is taken for the pre-swelling of the second copolymer incorporating 0.3 % DVB, an additional increase in the swelling of final materials can be registered. Because of the above discussed dilution factor, this is not surprising. But, again, crosslinking of linear polystyrene dissolved in an equal amount of dichloroethane, leads to polymers of markedly higher swelling ability (curves 3 and 4 in Fig. 3).

The copolymer beads containing 0.3 and 1.0 % DVB, like any other conventional styrene-DVB copolymers, were prepared in the absence of any diluent. When in dry state, networks of these copolymers are not strained. Internal strains arise and grow in such networks during their swelling. Swelling stops in the moment where strained elastically active chains do not yield any more to swelling pressure. On crosslinking additionally the swollen gel, all pre-strained structural fragments appear immobilized in the final framework. There is no possibility for the strains to relax or dissipate. They would always manifest themselves by limiting the ultimate solvent uptake value of products prepared by additional crosslinking. There is only one initial system possible which is free from pre-strained chains - that is a solution of linear polystyrene. This system always produces materials of highest swelling ability.

3. Influence of the Distribution Character of Crosslinks

Since the initial reaction mixtures in the crosslinking reaction of linear polystyrene with monochlorodimethyl ether represent homogeneous solutions of the polymer, reagent and catalyst, one can expect a more or less statistical distribution of cross bridges in the final homogeneous gel, irrespective of the fact that, in the first step of the reaction, a partial chloromethylation of polystyrene takes place and, in the second step, the cross bridges are formed by the chloromethyl groups interacting with unsubstituted phenyl rings.

However, the distribution of cross bridges would deviate from the statistical one in the case that partially chloromethylated polystyrene would be taken for crosslinking unsubstituted polymeric chains. The chloromethylated polystyrene coils could be expected to give rise to more densely crosslinked micro area in the final gel. The more chlorine atoms contained in the starting polymeric agent, the smaller are the quantities of the latter needed to attain a given crosslinking degree of the gel, and the less homogeneous should be the internal structure of the gel formed. Fig. 4 indicates that the inhomogeneous distribution of crosslinks manifests itself in the higher swelling ability of the networks.

4. The Role of the Thermodynamic Quality of the Solvent and the Structure of the Crosslinking Reagent

Current theories predict that polymeric networks swell with good solvents and shrink in liquids that are bad from the thermodynamic point of view. Contrary to this, hypercrosslinked polystyrene gels prepared in dichloroethane do not exhibit any noticeable change in their volume on a replacement of this thermodynamically good solvent by any other organic liquid, among them such typical precipitators for linear polystyrene

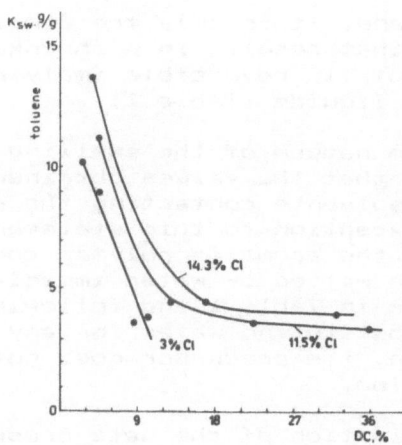

Fig. 4. Swelling in toluene of products of a reaction of polystyrene with partially chloromethylated polystyrene as a function of crosslinking degree. Content of chlorine in the chloromethylated polymer. 3.0 % (1), 11.5 % (2), 14.8 % (3).

Table 1. Swelling (ml/g) of Polystyrene Crosslinked with Different Reagents (measured by weights of dry and swollen products).

gree of ossnking	Toluene	Methanol	Hexane	Water	Toluene	Methanol	Hexane	Water
	Monochlorodimethyl ether				Dimethyl formal			
5	6.23	0.04	0.06	–	6.42	0.04	0.07	–
1	3.97	0.06	0.06	0.17	3.76	0.04	0.07	–
8	3.19	–	0.21	–	–	–	–	–
5	3.10	0.25	0.48	0.29	2.81	0.63	1.00	0.34
3	3.86	3.08	3.48	0.72	3.14	1.34	1.64	0.62
6	4.25	4.10	4.30	1.49	3.51	2.92	3.05	0.96
0	4.37	3.84	4.12	2.27	3.49	2.90	3.23	1.23
	Tris-(chloromethyl)-mesitylene				Bis-(chloromethyl-phenyl)-butane			
5	3.86	0.06	0.08	–	6.70	0.09	0.07	–
1	2.83	0.06	0.20	–	4.51	0.09	0.07	–
8	2.82	0.06	1.03	0.25	–	–	–	–
5	2.91	1.81	2.21	0.37	3.67	0.10	0.17	–
3	3.08	3.06	3.10	0.81	3.29	0.15	0.26	–
6	2.88	3.06	3.10	0.87	3.71	0.19	0.32	–
0	2.91	3.05	3.10	0.94	3.91	0.09	0.18	–
	Bis-chloromethyl-diphenyl				Xylilene dichloride			
5	5.51	–	–	–	–	–	–	–
1	3.90	–	–	–	3.69	0.15	0.15	0.15
8	3.21	–	–	–	–	–	–	–
5	2.98	0.20	0.12	0.14	2.46	0.32	0.58	0.10
3	3.17	1.09	1.00	0.27	2.19	1,13	1.70	0.30
6	3.38	2.17	2.48	0.77	2.19	1.71	2.00	0.80
0	3.60	2.91	2.68	1.23	1.92	1.67	1,81	0.84

as methanol and hexane. It is only the complete removal of the solvent on drying, that results in a shrinkage of the polymer. This shrinkage is totally reversible implying that dry polymers swell with all liquids (Table 1).

The equilibrium nature of the swelling is further corroborated by the fact that the values obtained are independent of the sequence of solvents contacting the polymer. Water represents the only exception to this statement, since the hydrophobic surface of the aromatic polymer contains no polar groups and cannot be wetted by water immediately. The values of swelling in water in Table 1 and following Figures 5, 7-10 were obtained by substituting water for any other water-miscible solvent. Again, the precursor does not affect the equilibrium swelling value.

A critical evaluation of the data presented in Table 1 reveals no suggestive correlations between the swelling properties of the networks and such structural parameters of the crosslinking agents as their length or number of aromatic phenylene rings. Quite evident is, however, the regularity that swelling in typical precipitators requires a certain rigidity of the network as a whole, ie, a sufficient degree of crosslinking and a definite conformational rigidity of the cross bridges. This threshold rigidity is already attained at a degree of crosslinking of about 25 % if a trifunctional reagent, tris-(chloromethyl)-mesitylene, is applied which ninds three polymeric chains in every junction knot. With bifunctional reagents, a more extensive crosslinking is required, that of about 40 %. Bis-(chloromethylphenyl)-butane containing a flexible tetramethylene fragment in the molecule does not yield a hypercrosslinked polymer that would swell with methanol or hexane, whatsoever.

The decisive role of the network rigidity finds a logical explanation as follows. The hypercrosslinked networks are prepared in the presence of an about 10-fold excess of a good solvent. The removal of this solvent from the polymer gels is, undoubtedly, associated with developing stresses within the shrinking networks, since large amounts of rigid spacers prevent polymeric chains from approaching each other. From this reason, dry polymers obtained must exhibit large internal stresses. Indeed, these stresses can be seen in dry beads of hypercrosslinked polystyrene in the form of a Maltese cross in a polarization microscope (Fig. 5). The stresses disappear on swelling the beads with any solvent, both good or bad for polystyrene. We consider the relaxation of internal stresses as the driving force for the swelling process with bad solvents.

An interesting conclusion from the data of Table 1 is that, unlike slightly crosslinked materials, the hypercrosslinked polymers swell with good and bad solvents to almost equal extents. (There are only two types of liquids that produce smaller values of swelling coefficients. These are perfluorocarbons and water).

This observation corresponds to the results of measuring the integral heats of interaction of dry hypercrosslinked polystyrene with good and bad solvents (Table 2) as well as with vapours of various liquids (Table 3). The former values appear unusually large: circa 130 J/g for toluene and 100 J/g for me-

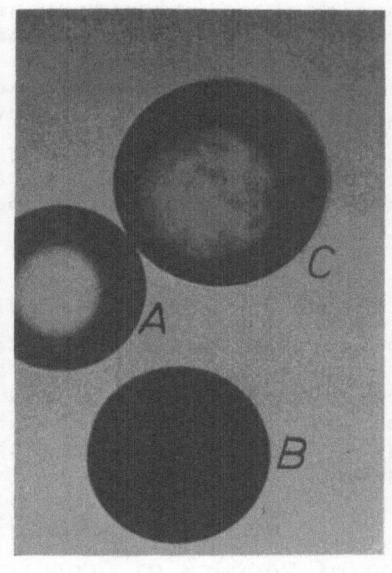
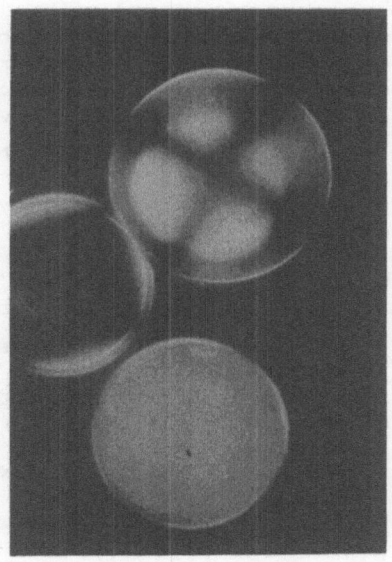

1   2

Fig. 5.   Polarization microscope photographs of beads of a
          conventional gel-type styrene-DVB copolymer (A), a
          macroporous copolymer (B), and a hypercrosslinked
          polymer (C) under normal (1) and cross-polarized
          (2) conditions.

thanol. Undoubtedly, these figures reflect additional significant contributions from the relaxation of the internal stresses, but they imply that methanol solvates accessible polystyrene chains almost as energetically as does toluene. The small
difference of 30 J/g may then correspond to the difference in
the enthalpy of interaction of polystyrene chains with the

Table 2.   Integral Heats of Interaction of Po-
           lystyrene Networks with Solvents and
           Precipitators.

Degree of cross-linking	Type* of networks	Interaction enthalpy, J/g	
		Toluene	Methanol
5	I	33.5	0
11	I	36.0	3.4
25	I	75.4	27.2
43	I	100.6	66.6
66	I	119.8	101.4
100	I	127.0	104.3
100	II	131.6	96.4
2	III	25.6	0
20	III	39.4	0

* I   - Linear polystyrene crosslinked with MCDE.
  II - Copolymer of styrene with 0.7 % DVB cross-
        linked with MCDE.
  III - Gel-type copolymers of styrene with DVB.

above good and bad solvents. Conventional gel-type styrene-DVB copolymers are not solvated by methanol. It is only a good solvent, toluene, that appears in position of disrupting strong intermolecular interactions between the polystyrene macromolecules and cause an exothermic swelling of the material. (In a good agreement with results by Soldatov[5], the data of Table 2 indicate that DVB molecules diminish the interaction energy and packing density of polystyrene chains in the copolymers.)

Similar conclusions can be drawn from the values of free enthalpy of adsorption of vapours (Table 3) which were calculated[6] from adsorption isotherms on various polystyrene samples. Whereas extremely high adsorption energies are characteristic of all hypercrosslinked materials, both for good and bad solvents, linear polystyrene as well as macroporous copolymers display lower surface energy values (which are proportional to the accessible surface area).

Table 3. Free Energy of Adsorption of Vapours of Organic Solvents (calculated from adsorption isotherms at a mole fraction of the solvent of 0.2).

Type of sample[*]	Surface area, m^2/g	Adsorption energy, J/mol				
		Benzene	Methanol	Cyclohexane	Hexane	Heptane
I	950	2135	–	1886	–	–
II	1200	–	2260	–	2721	2260
III-a	7.6	578	59	100	–	–
III-b	50.6	838	335	209	–	–
III-c	38.1	922	293	587	–	–

[*]I – Linear polystyrene crosslinked to 100 % with XDC.
 II – Copolymer of styrene with 0.7 % DVB crosslinked to 100 % with MCDE.
 III-a – Linear polystyrene.
 III-b – Porous linear polystyrene precipitated from a toluene solution into methanol.
 III-c – Macroporous copolymer of styrene with 20 % DVB.

However, a combination of methods reveals that the interaction energy of various compounds with hypercrosslinked polystyrene decreases in the following order:

| aromatic and aliphatic hydrocarbons, alcohols, ketones etc. | > | perfluoro- carbons | > | water | > | liquid argon at −196 °C | > | vapours of Ar, N_2, CO_2 at 20 °C |

It is remarkable that hypercrosslinked polystyrene swells during adsorption of argon vapours at the temperature of liquid nitrogen.

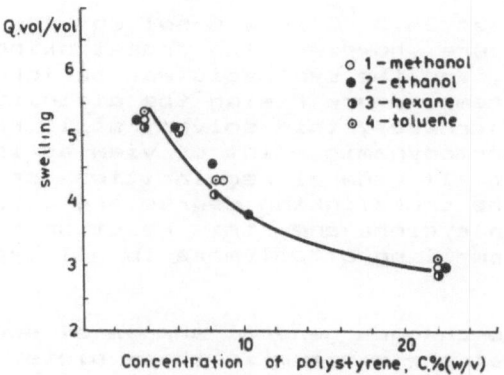

Fig. 6. Increase in the volume of polystyrene crosslinked to 66 % with MCDE on contacting methanol (1), ethanol (2), hexane (3), and toluene (4), as a function of the concentration of polystyrene solutions in dichloroethane during the synthesis.

It should be mentioned here, that swelling with precipitators depends on the concentration of the system that is subjected to crosslinking (Fig. 6). From a dilute solution, it is possible to obtain polymers which increase their volume in methanol, hexane or ethanol by a factor of 5, very similar to their swelling capacity in toluene.

5. Influence of the Reaction Media

Suitable solvents for conducting the Friedel-Crafts crosslinking reaction of polystyrene are dichloroethane, tetrachloroethane, nitrobenzene, and cyclohexane. Reaction products in the first two solvents are practically identical in all their properties (Fig. 7).

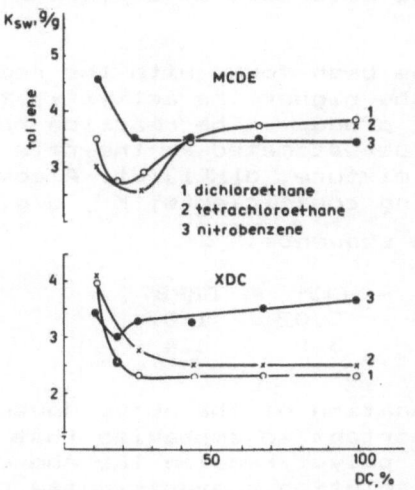

Fig. 7. Swelling in toluene of products of crosslinking with MCDE and XDC of linear polystyrene solutions as a function of crosslinking degree.

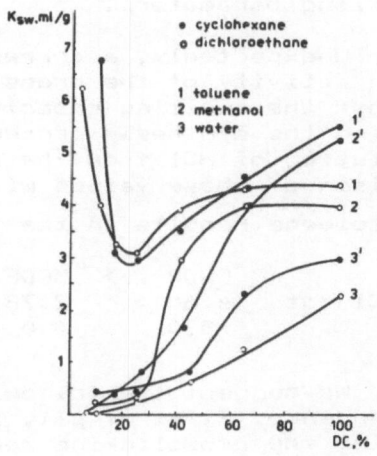

Fig. 8. Swelling of products of crosslinking with MCDE of polystyrene solutions in dichloroethane (1-3) and cyclohexane (1'-3') as a function of crosslinking degree.

Cyclohexane at 34.5 $^{\circ}$C is a $\ominus$-solvent for polystyrene. At that low temperature, however, the crosslinking reaction proceeds too slowly, and the synthesis was performed at 60 $^{\circ}$C. Though with the temperature rising the affinity of cyclohexane to polystyrene increases, this solvent still remains not as good from the thermodynamic point of view as is dichloroethane at 80 $^{\circ}$C. Whereas all general regularities for the dependence of swelling on the crosslinking degree are also valid for products prepared in cyclohexane, this reaction medium causes remarkably high swelling of polymers in all types of solvents (Fig. 8).

Nitrobenzene changes general shapes of swelling-crosslinking plots in that the minimum in these plots, that is very characteristic of products of crosslinking with MCDE in di- and tetrachloroethane, disappears in nitrobenzene (Fig. 7). Gel formation in nitrobenzene takes place too fast, so that there is no possibility for obtaining a homogeneous distribution of the catalyst in the initial reaction mixture. However, a similar, artificially created inhomogeneity at the onset of reaction with MCDE in dichloroethane did not cause any peculiar properties of the products. When crosslinked with xylylene dichloride in nitrobenzene, polymers of very high swelling capacity are obtained (Fig. 7). The special role of nitrobenzene is not well understood, thus far.

6. The Role of the Reaction Rate of Polystyrene with the Crosslinking Agent

As it was shown in Table 1, different crosslinking agents produce hypercrosslinked polymers of different swelling ability in toluene. There is no correlation between this property of the network and the structure of the cross bridge. Thus, monochlorodimethyl ether and dimethyl formal, while contributing identical fragments to the cross link, namely, a methylene group, result in hypercrosslinked materials of differing swelling parameters.

Unexpectedly, a correlation has been found with the reaction activity of the cross agent: the higher the activity, the higher the swelling capacity of the product. The reaction rate during the synthesis procedure can be estimated as the rate of evolution of HCl from the reaction mixture, $d[HCl]/dt$. A comparison of these values with swelling coefficients, K_w, g/g, in toluene results in the following sequence:

	CMDP	>>	MCDE	>	XDC	~	CMM	~	CMPB
$d[HCl]/dt$	6.60		2.78		1.90		2.02		1.87
K_{sw}	>5.0		4.0		3.8		3.5		3.4

We suggest the following explanation of the above unusual dependence. First of all, it is important to emphasize that during the crosslinking reaction of polystyrene in the above good solvents, the initial polymer solution converts quite rapidly into a gel. Then a relatively large portion of the solvent is observed to be gradually excluded (macro-syneresis) from the gel phase in the course of further heating the reaction mixture (heating is required for achieving a complete conversion of all reactive groups of the crosslinker). At the end, the amount of the solvent that happens to be included in

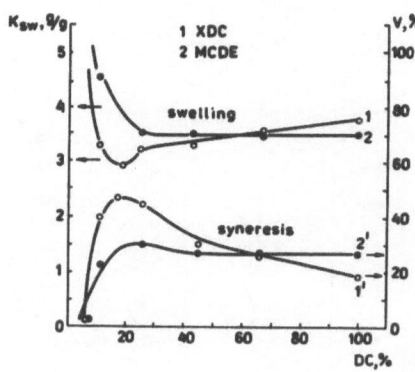

Fig. 9.   The portion of the solvent, nitrobenzene, excluded
          from the gel during the synthesis (1', 2') and the
          swelling ability in toluene (1, 2) of the polymer
          crosslinked with XDC (1, 1') and MCDE (2, 2'), as
          a function of the crosslinking degree.

the final gel corresponds to the equilibrium swelling ability
of the polymer in that solvent or other good solvents (Fig. 9).

     The above macro-syneresis was observed to be especially
intensive at the early stage of the reaction, but then it
stops almost completely, probably, due to the fact that, sooner or later, the network acquires a sufficient rigidity. Since
the macro-syneresis   implies transportation of solvent molecules through the gel phase toward its border, it must be a
relatively slow process. Therefore, the portion of the solvent
excluded from the gel was found to fall with the activity of
the crosslinking agent rising. In other words, the sooner the
polymer solution converts into a rigid gel, the larger the
quantities of the solvent which remain included in the latter.
This competition between the macro-syneresis process and the
crosslinking reaction is also evident from the fact that increasing the volume of the reaction mixture results in obtaining products of higher swelling ability.

     Many other evidences for the influence of the rate of
conversion of the initial solution into a gel can be presented. Thus, substituting nitrobenzene for dichloroethane as the
reaction media for crosslinking with XDC enhances considerably
both the relative rate of HCl evolution (from 1.90 to 4,25)
and the swelling capacity of the 100 %-crosslinked product in
toluene (from 2.3 to 3.6 g/g). Increasing the catalyst ($SnCl_4$)
concentration (from 0.5 to 2.0 mol catalyst per mole MCDE) during crosslinking linear polystyrene in cyclohexane results in
rising solvent quantities immobilized in the gel and enhanced
swelling properties of dry products in all solvents examined
(Fig. 10).

     The same reason, entrapment of the reaction media by the
gel phase, most probably, accounts for the surprising observation that, with the degree of crosslinking rising in the range
from 25 to 100 %, the swelling ability of the hypercrosslinked
polymer increases again, so that a totally crosslinked polystyrene accumulates 4 to 5 volumes of toluene, as does a styrene copolymer containing as little as 0.5 to 1.0 % DVB (eg,
see Table 1, curves 1 and 2 in Fig. 7a, curves 1 and 1' in

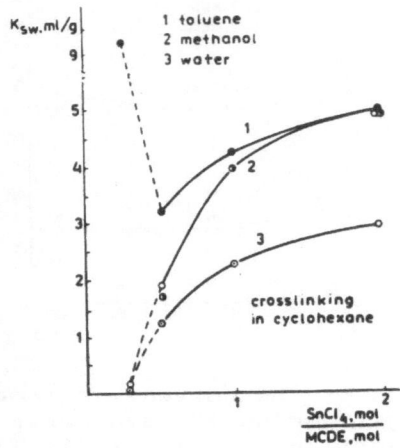

Fig. 10. Swelling in toluene (1), methanol or hexane (2),
and water (3) of products of crosslinking to 100 %
of linear polystyrene with MCDE in cyclohexane, as
a function of the catalyst (SnCl$_4$) amount applied.

(0.25 Mol of the catalyst are not sufficient for
the conversion to be complete).

Fig. 8). Obviously, with the high amounts of the crosslinking
agent and catalyst that are applied in such reaction batches,
formation of a rigid gel proceeds too fast, as compared with
the time required for the solvent to be expelled from the gel
during the macro-syneresis of the latter.

7. Influence of the Molecular Weight of Polystyrene

Fig. 11 illustrates an unexpected finding that using linear polystyrene of 8,800 Dalton molecular weight, instead of
a material of 300,000 Dalton, results in crosslinked products
of higher swelling ability, which holds for all crosslinking
degrees and all solvating media. We do not have any satisfying
explanation for this phenomenon, thus far.

8. General Remarks on Swelling Properties of Hypercrosslinked Polystyrene

High swelling ability of hypercrosslinked polystyrene implies high mobility of their networks. This property is not
trivial, as the distance between the neighbouring junctions in
highly crosslinked preparations approaches the length of a
single repeating monomer unit. Most important factors are the
special topology of networks considered, loose packing of polymeric chains, reduced portion of entrapped entanglements.
Another factor is the rigidity of the network and significant
internal stresses which emerge in the structure on removing
the solvating media.

POROSITY OF HYPERCROSSLINKED POLYMERS

An important consequence of the fact that rigid cross
bridges actively prevent the polymeric chains from attaining
dense packing is the abnormally low density of dry hypercross-

192

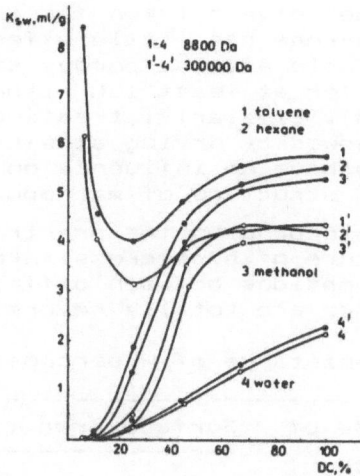

Fig. 11. Swelling in toluene (1. 1'), hexane (2. 2'), meth-
           anol (3.3'), and water (4, 4') of products of
           crosslinking polystyrene 8.800 Da (1-4) and
           300,000 Da (1'-4') with MCDE in dichloroethane,
           as a function of crosslinking degree.

linked polymers, which may correspond to an apparent porosity
of 0.2-0.3 cm^3/g (maximal. 0.7 cm^3/g). Though the polymer remains transparent and blemishless on drying, it readily ab-
sorbs gases. Surprisingly, the ultimate absorption capacity
was always observed to be definitely larger than the initial
volume of voids in the material. This implies that the adsorption process of gases, even at the low temperature of liquid
nitrogen, is accompanied by a swelling of the network in the
gas condensed in the polymer phase.

      The apparent inner surface area of hypercrosslinked poly-
mers can be formally calculated in accordance to the BET method which perfectly describes adsorption isotherms, in spite
of the fact that the theory should not apply to swelling systems[7]. Anyway, polystyrenes crosslinked to high extents with
rigid cross bridges behave as porous materials which internal
surface area amounts to extremely high values of about 1000
m^2/g (Table 4). Flexible cross agents (CMPB) are not qualified
for obtaining porous materials.

Table 4. Apparent Inner Surface Area, S_{sp}, m^2/g, of Products
          of Crosslinking Linear Polystyrene in Various Media
          by Various Cross Agents

Degree of cross-linking	Dichloroethane		Nitrobenzene		Tetrachloroethane	
	XDC	MCDE	XDC	MCDE	XDC	MCDE
11			0	60	0	0
18			–	150	–	–
25	0-100	240	300	140	100	90
43	530	640	560	640	300	800
66	820	1000	1100	860	900	930
100	1000	1000	900	1100	1200	1030

The nature of the solvent taken for the synthesis of hypercrosslinked polystyrene has little effect on the value of inner surface area (Table 4). The porous structure tolerates heating up to 150 $^{\circ}$C for at least 1 h without any loss in the surface area (Table 5). Similarly, treating the polymer with various solvents followed by drying at elevated temperatures (Table 6), appears to have no influence on the porous structure. Contrary to this, structure of macroporous styrene-DVB copolymers is known[8] to depend on the pre-treatment protocol. Obviously, the structure of hypercrosslinked polymers is stable, and all transformations between different equilibrium states of the structure are totally reversible.

Table 5.　Heat Resistance of Hypercrosslinked Polymers.

Polymer type*	Cross agent	Degree of cross-linking	Surface area (m^2/g) after heating		
			80°C, 5h	100°C, 1h	150°C, 1h
I	XDC	43	530	510	133
		66	820	830	700
		100	1000	1000	1050
I	MCDE	25	240	220	67
		43	640	650	460
		66	1000	1000	800
		100	1000	920	900
II	MCDE	100	1010	1040	980

*I　- Linear polystyrene crosslinked to 100 % with MCDE.
II - Styrene copolymer with 0.7 % crosslinked to 100 % with MCDE.

Table 6. Influence of Precursor Solvents on the Surface Area of Hypercrosslinked Polymers

Polymer type*	Precursor	Temperature of drying	Surface area m^2/g
I	Toluene	125	700
	Hexane	125	840
	Water	125	950
	Water	180	720
	Toluene	125	940
	Hexane	125	980
	Water	125	1000
II	Water	80	1000

*I　- Linear polystyrene crosslinked to 100 % with MCDE.
II - Styrene copolymer with 0.7 % crosslinked to 100 % with MCDE.

The porous structure of polymers prepared in cyclohexane appears to be more sensitive to synthesis conditions. Thus, reducing the quantity of the catalyst applied results in the drop of both swelling ability (in precipitating media) and porosity of the material:

Mol $SnCl_4$, per mol MCDE	2.0	1.0	0.5	0.25
Surface area, m^2/g	1060	960	260	0

Pore diameters of swollen polymers can be easily estimated be using inverted gel permeation chromatography. This method gives a very narrow pore size distribution with a maximum in the range of 15 A^o. There are no pores accessible to polystyrene standard coils of larger than 50 A^o in diameter. Contrary to these findings, classical mercury porosimetry seems to reveal much larger pores - from 60 to 600 A^o in diameter. Pores of that size would make the polymer opaque, which is not the case. In addition, large inhomogeneities would cause strong small-angle X-ray scattering, whereas the scattering from hypercrosslinked spherical styrene-DVB copolymers was observed to be rather weak. Obviously, the low-density polymer just undergoes compression under high pressure applied through mercury, this compression being experimentally indistinguishable from intrusion of mercury into large pores. This behavior points out a definite flexibility of the network exposed to mechanical forces.

Products of crosslinking polystyrene in cyclohexane appear to have bimodal pore size distribution with two maxima at about 15 and 60 A^o. These materials are non-transparent.

MOBILITY OF HYPERCROSSLINKED NETWORKS

If the polymer can be compressed, several methods must be in position to reveal mobility of its network, as well. Indeed, precise dilatometric measurements reveal several phase transitions with changing slopes in the plot of the volume of polymer particles versus temperature, whereas, in the temperature range of 150 to 450 K examined, linear polystyrene only exhibits a β-transition at -71 ^{o}C and glass transition close to 100 ^{o}C (Fig. 12).

Even more impressive are the results of measuring dynamic mechanical losses by a vibration technique (Fig. 13) on a bun-

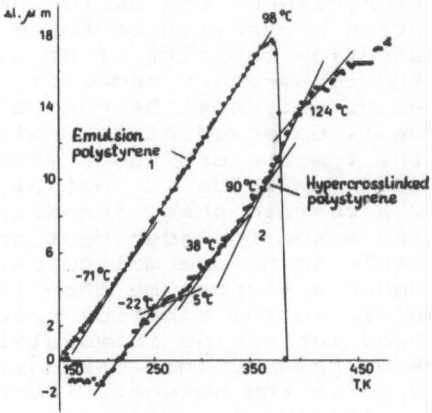

Fig. 12. Dilatometric examination of a linear polystyrene (1) and a product of crosslinking linear polystyrene to 100 % with MCDE.

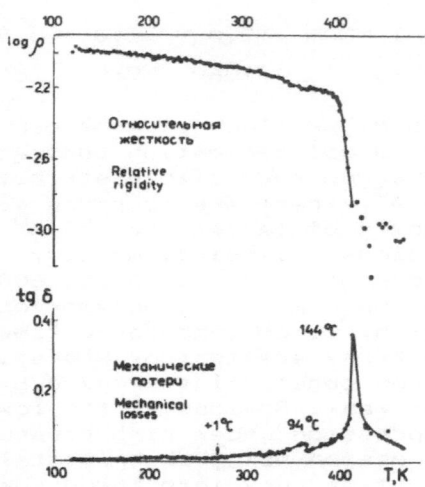

Fig. 13.   Temperature dependence of the relative rigidity
and mechanical losses of polystyrene crosslinked
with XDC.

dle of polyamide fibers that is cowered with a XDC-crosslinked
polystyrene. Though the porosity of the polymer film is rather
low (inner surface area. 360 m^2/g), an intensive transition
effect can be observed at 144 $^\circ$C, as well as a weak conformational transition at 98 $^\circ$C. It is not possible thus far to as-
cribe the intensive effect to glass transition, since, according to theories, it could be expected to lie at much higher
temperatures. However, it should be taken into account that,
due to the low density of the material and the rigid spacers
preventing the polystyrene chains from approaching each other
closely, the total energy of the between-chain interactions in
the hypercrosslinked polymer must be reduced quite considerably. In its turn, this factor must reduce the value of T_g.

   The above suggestion that certain important conformational transitions in a hypercrosslinked network may take place
at relatively low temperatures can be further strengthened by
anomalies of adsorption properties of these materials (Fig.
14). Generally, adsorption capacity of porous adsorbents gradually rises with the temperature decreasing. However, with
the hypercrosslinked polystyrene "Styrosorb", an inversion of
adsorption isotherms is observed at low relative pressures of
hexane vapours in the temperature range from 0 to 20 $^\circ$C: adsorption capacity at 0 $^\circ$C yields to that at higher temperatures. Most probably, a certain phase transition takes place
around 0-10 $^\circ$C, which makes the adsorbent network more flexible and more accessible to hexane molecules. Special experiments carried out under a microscope show the adsorbent network to respond rapidly to the sorption process with the volume of the polymer bead increasing immediately on sorption of
very first portions of hexane. This response gradually provides new sorption space in the network. Therefore, in spite of
a relatively small initial volume of voids in the material
(circa 0.2 cm^3/g), the ultimate adsorption capacity amounts to
extremely high values, eg, 1.2 ml hexane per gram of the polymer.

196

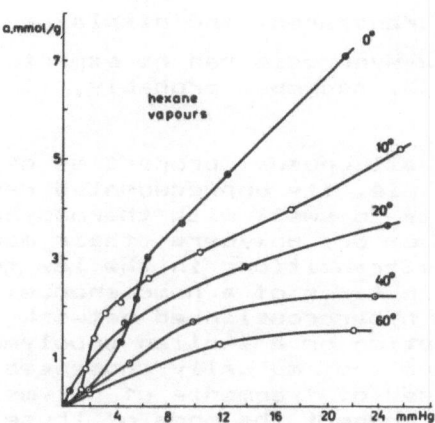

Fig. 14. Adsorption isotherms of hexane vapours at several temperatures on a styrene-DVB (0.7 %) copolymer crosslinked additionally with MCDE to 100 %.

STRUCTURE OF THE HYPERCROSSLINKED POLYSTYRENE

The above presented examples of a remarkable ability of a
densely crosslinked polymer to easily change its volume and
flexibly respond to mechanical strains requires a deeper understanding of the relation structure/properties than usually
exhibited in classical theories of elasticity of a network.

From the structural point of view, most important is the
question of homogeneity or heterogeneity of hypercrosslinked
polystyrene. Properties of homogeneous networks have to be
explained on a molecular level, whereas those of heterogeneous
materials can partially result from specific inter-phase phenomena. Generally, heterogeneous networks are formed under
conditions of a micro-phase segregation during the synthesis
procedure, which can be caused either by (i) poor solvation of
the polymer formed by the low-molecular-weight diluent of the
system (x-syneresis) (ii) an extremely inhomogeneous distribution of the cross bridges (ν-syneresis) or (iii) presence of
polymeric coils that are poorly compatible with the network
chains. Non of the above factors operates during the synthesis
of hypercrosslinked polystyrene in thermodynamically good solvents (di- or tetrachloroethane, nitrobenzene). By using electron microscopy and small-angle X-ray scattering techniques,
as well as size exclusion chromatography technique, we failed
revealing inhomogeneities on a macromolecular level. Indeed,
when crosslinking swollen copolymers of styrene with, eg, 2 %
DVB, inhomogeneities of that size should be excluded from a
purely theoretical point of view. (Crosslinking of linear po-
polymer solutions requires a more detailed consideration[10]).

Finally, we were able to show more recently[11] that all the unusual properties of hypercrosslinked polystyrene networks can
be simulated in a single polystyrene macromolecule.

Contrary to this, polystyrene crosslinking products in
cyclohexane, a poor solvating media, appear to be heterogene-

ous. They are non-transparent and display a bimodal pore size distribution[12]. A $\varkappa$-syneresis can be expected to proceed in this synthesis media, and most probably, it takes place in reality.

Nevertheless, all unusual properties of the hypercross-linked polystyrene, ie, its unprecedented range of volume changes, the ability to swell with thermodynamically poor solvents, low density of dry polymers, their marked ability to absorb gases, phase transitions in the low temperature range, must be explained in terms of a homogeneous, ie, one-phase-structure model. A hypercrosslinked network that was obtained from a polymer solution or a swollen copolymer, should be considered as an ensemble of mutually condensed and interpenetrating cycles comprised of fragments of polymeric chains and cross bridges that connect the ends of these fragments. According to calculations by Flory[13], two neighbouring phenyl rings in a polystyrene chain, in order to avoid close contacts, cause a twist of the chain by an angle of about 20°. This appreciably lowers the probability of closing the smallest thinkable network mesh by forming two neighbouring cross bridges (Fig. 15 A). The smallest unstressed cycle comprises three pairs of phenyl groups belonging to three different chains (or chain segments) and combined by three bridges (Fig. 15 B). Larger cycles may also form with high probability.

An important statement is that each cycle in a polymer crosslinked to 100 % has a sufficient contour length and should be conformationally flexible. Certainly, all cycles are condensed with many neighbor cycles and cannot behave independently of them. There must be a cooperative rearrangement of conformations of large assembles of cycles that enables the network to change its volume. To our opinion, the decisive factors which make possible this cooperative rearrangement are the low degree of interpenetration of the network meshes and low number of entrapped entanglements, which, in their turn, are caused by the strong dilution of the polymeric system during the formation of cross bridges.

In a sufficiently rigid network, the above cooperative reconstruction requires strong forces to be involved. On changing one solvent in a swollen gel for another, eg, a precipitator for a solvent, one would cause a change in the solvation

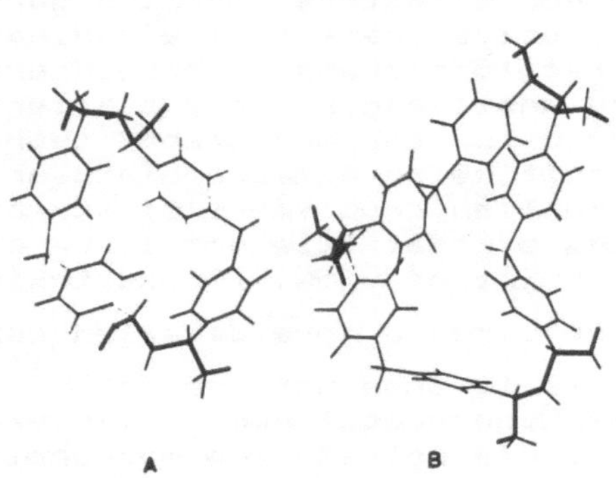

A   B

Fig. 15.   Computer simulation of smallest possible network
           meshes, a conformationally unfavorable (A) and a
           more probable (B) one.

energy of the polymeric network. This difference, however, is relatively small and cannot reduce significantly the total volume of the rigid gel, with the exception of perfluorocarbons and water which have an exceptionally low affinity to polystyrene. Therefore, aside from these two types of media, the equilibrium swelling capacities of rigid hypercrosslinked polystyrenes, in contrast to flexible or slightly crosslinked structures, are almost identical in all solvents and precipitants.

Strong forces arise on removing the solvent from the rigid network. It would be an extremely unfavored situation if polymeric chains would remain naked. It is here that cooperative changes in the conformation of meshes takes place, in order to achieve as much of close intramolecular contacts and gain as much of dispersion energy as possible. As shown above, the contraction of the network on drying may exceed a factor of 5, still giving a product of low density and large internal stresses. Naturally, these peculiarities are characteristic of rigid networks, only, and a hypercrosslinked polystyrene prepared by crosslinking with the flexible 1,4-bis(p-chloromethylphenyl)-butane behaves normally.

CONCLUSION

Spacious and rigid networks of hypercrosslinked polystyrene that are prepared by an intensive crosslinking of a dilute polystyrene-solvent system represent a new type of polymeric materials. Besides the above discussed new theoretical knowledge they allow to gain, these materials are of great practical interest. Their unprecedented sorption capacity to all kinds of organic compounds opens new possibilities for developing progressive technologies of recovery and purification of important chemicals, environment protection, solid phase extraction, exclusion chromatography, etc. A whole palette of adsorbents of the "Styrosorb" series is being successfully developed.

ACKNOWLEDGEMENT

We greatly acknowledge the support of this work by Purolite International Ltd. (Pontyclun, UK).

REFERENCES

1. T. R. E. Kressman and J. R. Millar, Relationship between swelling and crosslinking of sulfonated polystyrene resins, Chem. and Ind., 45:1833 (1961).
2. J. H. Glans and D. T. Turner, Glass transition elevation of polystyrene by crosslinks, Polymer, 22:1540 (1981).
3. M. Seno and T. Yamabe, A note on the characterization of the network structure of the ion exchange resins, Bull. Chem. Soc. Japan, 37:754 (1964).
4. K. Dušek and W. Prins, Structure and elasticity of non-crystalline polymer networks, Adv. Polym. Sci., 6:1 (1969).
5. V. A. Artamonov and V. S. Soldatov, Study into permeability of copolymers of styrene with divinylbenzene, Izvestia Akad. Nauk BSSR, Ser. Khim., N 3:34 (1974).

6. D. A. Reshetko, On relations between processes of physical adsorption and dissolution on interaction of polymers with vapours of low-molecular-weight liquids. Ph D Thesis, Sverdlowsk, (1976), p. 176.

7. I. F. Khirsanova, V. S. Soldatov, R. V, Marzinkevich, M. P. Tsyurupa, and V. A. Davankov, Sorption properties of polystyrene gels crosslinked with p-xylilene dichlorde, Kolloid. Zhurnal, 40: 1025 (1978).

8. J. Baldrian, B. N. Kolarz, and H. Galina, Small angle x-ray scattering study of porosity variation in styrene-divinylbenzene copolymers, Collect. Czech. Chem. Commun., 46: 1675 (1981).

9. G.I. Rozenberg, A. S. Shabaeva, V. S. Moryakov, T. G. Musin, M. P. Tsyurupa, and V. A. Davankov, Sorption properrties of hypercrosslinked polystyrene sorbents, Reactive Polymers, 1: 175 (1983).

10. M. P Tsyurupa, E. A. Pankratov, D. Ya. Tswankin, V. P. Zhukov, and V. A. Davankov, Morphology of macronet isoporous styrene polymers of the "Styrosorb"-type, Vysokomolek. Soed., A 27: 339 (1985).

11. M. P. Tsyurupa, T. A. Mrachkovskaya, L. A. Maslova, G. I. Timofeeva, L. V. Dubrovina, E. F. Titova, and V. A. Davankov, Soluble intramolecularly-hypercrosslinked polystyrene, Reactive Polymers, in press.

12. L. D. Belyakova, T. I. Shevchenko, V. A. Davankov, and M. P. Tsyurupa, Sorption of vapours of various substances by hypercrosslinked "Styrosorb" polystyrene, Adv. in Colloid and Interface Sci., 25: 249 (1986).

13. J. D. Joon, P. R. Sandararajan, and P. J. Flory, Conformational characterization of polystyrene, Macromolecules, 8: 776 (1975).

MOLECULAR MODELING OF THE MECHANICAL PROPERTIES
OF CROSSLINKED NETWORKS

Yves Termonia

Central Research and Development
Experimental Station
E.I. du Pont de Nemours, Inc.
Wilmington, DE 19880-0356

INTRODUCTION

It is now well accepted that entanglements have a strong effect on the mechanical properties of crosslinked networks. The relationship[1]

$$G = \nu \, kt \qquad\qquad\qquad (1)$$

in which ν denotes the number of starting polymer chains per unit volume has been shown[2-3] to greatly underestimate the value of the shear modulus G. Entanglements, which are always present in polymeric systems prior to crosslinking, indeed add a significant contribution to the modulus value given by Eq.1.

In a recent series of publications[4-8], we have introduced a Monte-Carlo model for the study of the effects of entanglements on the deformation behavior of crosslinked systems. In our approach, a regular lattice is first filled-in with an array of linear macromolecules having a prescribed molecular weight distribution. The resulting entangled network is then crosslinked in situ and its mechanical properties are tested as follows. The network is strained in small increments on the computer and, for each increment, the following processes are allowed to occur: (i) extension and orientation of chain strands between junctions (entanglements and crosslinks); (ii) slippage of chains through entanglements; (iii) breakage of strands at maximum extension. At each value of the external strain, the network is relaxed towards mechanical equilibrium by a series of fast computer algorithms which steadily reduce the net residual force acting on each network junction.

The model has been applied to crosslinking of difunctional polymeric chains with particular emphasis on poly(dimethylsiloxane) (PDMS) networks. The model is described in detail in Ref.4. The importance of the initial molecular weight distribution is studied in Ref.5 in an attempt to elucidate some unusual toughness properties found experimentally in bimodal PDMS networks. The approach has been also applied to networks crosslinked in a state of strain[6] as well as to interpenetrating networks[7] made by crosslinking one polymer in the immediate presence of another. The model has been recently extended to multiaxial deformation[8] in an attempt to evaluate the strain energy density function. Here, we shall limit ourselves to review some of our results on the effects of entanglements on the mechanical properties of crosslinked networks.

Synthesis, Characterization, and Theory of Polymeric Networks and Gels
Edited by S.M. Aharoni, Plenum Press, New York, 1992

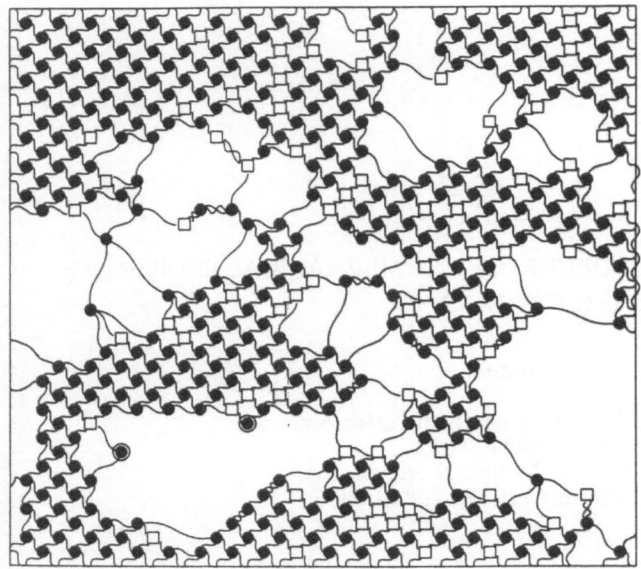

Fig.1. Actual network prior to deformation, as obtained from the computer model for a polymer with M=4M$_e$ and crosslinked with tetra-functional monomers. The fractional conversion of crosslinker groups is p=0.96. The figure shows only the gel fraction with its elastically active junctions (crosslinks $\square$ and entanglements $\bullet$) and their connecting strands. Dangling ends have not been represented.

MODEL

Figure 1 shows a typical network structure, as obtained from our computer model, prior to deformation. The network was obtained through end-linking difunctional chains with tetrafunctional monomers (symbol $\square$). The starting molecular weight of the chains equals 4 times the molecular weight between entanglements, M=4M$_e$, and the degree of advancement of the reaction p=0.96. Trapped entanglements are denoted by symbol $\bullet$. The figure shows only the gel fraction with its elastically active junctions and their connecting strands. Dangling ends have not been represented.

The network of Fig.1 is strained in small increments either uniaxially or biaxially and, for each value of the external strain, each junction (entanglement and crosslink) is relaxed towards mechanical equilibrium with its neighbors. The state of stress of a particular chain strand between junctions is obtained as follows. Let r and w(r) denote respectively the end-to-end vector length and free energy of deformation for that chain. The engineering stresses σ_x and σ_y along the x and y directions, are given by

$$\sigma_x = \partial w(r) / \partial \lambda_x = [dw(r) / dr] [\partial r/ \partial \lambda_x]$$
$$\sigma_y = \partial w(r) / \partial \lambda_y = [dw(r) / dr] [\partial r/ \partial \lambda_y]$$

(2)

in which λ_x and λ_y are the draw ratios along the x- and y-directions, respectively. Eqs. 2 can be rewritten in simpler form

$$\sigma_x = f [\partial r/ \partial \lambda_x]$$
$$\sigma_y = f [\partial r/ \partial \lambda_y]$$

(3)

in which

$$f = dw(r) / dr \qquad (4)$$

is the force on the chain. The latter is obtained from

$$f = (kT/\ell) \, \mathcal{L}^{-1} \, (r/n\ell) \qquad (5)$$

In Eq.5, $\mathcal{L}^{-1}$ represents the inverse Langevin function whereas n and ℓ denote respectively the number and length of the statistical segments for the chain. The non-affine displacements of the various junctions, obtained through network relaxation, create large force gradients on chains passing through trapped entanglements. These gradients lead to the possibility of chain slippage which is allowed in the model until the difference in force in the two strands of a chain separated by an entanglement falls below an entanglement friction force denoted by f_e. The latter, which is left as a free parameter, thus effectively prevents an entanglement from moving freely along the chain contour.

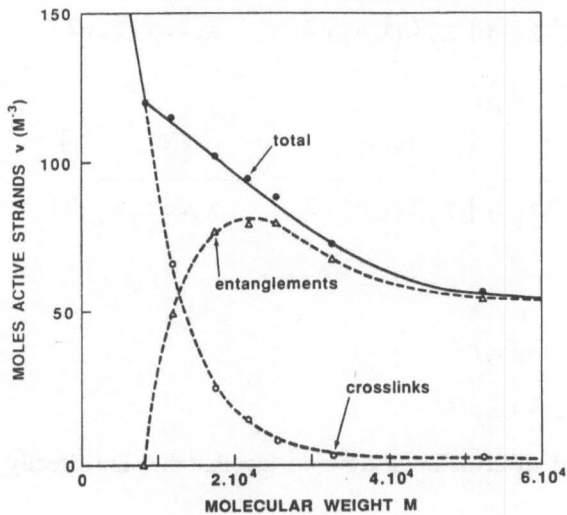

Fig.2. Dependence of the number of elastically active strands ν on the initial molecular weight for (close to) monodisperse PDMS (ρ=1,000kg/m^3; M_e=8,100g/mol). The results are for a tetrafunctional crosslinker with p=0.98. The notation is as follows:
●: total number of active strands; Δ: contribution from trapped entanglements; ○: contribution from crosslinks.

RESULTS AND DISCUSSION

We start by studying in detail the relative contributions of entanglements and crosslinks to the total number of elastically active junctions. The results (Fig.2) are presented as a function of the starting molecular weight M for (close to) monodisperse PDMS for which the molecular weight between entanglements, M_e, equals 8,100. The figure is for a tetrafunctional crosslinker with p=0.98. As a general result, the total number of elastically active strands decreases with the molecular weight and follows different regimes below and above M_e. A detailed analysis of the relative contributions of crosslinks and entanglements gives the following insight. At low $M<M_e$, entanglements are absent and all the active strands originate from crosslinks. Within that regime, our results follow the expected relation $\nu \sim 1/M_e$. At $M>M_e$, entanglements trapped between crosslinks start to play a contributing role to ν.

Inspection of the figure shows that the latter contribution already exceeds that from crosslinks at rather low values of M of the order of 2 to 3 times M_e. That finding is of the uttermost importance since those molecular weight values fall within the range of those usually reported for PDMS and many commercial rubbers. These results thus point to the relevance of the present approach which takes full account of the effects of trapped entanglements on mechanical properties.

We now turn to a study of the effect of entanglements on deformation. Of particular importance in that respect is the behavior of the strain energy density function W. That study also constitutes an excellent test of the validity of our model since none of the previous existing theories has been successful in fully describing the dependence of W over the whole range of deformation[9]. In theory, W can be directly obtained from numerical integration of stress with respect to strain (see Eqs.2-5). Of more importance, however, is the determination of its partial derivatives[10]

$$\frac{\partial W(I_1, I_2)}{\partial I_1} = \frac{1}{2(\lambda_y{}^2 - \lambda_x{}^2)} \left[\frac{\lambda_y{}^3 \sigma_y}{\lambda_y{}^2 - (\lambda_y \lambda_x)^{-2}} - \frac{\lambda_x{}^3 \sigma_x}{\lambda_x{}^2 - (\lambda_y \lambda_x)^{-2}} \right]$$

(6)

$$\frac{\partial W(I_1, I_2)}{\partial I_2} = \frac{1}{2(\lambda_x{}^2 - \lambda_y{}^2)} \left[\frac{\lambda_y \sigma_y}{\lambda_y{}^2 - (\lambda_y \lambda_x)^{-2}} - \frac{\lambda_x \sigma_x}{\lambda_x{}^2 - (\lambda_y \lambda_x)^{-2}} \right]$$

in which

$$I_1 = \lambda_x{}^2 + \lambda_y{}^2 + (\lambda_x \lambda_y)^{-2}$$

$$I_2 = (\lambda_x \lambda_y)^2 + \lambda_x{}^{-2} + \lambda_y{}^{-2}$$

(7)

The interest in those derivatives stems from the fact that they are directly related to the stress-strain curves:

(i) simple uniaxial extension ($\lambda \equiv \lambda_y = 1/\lambda_x{}^2$)

$$\sigma_y = 2 \left(\lambda - \frac{1}{\lambda^2} \right) \left[\frac{\partial W(I_1, I_2)}{\partial I_1} + \frac{1}{\lambda} \frac{\partial W(I_1, I_2)}{\partial I_2} \right]$$

(8)

(ii) equi-biaxial deformation ($\lambda \equiv \lambda_y = \lambda_x$)

$$\sigma_y = 2 \left(\lambda - \frac{1}{\lambda^5} \right) \left[\frac{\partial W(I_1, I_2)}{\partial I_1} + \lambda^2 \frac{\partial W(I_1, I_2)}{\partial I_2} \right]$$

(9)

A full report on our results for the dependence of the $\partial W / \partial I$'s for general biaxial deformation can be found in Ref.8. Here, we shall limit ourselves to present our data for the two limiting cases of uniaxial and equibiaxial extensions.

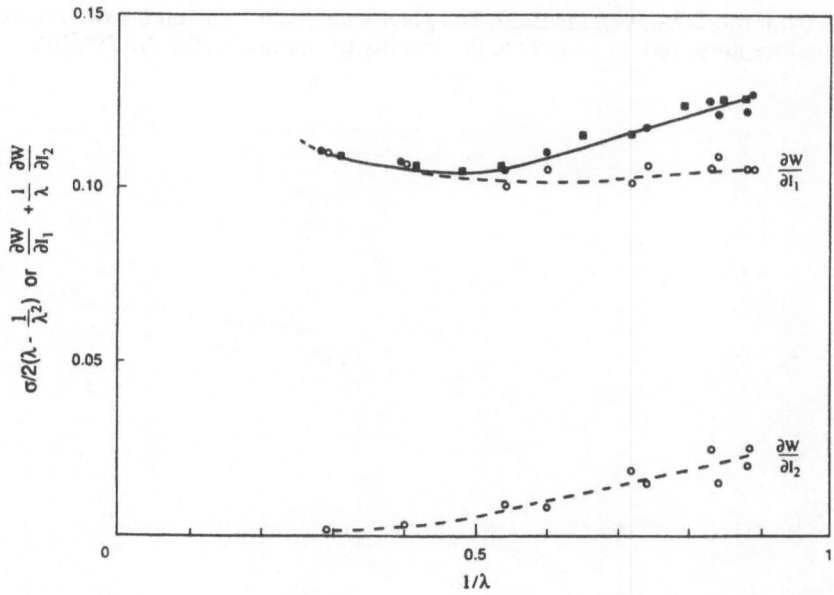

Fig.3. Mooney-Rivlin plot of modulus data for uniaxial extension. Symbol ● represents our calculated values for the sum $[\partial W/\partial I_1 + (1/\lambda)\partial W/\partial I_2]$ in which the partial derivatives (symbol ○) have been obtained by extrapolating our results from biaxial[8] to uniaxial extension. Symbol ■ denotes our modulus values directly calculated from $\sigma/2(\lambda-1/\lambda^2)$.

Fig.3 shows our estimations of the $\partial W/\partial I_i$'s in pure extension (symbol ○) as a function of $1/\lambda$ ($\equiv 1/\lambda_y$), together with our calculated values (symbol ●) of the modulus $G/2 = [\partial W/\partial I_1 + (1/\lambda)\partial W/\partial I_2]$, see Eq.8. A linear decrease of the modulus with $1/\lambda_y$ is observed, in agreement with the Mooney-Rivlin prediction $G/2 = C_1 + (1/\lambda)C_2$. Inspection of the figure however also reveals that $\partial W/\partial I_1$ and $\partial W/\partial I_2$ are not constant and therefore cannot be identified with the Mooney-Rivlin constants C_1 and C_2. Similar conclusions have been reached in previous experimental studies[10]. Also represented in Fig.3 (symbol ■) are our modulus values directly calculated from $\sigma_y/2(\lambda-1/\lambda^2)$, see Eq.8. The two sets of $G/2$ data are in excellent agreement with each other. This, in turn, leads to strong confidence in our results for the partial derivatives $\partial W/\partial I_i$. Further investigation reveals that the strong dependence of the $\partial W/\partial I_i$'s on strain is uniquely due to slippage of chains through entanglements[8].

The case of equi-biaxial deformation is described in Fig.4. Here again we observe that, due to chain slippage, the values of the strain derivatives (symbol ○) are not constant and exhibit a very complex behavior. At very low strains, $\partial W/\partial I_1$ shows a sharp upturn, whereas $\partial W/\partial I_2$ diverges towards large negative values. In spite of those dramatic changes, however, the expression for the modulus $G/2 = [\partial W/\partial I_1 + \lambda^2 \partial W/\partial I_2]$ (symbol ●), see Eq.9, stays at a constant value at low strains. Note that the peculiar behavior of the strain derivatives obtained here for equi-biaxial deformation is quite similar to that observed experimentally[10] for the case of uniaxial extension. The reason for that similarity in extreme types of deformation is not clear to us. Also represented in Fig.4 (symbol ■) are our modulus values estimated from $\sigma_y /2(\lambda-1/\lambda^5)$, see Eq.9. The good agreement between the two sets of modulus values gives further support to the validity of our results for the $\partial W/\partial I_i$'s.

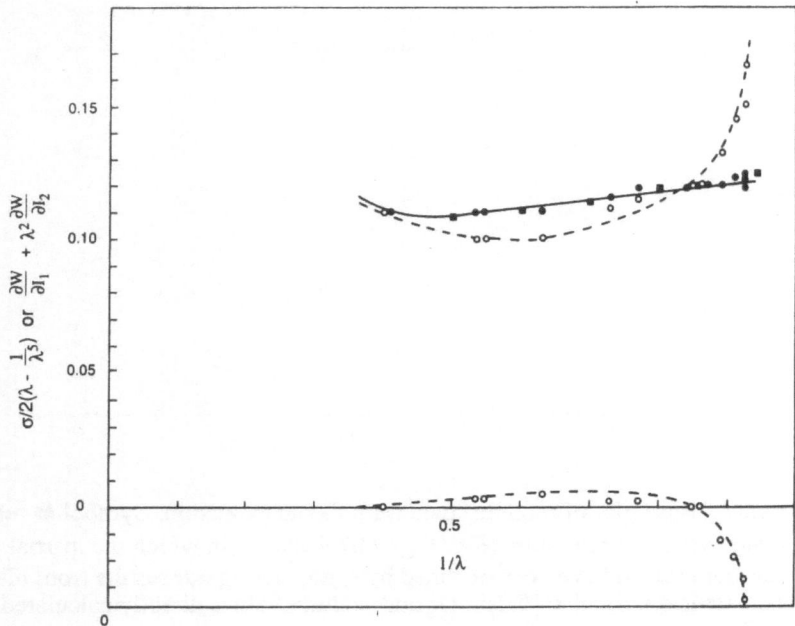

Fig.4. Mooney-Rivlin plot of modulus data for equi-biaxial extension. Symbol ● represents our calculated values for the sum $[\partial W/\partial I_1 + \lambda^2\, \partial W/\partial I_2]$ in which the partial derivatives (symbol ○) have been obtained by extrapolating our results from biaxial[8] to equi-biaxial extension. Symbol ■ denotes our modulus values directly calculated from $\sigma /2(\lambda-1/\lambda^5)$.

CONCLUSIONS

We have introduced a Monte-Carlo lattice model for the description of the factors controlling the mechanical properties of crosslinked networks. The approach takes explicitly into account the molecular weight distribution of the starting polymer as well as the role of entanglements latent in the polymer prior to crosslinking. Our results clearly demonstrate the important role of entanglements in controlling the deformation behavior.

REFERENCES

1. P.J. Flory, Theory of Elasticity of Polymer Networks. The Effect of Local Constraints on Junctions, J. Chem. Phys., 66:5720 (1977)
2. N.R. Langley, Elastically Effective Strand Density in Polymer Networks, Macromolecules, 1:348 (1968)
3. L.M. Dossin and W.W. Graessley, Rubber Elasticity of Well-Characterized Polybutadiene Networks, Macromolecules, 12:123 (1979)
4. Y. Termonia, Molecular Model for the Mechanical Properties of Elastomers. 1. Network Formation and Role of Entanglements, Macromolecules, 22:3633 (1989)
5. Y. Termonia, Molecular Model for the Mechanical Properties of Elastomers. 2. Synergetic Effects in Bimodal Crosslinked Networks, Macromolecules, 23:1481 (1990)
6. Y. Termonia, Molecular Model for the Mechanical Properties of Elastomers. 3. Networks Crosslinked in a State of Strain, Macromolecules, 23:1976 (1990)

7. Y. Termonia, Molecular Model for the Mechanical Properties of Elastomers. 4. Interpenetrating Networks, Macromolecules, 24:1392 (1991)

8. Y. Termonia, Multiaxial Deformation of Polymer Networks, Macromolecules, 24:1128 (1991)

9. M. Gottlieb and R.J. Gaylord, Experimental Tests of Entanglement Models of Rubber Elasticity. 3. Biaxial Deformations, Macromolecules, 20:130 (1987)

10. S. Kawabata and H. Kawai, Strain Energy Density Functions of Rubber Vulcanizates from Biaxial Extension, Adv. Polym. Sci., 24:89 (1977)

ATOMISTIC NATURE OF STRESS IN POLYMER NETWORKS

J. H. Weiner and J. Gao

Division of Engineering and Department of Physics
Brown University
Providence, Rhode Island 02912

INTRODUCTION

The principles of macroscopic thermodynamics applied to experimental observations of the thermomechanical behavior of rubber clearly demonstrate the entropic character of rubber elasticity. That is, the restoring force acting on a piece of stretched rubber is primarily due to its change in entropy, rather than to its change in internal energy. This was established by the work of Kelvin[1] and Joule[2] in the middle of the last century. After the nature of macromolecules was recognized in the 1920's, primarily through the work of Staudinger[3], it became clear that an isolated long-chain molecule in thermal motion would behave as an entropic spring; a tensile applied force is required to maintain a given end-to-end distance of the chain since its configurational entropy increases as this distance decreases[4]. It appeared to be natural, then, to regard the entropic spring concept as the basis for the observed macroscopic entropic behavior of rubber and rubber-like solids. In fact, this concept has played the central role in the development of the classical molecular theory of rubber elasticity and continues to play a central role in current theories.

An important conceptual difficulty with a theory that regards a polymer network in its rubbery state as a collection of cross-linked entropic springs became apparent early.[5] Since all of the chains act as entropic springs in tension, such a network, unopposed by other forces, would collapse to a point. Here, the fact that atoms have finite volumes and therefore interact with a repulsive excluded-volume potential was invoked. It was assumed that, as in a simple liquid, these interactions produced a hydrostatic pressure and that this pressure would prevent the collapse of the systems of chains all in tension. Furthermore, it was assumed that when the network was defo. ned, e.g. subjected to a uniform extension, the excluded volume interactions continued to make only an isotropic contribution to the stress. Since the isotropic portion of the stress

Synthesis, Characterization, and Theory of Polymeric Networks and Gels
Edited by S.M. Aharoni, Plenum Press, New York, 1992

is usually determined by the boundary conditions on the problem, little more is said about the role of excluded volume interactions in most theories.

We have been re-studying this question by means of the computer simulation of idealized models of polymer systems, Refs (6)-(9). These simulations lead to a physical picture that is totally at variance with that of the classical theory, one that gives a central role to the excluded volume interactions in the development of the anisotropic stress in a stretched rubber-like solid. In what follows we briefly describe these results.

CHAIN MODEL

The idealized chain model employed approximates the hard-sphere, freely jointed chain and is the simplest chain model that has both the covalent bonding characteristic of macromolecules and the attribute of excluded volume. For computational and conceptual convenience the covalent potential is represented by a stiff linear spring, and the hard-sphere potential is replaced by the repulsive part of the Lennard-Jones potential. That is, the covalent potential $u_c(r)$ is

$$u_c(r) = \frac{1}{2} \kappa (r - a)^2 \tag{1}$$

where r is the distance between adjacent atoms on a given chain and a is the zero-force bond length; the noncovalent or nonbonded potential is

$$u_{nc}(r) = 4\varepsilon[(\sigma/r)^{12} - (\sigma/r)^6] + \varepsilon \text{ for } r \leq r_0,$$
$$= 0 \text{ for } r \geq r_0 \tag{2}$$

where $r_0 = 2^{1/6}\sigma$. The latter potential applies between all pairs of atoms of the system, inter- as well as intra-chain, except for neighboring pairs of atoms along a chain.

The values of parameters employed are such that σ may be regarded as an effective hard-sphere diameter. An important parameter in a discussion of systems of such chains is the reduced density ρ where

$$\rho = \frac{n\sigma^3}{v} \tag{3}$$

with n/v the atom number density. The packing fraction $\phi = \frac{\pi}{6} \rho$. For dense polymer systems, as for liquids, $\rho \sim 1$.

COVALENT BOND FORCE

Computer simulations by the method of molecular dynamics of systems of chains forming either polymer melts or model networks have been performed; details may be found in Refs. (6)-(9). These programs include the computation of $<u'_c(r)>$, the time average of the force in the covalent bonds, averaged over all of the bonds of the system. For low values of reduced density, this average covalent bond force is found to be positive, i.e. the bonds are in tension. This result may be understood as a consequence of the "centrifugal force" acting on the bond due to the thermal motion of the constituent atoms. However, as the value of the reduced density is increased, the covalent bond force decreases, and it becomes negative for values of $\rho > \sim 0.6$.

EXCLUDED VOLUME STRESS

It is difficult to reconcile the fact that the covalent bond force in a dense polymer system is negative (compressive) at realistic packing fractions, with the usual picture of the tensile stress in a stretched network arising from the tensile forces in the chains acting as entropic springs. Simulations of various model systems showed, in fact, that the covalent contribution to the anisotropic part of the stress in an extended network is a compressive stress. It is, rather, the excluded volume contribution to this stress which is tensile, and it is sufficiently greater in magnitude than the covalent contribution so that the overall stress is, indeed, tensile.

When this result is first encountered, it appears highly counterintuitive that a two-body, spherically symmetric repulsive potential modelling excluded volume interactions, such as $u_{nc}(r)$, eq 2, could give rise to an anisotropic tensile stress. However, the result has been traced to the screening of the excluded volume interactions by the covalent structure linking the atoms of the system. In the reference, stress-free, state this covalent structure is isotropic, and the screening leaves the excluded volume contribution to the stress isotropic. However, when the system is deformed, the covalent structure becomes anisotropic, and its screening action results in the observed anisotropic contribution of the excluded volume interactions.

ENTROPIC ELASTICITY

As noted previously, one of the principal reasons the concept of the entropic spring was made central to the classical theory of rubber elasticity was that it appeared to provide a natural molecular explanation for the macroscopic observation and deduction that rubber elasticity was primarily entropic. It is important to emphasize, in this connection, that a soft sphere model of a gas is also a primarily entropic system, and the same is the case for a model of a polymer system consisting of hard spheres, interconnected by covalent bonds. The view of the excluded volume interaction basis of the stress in polymer systems is therefore fully consistent with the known entropic character of rubber elasticity.

CHAIN FORCE REINTERPRETATION

The new physical picture revealed by the computer simulations appears totally different from the classical picture based on entropic springs. Yet the simple classical theory based on this picture predicts values of the stress for a rubberlike solid in simple extension that is in excellent agreement with experiment, although there are small, systematic discrepancies sometimes referred to[10] as the Mooney effect, that still require explanation. How is the excellent agreement of the theory based on the entropic spring concept to be reconciled with the new physical picture based on excluded volume interactions?

To answer this question, we have re-examined the concept of the chain force. For an isolated chain, the chain force is that force that must be applied to its end atoms in order to maintain them fixed at a given separating distance while the remaining atoms of the chain are free to undergo thermal motion in a manner consistent with the covalent connectivity of the chain.

We have extended[9] this concept of chain force to apply to a system of chains whose atoms are subject to both intrachain and interchain interactions. Based on the principles of equilibrium statistical mechanics, equivalent expressions for this chain force were derived for this case; they reduce to the conventional expressions when interchain interactions are absent.

We then conducted[9] computer simulations of model chain systems to explore this extended concept. The systems consisted of a single chain with end atoms fixed (tethered chain) in interaction with a surrounding dense system of chains. Two cases were considered: one in which the surrounding system represented a free polymer melt, and the second in which it modelled a deformable network.

In the first case it was found that the numerical values of the chain force acting on the tethered chain were quite close to those acting on the corresponding isolated ideal chain. However, the physical picture underlying the chain force is quite different in the two cases. For the isolated chain, the end atoms of the chain are being <u>pulled</u> inward by the adjacent covalent bonds of the chain, and the applied tensile chain force is necessary to counter this inward force. On the other hand, for the tethered chain, the covalent bonds are in compression, and they cannot pull inward. Rather the end atoms of the tethered chain are being <u>pushed</u> inward by the interchain excluded volume interactions that are screened by the covalent structure of the tethered chain.

The situation is quite similar when the tethered chain is in interaction with a model of a deformable network, particularly when the latter is in the isotropic condition corresponding to its unstressed reference state. However, when the deformed network model is subjected to a uniaxial extension, it becomes anisotropic and its interactions with the tethered chain then exhibit a new feature. The chain force on the tethered chain due to these interactions is no longer axial--that is, it is no longer parallel to its end-to-end vector. The chain force on the tethered chain can now be represented as an axial force plus an applied moment.

The approximate numerical agreement of the chain force on an isolated ideal chain with that acting on a tethered chain provides an explanation of the great predictive success of the classical theory that is based on the entropic spring concept with neglect of excluded volume interactions. On the other hand, the additional moments found operative when the tethered chain is in interaction with a deformed system may provide a partial explanation for the observed deficiency in the classical theory called the Mooney effect. We turn next to this question.

MOONEY EFFECT

This terminology has been used[10] to denote the softening of the experimentally observed tensile stress in rubberlike solids in uniaxial extension relative to the value predicted by the classical theory. Typically, this softening is observed only for extension ratios $\lambda > \sim 2$ and is $\sim 20\%$ in the range $\lambda = 3$.

Numerous attempts have been made to provide a molecular explanation for the Mooney effect. One class of approaches[11-13] that relate to excluded volume effects seeks the explanation in packing effects. A second, more recent, line of inquiry[14-16] deals with nematic interactions between chains which results in new ordering effects not present in non-interacting systems. Our simulations show that it is not necessary to invoke a phenomenological nematic potential; these additional ordering effects occur naturally in the presence of excluded volume effects even when, as in our models, the latter are represented by spherically symmetric two-body potentials. Therefore, theories based on packing effects and on phenomenological nematic interactions are closely related. Indeed, both classes of theories lead to corrections to the classical theory of similar form. Most of these theories (an exception is the work of Warner and Wang[16]) come to the conclusion that the softening that they predict is too small to justify regarding the new effects they include as major contributors to the Mooney effect.

The results of our simulations indicate that a network of chains with both intrachain and interchain excluded volume interactions can be modelled as a set of non-interacting ideal chains with, however, the addition of applied moments in addition to axial chain forces. Furthermore, our simulations show that these moments have a sense such that they have a softening effect on the tensile stress in extension relative to the predicted stress in the absence of moments.

We have constructed a theory of rubber elasticity[9] which includes the effects of the applied moments. The form of this theory is quite similar in form to the generalizations of the classical theory based on either packing or nematic interaction effects. The magnitude of the softening effect it predicts depends critically on that of the chain moments. We turn next, therefore, to a consideration of this question.

ENHANCEMENT FACTOR

One of three equivalent methods we have found[9] for the calculation of the chain force $\underline{f}$ acting on a chain of a network with intrachain and interchain interactions is through the relation

$$\underline{f} = -kT\frac{\partial}{\partial \underline{r}} \log p(\underline{r};\lambda) \tag{4}$$

where $p(\underline{r};\lambda)$ is the probability density for the end-to-end chain displacement vector $\underline{r}$ for a <u>free</u> chain immersed and interacting with the network. The presence of the extension λ as parameter in this function is required since the extension causes the network to be anisotropic, and the resulting interactions affect the distribution $p(\underline{r};\lambda)$ and lead to its anisotropy. It is clear from eq 4 that an anisotropic $p(\underline{r};\lambda)$ will lead to a non-axial chain force, or equivalently, to an axial force plus a moment. It is, therefore, important for the assessment of the magnitude of the moments to study the degree of anisotropy of the function $p(\underline{r};\lambda)$ describing the chain vector distribution of a free chain embedded in the stretched network.

It is of interest that this mathematical requirement of considering the chain vector distribution of a free chain embedded in a stretched network parallels the physical system considered experimentally by Sotta et al[17] with the aid of ^{2}H NMR. By this technique they have determined the segment orientation $<P_2>_s$ acquired by the embedded free chain when the host network is stretched. The question then before us is how this segment orientation $<P_2>_s$ is translated in the presence of segment-segment correlation to the overall chain vector $\underline{r}$ orientation $<P_2>_v$, an orientation which, of course, implies anisotropy of the function $p(\underline{r};\lambda)$. By extensive Monte Carlo simulations of the rotational isomeric model for polyethylene of Jernigan and Flory[18], we have found that

$$<P_2>_v = \xi <P_2>_s \tag{5}$$

where $\xi \cong 4$.

We term the parameter ξ the enhancement factor. The softening effect of the chain moments is directly proportional to this enhancement factor. With the value $\xi \cong 4$ that we observe, we find a softening effect of ~10% at $\lambda=3$. We tentatively conclude, therefore, that the excluded volume interactions and the chain moments they produce play a substantial role in the production of the Mooney effect.

ACKNOWLEDGEMENT

This work has been supported by the Gas Research Institute (Contract 5085-260-1152). The computations were performed on the Cray Y-MP at the Pittsburgh Supercomputing Center.

REFERENCES

1. W. Thomson (later Lord Kelvin), Q. Journal of Pure and Applied Math., 1:57 (1857).
2. J. P. Joule, Phil. Trans.-London, 149:91 (1859).
3. For an excellent historical account, see H. Morawetz, Polymers: The Origins and Growth of a Science; John Wiley, New York, 1983.
4. E. Guth and H. Mark, Monats. f. Chemie, 65:93 (1934).
5. H. M. James and E. Guth, J. Polymer Science, 4:153 (1949).
6. J. Gao and J. H. Weiner, Macromolecules, 20:2525 (1987).
7. J. Gao and J. H. Weiner, Macromolecules, 22:979 (1989).
8. J. Gao and J. H. Weiner, Macromolecules, 24:1519 (1991).
9. J. Gao and J. H. Weiner, Macromolecules, 24:5179 (1991).
10. G. Ronca and G. Allegra, J. Chem. Phys., 63:4990 (1975).
11. E. A. DiMarzio, J. Chem. Phys., 36:1563 (1962).

12. J. L. Jackson, M. C. Shen and D. A. McQuarrie, J. Chem. Phys., 44:2388, (1966).
13. T. Tanaka and G. Allen, Macromolecules, 10:426 (1977).
14. J.-P. Jarry and L. Monnerie, Macromolecules, 12:316 (1979).
15. B. Deloche and E.T. Samulski, Macromolecules, 21:3107 (1989).
16. M. Warner and X.J. Wang, Macromolecules, 24:4932 (1991).
17. P. Sotta et al, Macromolecules, 20:2769 (1987).
18. R. L. Jernigan and P. J. Flory, J. Chem. Phys., 50:4165 (1969).

INVESTIGATION OF GELATION PROCESSES AND GEL STRUCTURES BY MEANS OF MECHANICAL PROPERTY MEASUREMENTS

N. Ichise, Y. Yang, Z. Li, Q. Yuan and J. E. Mark

Department of Chemistry and Polymer Research Center
University of Cincinnati, Cincinnati, Ohio 45221-0172

E. K. M. Chan, R. G. Alamo, and L. Mandelkern

Department of Chemistry, Florida State University
Tallahassee, FL, 32306

INTRODUCTION

There are a variety of gels (highly swollen solids) that are of considerable interest to polymer scientists, materials scientists, and ceramists.[1-42] One type consists of typical organic polymers such as polyethylene or polystyrene, in networks which are formed by means of physical cross links, such as crystallites or physical aggregates. Such gels are thermoreversible in that liquefaction occurs upon heating. Another type consists of chain-like structures permanently bonded into covalent networks. These permanently branched and cross-linked chains can be either inorganic [silica (SiO_2), titania (TiO_2), zirconia (ZrO_2), etc.] or organic (phenol-formaldehyde resins, epoxies, etc.). Both the inorganic and iorganic covalent types have been used to prepare aerogels, and the inorganic ones are now much used to prepare high-tech ceramics by the new sol-gel route.

In the case of the thermoreversible, organic polymer gels, moduli can be measured as a function of concentration, temperature, and structural characteristics of the polymer (molecular weight, molecular weight distribution, and nature and degree of any chain branching). Such equilibrium results give information on the nature of the gels, including the influence of morphology and the presence of dangling-chain irregularities. Measurements carried out as a function of time, for example, on polyethylene homopolymers and copolymers, can give information about their gelation (crystallization) kinetics.[42]

In the case of the ceramic materials, the evolution of the shear modulus with time is very useful in establishing induction times, rates of gelation, and aging effects. Correlation of such information with results of scattering studies can give much insight into the nature of the sol-gel process.

Gels are solids but, because of their highly swollen nature, are generally very weak. This effect can be very marked in the case of the "thermoreversible"

Synthesis, Characterization, and Theory of Polymeric Networks and Gels
Edited by S.M. Aharoni, Plenum Press, New York, 1992

gels,[1-6,10,13,16,17,28,30,32-37] which reliquefy upon increase in temperature. Here the linkges are physical associations, for example the very small crystallites or morphological structures present in polyethylene gels,[10,13,17,43] or the ionic aggregates present in ionomers.[44] If the linkages between the units which give rise to the required network structure are covalent bonds, then the structure is more permanent.[1,2,45,46] Nonetheless, large stresses or deformations do cause creep (irreversible flow) even in these systems, because of the dilute nature of the system.

As can be concluded from the above remarks, the mechanical properties of the gels are some of their most important characteristics. In the case of the thermoreversible gels, one aspect of obvious interest is the way in which the modulus of the gel, in compression or shear, develops with time, particularly as a function of polymer molecular weight, concentration, and temperature. If the gelation is due to crystallization, then this information could be correlated with independently-obtained results on crystallization kinetics and polymer morphology. Also, structural information could be obtained by quantitative analysis of the moduli using the molecular theories of rubberlike elasticity.[46] In the case of permanent gels, it is of particular interest to determine how the modulus depends on composition of the reacting system, pH, temperature, and the presence of catalysts, additives, and sonication fields.[1-3,13,16,21,39,47] Such kinetic data should help elucidate the mechanisms of the reactions used to form the gels of interest. Features of interest would include (i) the lengths of induction periods associated with the formation of precursor tree-like structures and possibly with the presence of inhibitors, (ii) the kinetic order of the reaction, with which to evaluate postulated growth mechanisms, and (iii) the amount of time required for the modulus to reach its maximum value.[42]

Some of the earliest studies of the mechanical properties of gels were carried out in the rubber industry, on commercial elastomers which had been prepared in the dry state and then swollen with various amounts and types of diluents.[1,45,46] Such gels were typically robust enough for elongation deformations to be used in stress-strain and stress-temperature investigations. More recent work has involved much more fragile gels, for example the thermoreversible ones obtained by cooling relatively dilute solutions of crystallizable polymers such as polyethylene.[43] Because of their fragility, these materials are better studied in compression or shear.

The present report summarizes some work being carried out in this area by the authors, and shows how the results can yield useful information on both the gelation process itself and the structure of the resulting gels.

PREPARATION

Thermoreversible Polyethylene Gels

One study of polyethylene gels emphasized homopolymeric chains, with both whole polymers of relatively broad molecular weight distributions and relatively narrow fractions being employed.[43] More recent, on-going work focuses on copolymers of ethylene and n-hexene-1, one of a class of materials known commercially as "linear low-density polyethylene".

In all cases, the polyethylene was dissolved in a high-boiling solvent and then cooled to the temperature chosen for the gelation (crystallization) process. The gelation point was determined visually, as the point at which the solution no longer flowed. Measurements carried out as a function of concentration were used to determine the minimum (critical) concentration c^* below which gelation does not occur.

Permanent Silical Gels

In a typical chemical reaction of this type, tetraorthosilicate (TEOS) [Si(OEt)$_4$] was hydrolyzed and polycondensed into a silica-like network structure.[21,39,48,49] Both acids and bases can be used as catalysts, ethanol is frequently present as a solubilizing agent, and a pre-hydrolysis step is sometimes included to decrease the likelihood of premature phase separation.

Precursors for Organic Aerogels

The reaction chosen to illustrate this type of system involves the reaction of resorcinol with formaldehyde to give a tightly cross-linked gel which can subsequently be converted into an aerogel by supercritical drying.[50,51] Typically, the two reactants are mixed with a catalyst, the pH adjusted if necessary, and the mixture is allowed to react at an elevated temperature until gelation occcurs.

MECHANICAL PROPERTY MEASUREMENTS

The earlier studies on polyethylene gels involved a rather rudimentary device for estimating moduli.[43] Specifically, weights were placed on the surface of the gel, and the extent of compression measured with a cathetometer. This method has two main disadvantages: the type of deformation is rather ill-defined, and it is difficult to obtain values of the modulus as a function of time, particularly in the case of relatively fast reactions.

Subsequent studies were carried in a device constructed like a dilatometer except that pressure could be applied to the gel in the sample compartment.[52] This caused the gel to bulge slightly at its lower end, and this change in volume was monitored by the change in height of a mercury column moving in a vertical capillary attached by a U-shaped glass tube to the bottom of the sample chamber. In this way very rapid measurements could be made, with the sample remaining undeformed between pressure pulses. With this design, it should also be possible in future work to carry out scattering measurements and mechanical property measurements at the same time.[42]

RESULTS AND DISCUSSION

Thermoreversible Polyethylene Gels

In the case of the polyethylene homopolymers,[43] the moduli of the gels were found to increase markedly with increase in concentration, with the largest changes occurring at the highest values of the polymer molecular weight, presumably because of the diminished importance of dangling chains. The moduli were found to increase approximately linearly with the square of the concentration, suggesting the importance of pairwise encounters between chain segments in the gelation process. The effect of the molecular weight M on the modulus in the limits of small deformations and at the critical concentration c^* were the same for the whole polymers and the fractions at the same number-average molecular weight M_n, again underscoring the importance of dangling ends on the gel structure and its mechanical properties. The absolute values of the moduli indicate that chains in a gel which was formed near c^* consist largely of elastically ineffective dangling ends.

The hexene-1 copolymers exhibited values of c^* which were considerably higher than those of the homopolymers of the same molecular weight, except at very high molecular weights. Similar changes were observed in some more limited results on other polyethylene copolymers.[43] Typical results are shown in Figure 1. At each molecular weight, c^* increases with increase in temperature, as was also found for the linear homopolymer.[43] At each temperature, c^* increases with decrease in molecular weight (as was also found for the linear homopolymer) because more cross links are required for network formation in the case of shorter chains. Plots of the low-deformation modulus as a function of time showed the modulus to reach an asymptotic limit as the crystallization approaches its equilibrium value.[42] The evolution of the modulus, at fixed temperature, reflects the nature of the crystallization process. The modulus was found to increase with increase in temperature, presumably because of an increase in number of cross links (crystallites).

Figure 2 shows the temperature dependence of the low-deformation modulus for the copolymer having 10^{-3} M_n = 43.5 g mol^{-1}. At each concentration, the increase in modulus with temperature is much larger than that expected from the molecular theories of rubberlike elasticity,[46] indicating a larger

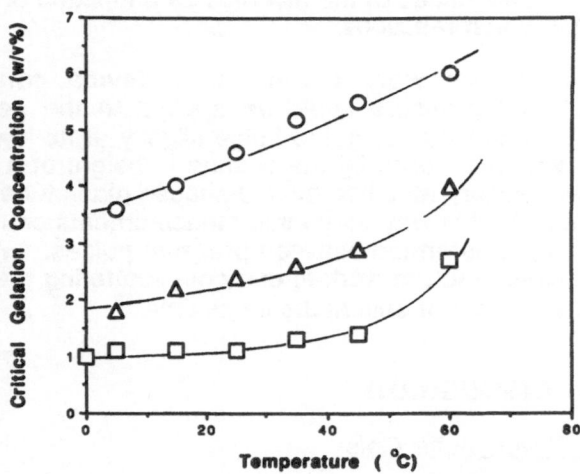

Fig. 1. Critical gelation concentrations for ethylene hexene-1 copolymers as a function of temperature. The values of 10^{-3} M_n are 26.1 (O), 43.5 ($\triangle$), and 111. ($\square$) g mol^{-1}, and values of the mol % branches are 1.21, 1.21, and 1.47, respectively.

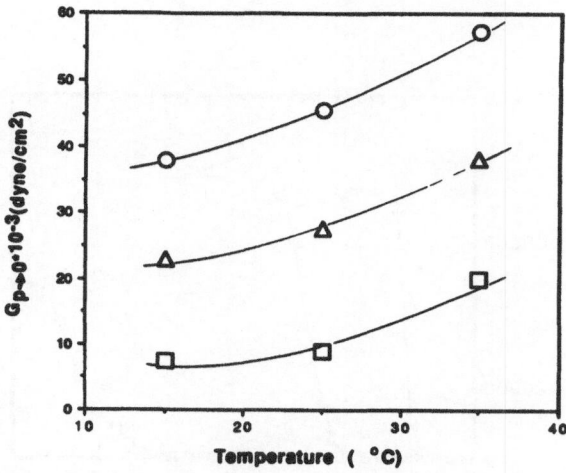

Fig. 2. Temperature dependence of the modulus in the limit of zero deformation pressure for the ethylene hexene-1 copolymer having $10^{-3} M_n = 43.5$ g mol^{-1} and 1.21 mol % branches. The concentrations are 3.66 ($\square$), 4.47 ($\triangle$), and 5.36 (O) g ml^{-1}, and the critical concentration is approximately 2.4 g ml^{-1}.

number of crystallites. At constant temperature, increase in concentration increases the modulus, also through increase in the number of crystallites, as expected.

Permanent Silical Gels

Some typical results obtained on the formation of a ceramic gel are shown in Figure 3. The period of time over which the modulus is approximately zero corresponds to the induction time of the gelation process. During this period, hydrolysis, polycondensation, and branching are occurring but a gel has not yet been formed. As is shown in Figure 4, this induction time decreases with increase in the amount of the NH$_4$OH catalyst present, as expected. Figure 5 shows the corresponding increase in the rate of reaction, as gauged by the rate of increase in modulus. The dependence is seen to be stronger than linear, and its quantitative interpretation will be used to obtain insight into the gelation mechanism.

Precursors for Organic Aerogels

The precursors for the resorcinol-formaldehyde aerogels not only underwent the required gelation, as expected and required, but also exhibited several color changes, going from colorless to yellow to orange to dark reddish-orange. Relatively long reaction times were required, and conditions are being modified to accelerate the reaction. Kinetic data obtained on the gelation process is being correlated with the reaction conditions, the color and mechanical properties of the resulting materials, and their X-ray scattering profiles.

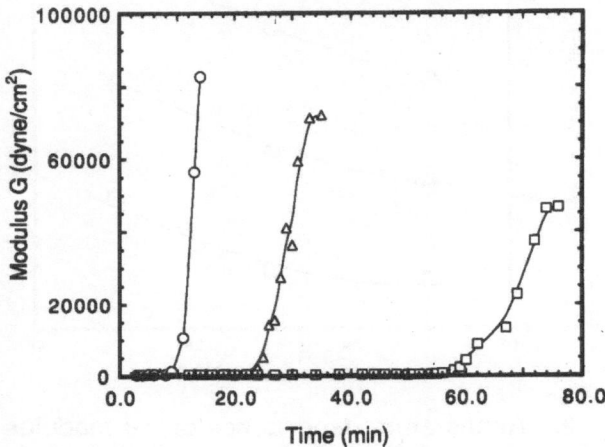

Fig. 3. Time dependence of the modulus of a silica-like gel prepared by hydrolyzing TEOS in the presence of ethanol, with ammonium hydroxide as catalyst. The values of the molar ratios were $[H_2O]/[TEOS] = 4.0$, $[C_2H_5OH]/[TEOS] = 3.0$, and $[NH_4OH]/[TEOS] = 0.0125$ (□), 0.025 (O), and 0.05 (Δ).

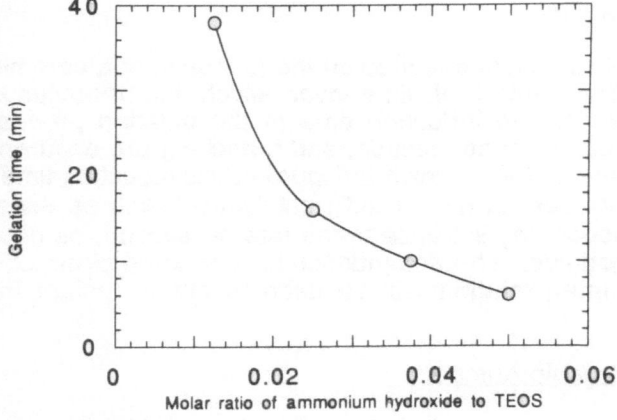

Fig. 4. The effect of catalyst concentration on the gelation time required in the hydrolysis of TEOS in the presence of ethanol, with ammonium hydroxide as catalyst.

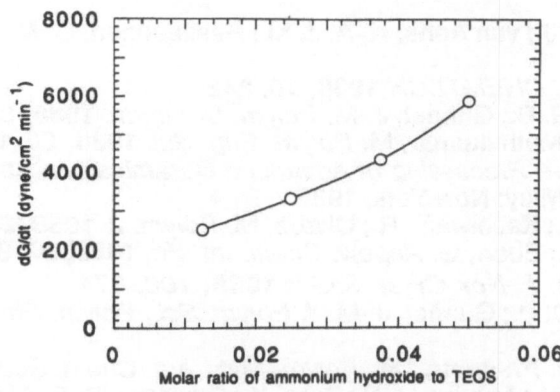

Fig. 5. The effect of catalyst concentration on the rate of reaction in the hydrolysis of TEOS in the presence of ethanol, with ammonium hydroxide as catalyst.

ACKNOWLEDGEMENTS

It is pleasure to acknowledge the financial support provided by the Army Research Office through Grant DAAL03-90-G-0131. The work carried out at Florida State University was supported by the Petroleum Research Fund, and the Polymers Program of the National Science Foundation (Grant DMR 89-14167).

REFERENCES

(1)   Flory, P. J. *Principles of Polymer Chemistry*; Cornell University Press: Ithaca NY, 1953.
(2)   Flory, P. J. *Faraday Disc. Chem. Soc.* **1974**, *57*, 7.
(3)   Lemstra, P. J.; Smith, P. *Br. Polym. J.* **1980**, *12*, 212.
(4)   Smith, P.; Lemstra, P. J.; Pijpers, J. P. L.; Kiel, A. M. *Coll. Polym. Sci.* **1981**, *251*, 1070.
(5)   Smith, P.; Lemstra, P. J.; Booij, H. C. *J. Polym. Sci., Polym. Phys. Ed.* **1981**, *19*, 877.
(6)   Edwards, C. O.; Mandelkern, L. *J. Polym. Sci., Lett. Ed.* **1982**, *20*, 355.
(7)   Okabe, M.; Isayama, M.; Matsuda, H. *J. Appl. Polym. Sci.* **1985**, *30*, 4735.
(8)   Klein, L. C. *Ann. Rev. Mat. Sci.* **1985**, *15*, 227.
(9)   Russo, P. S.; Siripanyo, S.; Saunders, M. J.; Karasz, F. E. *Macromolecules* **1986**, *19*, 2856.
(10)   Domszy, R. C.; Alamo, R.; Edwards, C. O.; Mandelkern, L. *Macromolecules* **1986**, *19*, 310.
(11)   Koltisko, B.; Keller, A.; Litt, M.; Baer, E.; Hiltner, A. *Macromolecules* **1986**, *19*, 1207.
(12)   Sawatari, E.; Okumura, T.; Matsuo, M. *Polym. J.* **1986**, *18*, 741.
(13)   *Reversible Polymeric Gels and Related Systems*; Russo, P. S., Ed.; American Chemical Society: Washington, DC, 1987.
(14)   Hiltner, A. In *Order in the Amorphous "State" of Polymers*; S. E. Keinath, R. L. Miller and J. K. Rieke, Ed.; Plenum Press: New York, 1987.
(15)   Clark, A. H.; Ross-Murphy, S. B. *Adv. Polym. Sci.* **1987**, *83*, 57.
(16)   Chan, E. K. M.; Mandelkern, L. *Preprints, Div. Polym. Chem., Inc., Am. Chem. Soc.* **1987**, *28(1)*, 130.

(17) Lemstra, P. J.; van Aerle, N. A. J. M.; Bastiaansen, C. M. *Polym. J.* **1987**, *19*, 85.
(18) Ulrich, D. R. *CHEMTECH* **1988**, *18*, 242.
(19) McKenna, G. B.; Guenet, J.-M. *Polym. Commun.* **1988**, *29*, 58.
(20) Skukla, P.; Muthukumar, M. *Polym. Eng. Sci.* **1988**, *28*, 1304.
(21) *Ultrastructure Processing of Advanced Ceramics*; Mackenzie, J. D.; Ulrich, D. R., Ed.; Wiley: New York, 1988.
(22) Matsuda, H.; Kashiwagi, R.; Okabe, M. *Polym. J.* **1988**, *20*, 189.
(23) Hoffman, H.; Ebert, G. *Angew. Chem. Int. Ed.* **1988**, *27*, 902.
(24) Ulrich, D. R. *J. Non-Cryst. Solids* **1988**, *100*, 174.
(25) McKenna, G. B.; Guenet, J.-M. *J. Polym. Sci., Polym. Phys. Ed.* **1988**, *26*, 267.
(26) LeMay, J. D. *Preprints, Div. Polym. Mat., Am. Chem. Soc.* **1989**, *60*, 695.
(27) Aharoni, S. M.; Murthy, N. S.; Zero, K.; Edwards, S. F. *Macromolecules* **1990**, *23*, 2533.
(28) *Physical Networks. Polymers and Gels*; Burchard, W.; Ross-Murphy, S. B., Ed.; Elsevier: London, 1990.
(29) Djabourov, M. *Polym. Int.* **1991**, *25*, 135.
(30) Mandelkern, L.; Prasad, A. *Preprints, Div. Polym. Chem., Inc., Am. Chem. Soc.* **1991**, *32(3)*, 420.
(31) Aharoni, S. M. *Macromolecules* **1991**, *24*, 235.
(32) Nguyen, H. P.; Delmas, G. *Preprints, Div. Polym. Chem., Inc., Am. Chem. Soc* **1991**, *32(3)*, 421.
(33) Plazek, D. J.; Chay, I.-C. *Preprints, Div. Polym. Chem., Inc., Am. Chem. Soc* **1991**, *32(3)*, 433.
(34) Jackson, C. J.; McKenna, G. B. *Preprints, Div. Polym. Chem., Inc., Am. Chem. Soc* **1991**, *32(3)*, 439.
(35) Amis, E. J.; Hodgson, D. F.; Yu, Q. *Preprints, Div. Polym. Chem., Inc., Am. Chem. Soc.* **1991**, *32(3)*, 447.
(36) Bucci, S.; Gallino, G.; Lockhart, T. P. *Preprints, Div. Polym. Chem., Inc., Am. Chem. Soc.* **1991**, *32(3)*, 457.
(37) Dagan, A.; Avichai, M.; Garstein, E.; Cohen, Y. *Preprints, Div. Polym. Chem., Inc., Am. Chem. Soc.* **1991**, *32(3)*, 459.
(38) Dickinson, E. *CHEMTECH* **1991**, *21*, 665.
(39) *Proceedings of the Fourth International Conference on Ultrastructure Processing (Tucson)*; Uhlmann, D. R., Ed.; Wiley: New York, 1992.
(40) Osada, Y.; Okuzaki, H.; Hori, H. *Nature* **1992**, *355*, 242.
(41) Aharoni, S. M. *Macromolecules* **1992**, *25*, 1510.
(42) Yang, Y.; Ichise, N.; Li, Z.; Yuan, Q.; Mark, J. E.; Chan, E. K. M.; Alamo, R. G.; Mandelkern, L. In *Complex Fluids*; E. B. Sirota, D. Weitz, T. Witten and J. Israelachvili, Ed.; Materials Research Society: Pittsburgh, PA, 1992.
(43) Li, Z.; Mark, J. E.; Chan, E. K. M.; Mandelkern, L. *Macromolecules* **1989**, *22*, 4273.
(44) Eisenberg, A.; King, M. *Ion-Containing Polymers*; Academic Press: New York, 1977.
(45) Treloar, L. R. G. *The Physics of Rubber Elasticity*; Clarendon Press: 1975.
(46) Mark, J. E.; Erman, B. *Rubberlike Elasticity. A Molecular Primer*; Wiley-Interscience: New York, 1988.
(47) Esquivias, L.; Zarzycki, J. In ; D. R. Uhlmann, Ed.; Wiley: New York, 1992; in press.
(48) *Better Ceramics Through Chemistry*; Brinker, C. J.; Clark, D. E.; Ulrich, D. R., Ed.; North Holland: New York, 1984.
(49) *Better Ceramics Through Chemistry IV*; Zelinski, B. J. J.; Brinker, C. J.; Clark, D. E.; Ulrich, D. R., Ed.; Materials Research Society: Pittsburgh, 1990.

(50)   LeMay, J. D.; Hopper, R. W.; Hrubesh, L. W.; Pekala, R. W. *MRS Bulletin* **1990**, *15*, 19.
(51)   LeMay, J. D.; Hopper, R. W.; Hrubesh, L. W.; Pekala, R. W. *MRS Bulletin* **1990**, *15*, 30.
(52)   Saunders, P. R.; Ward, A. G. In *Proceedings of the Second International Congress of Rheology* Butterworths Sci. Pub.: London, 1953.

THEORY OF STRAIN-INDUCED CRYSTALLIZATION IN REAL ELASTOMERIC NETWORKS

A. Kloczkowski and J. E. Mark

Department of Chemistry and Polymer Research Center
University of Cincinnati, Cincinnati, Ohio 45221-0172

M. A. Sharaf

Department of Chemistry, United Arab Emirates University
P.O. Box 17551, Al Ain, United Arab Emirates

B. Erman

School of Engineering, Bogazici University, Bebek 80815
Istanbul, Turkey

INTRODUCTION

It has frequently been observed that some elastomeric networks show an abrupt increase in the nominal stress, and consequently the modulus [f*], at high elongations.[1] Molecular interpretation of this upturn has generally been attributed to strain-induced crystallization.

A thermodynamic theory for such strain-induced crystallization was first developed by Flory.[2] His theory is most applicable under conditions of incipient equilibrium crystallization, and is based on many simplifying assumptions. Specifically, it assumed that the crystallites form parallel to the direction of stretch. In addition, the model ignored entropy changes associated with formation of crystalline nuclei, and assumed that the chains were Gaussian and deformed affinely. The theory explicitly expressed the effect of the strain on the elevation of the melting point, on the degree of crystallinity ω, and on the elastic force exhibited by the network. These relations are

$$\omega = 1 - \{ (3/2 - \phi(\lambda))/(3/2 - \theta) \}^{1/2} \tag{1}$$

with

$$\phi(\lambda) = (6/\pi n)^{1/2} \lambda - (\lambda^2/2 + \lambda^{-1})/n \tag{2}$$

$$\theta = (H_f/R) (1/T_m^0 - 1/T) \tag{3}$$

Synthesis, Characterization, and Theory of Polymeric Networks and Gels
Edited by S.M. Aharoni, Plenum Press, New York, 1992

227

Here, λ is the elongation in simple tension defined as the ratio of the deformed length to the length at the state of the formation of the network, n is the number of segments (statistical links) per chain, H_f is the molar heat of fusion per segment, and T_m^o is the incipient crystallization temperature of the undeformed polymer. The melting point elevation with strain is

$$\frac{1}{T_m} = \frac{1}{T_m^o} - \frac{R\,\phi(\lambda)}{H_f}$$

(4)

The retractive force f at equilibrium in a stretched polymer network according to the theory is

$$f = \nu RT\,[(\lambda - \lambda^{-2}) - (6n/\pi)^{1/2}\omega]/(1 - \omega)$$

(5)

where ν is the number of chains per unit volume.

The assumption that network chains deform affinely with the applied strain is very inaccurate since chains and junctions in the polymer network exhibit large fluctuations around their mean positions. A "phantom" model of the network which takes this into account but neglects all topological and excluded volume constraints on these fluctuations was formulated by James and Guth.[3] This is an idealized model of a network which is similar to the model of the ideal gas.

A molecular theory of rubberlike elasticity of real networks was developed by Flory.[4,5] This theory is based on a Gaussian distribution of end-to-end vectors of chains, and extends the earlier theory of phantom networks by taking into account constraints on the fluctuations of the junctions in real networks. This constrained-junction theory was later extensively elaborated by Flory and Erman[6,7] and applied with success to many problems in the area of rubberlike elasticity. One of the major premises of the theory is that local intermolecular entanglements and steric constraints restrict the fluctuations of the junctions and thus contribute to the modulus. Extension of these ideas to application of the constraints to the mass center of each chain, instead of the junctions, was developed by Erman and Monnerie in an analogous constrained-chain theory.[8]

In the present study we apply the constrained-junction theory and the constrained-chain theory of rubberlike elasticity to polymer networks undergoing strain-induced crystallization. Illustrative computations are performed for uniaxial extension of networks at constant temperature. The stress-strain isotherms thus obtained are compared qualitatively with relevant experimental results.

THEORY

The proposed treatment of a network undergoing crystallization by stretching is similar to Flory's.[2] We assume that crystallites in a stretched network form parallel to the direction of strain and there is no significant change in entropy associated with the formation of the nuclei. The configurational probability of an amorphous chain is usually expressed as a Gaussian function of the distance between its ends[1]

$$W(x, y, z) = (\beta/\pi^{1/2})^3 \exp[-\beta^2(x^2 + y^2 + z^2)]$$

(6)

with $\beta = (3/2n)^{1/2}/l$, where n is the number of statistical segments per chain, l is the length of a segment and x, y and z are cartesian components of the end-to-end vector r. If the deformation is affine, then after stretching along the z axis by a factor λ, the distribution of chain coordinates becomes

$$\Omega(x, y, z) = v(\beta/\pi^{1/2})^3 \exp[-\beta^2(\lambda x^2 + \lambda y^2 + \frac{z^2}{\lambda^2})]$$

(7)

with v being the total number of chains considered.

When η of the n segments of a chain crystallize, the relative number of configurations available to the n - η segments becomes[2]

$$W'(x, y, z') = (\beta'/\pi^{1/2})^3 \exp[-\beta'^2(x^2 + y^2 + z'^2)]$$

(8)

with $\beta' = \beta[n/(n - \eta)]^{1/2}$ and $z' = \pm(|z| - \eta l)$, where z' is the algebric sum of the displacement lengths of the two amorphous portions of the chain, with the plus sign for z > 0 and the minus sign for z < 0. The x and y displacements are unaffected. The distribution of the coordinates of v chains becomes

$$\Omega'(x, y, z') = v W'(x, y, z')$$

(9)

The constrained-junction model takes into account the constraints imposed on the fluctuations of junctions by neighboring chains. The Gaussian distribution function of the end-to-end vector may be expressed in term of the molecular deformation tensor Λ_t (t=x,y,z) as

$$\Omega(x, y, z) = \frac{v}{\Lambda_x \Lambda_y \Lambda_z}(\frac{\beta}{\pi^{1/2}})^3 \exp[-\beta^2(\frac{x^2}{\Lambda_x^2} + \frac{y^2}{\Lambda_y^2} + \frac{z^2}{\Lambda_z^2})]$$

(10)

where v is a the number of chains in the network. For a phantom network the molecular deformation tensor Λ_{ph} is related to the macroscopic deformation tensor λ through

$$\Lambda_{t,ph}^2 = (1 - \frac{2}{\varphi})\lambda_t^2 + \frac{2}{\varphi}$$

(11)

Erman and Flory[9] derived the following expression for the molecular deformation tensor in the constrained-junction theory

$$\Lambda_t^2 = (1 - \frac{2}{\varphi})\lambda_t^2 + \frac{2}{\varphi}(1 + B_t) \qquad (t = x, y, z)$$

(12)

with

$$B_t = (\lambda_t - 1)(\lambda_t + 1 - \zeta\lambda_t^2)/(1 + g_t)^2 \qquad (13)$$

and

$$g_t = \lambda_t^2[\kappa^{-1} + \zeta(\lambda_t - 1)] \qquad (14)$$

In these equations, Λ_t represents the element of the molecular deformation tensor in the direction t, λ_t the principal extension relative to the state of reference in the t direction, and φ the network functionality. The parameter κ is a measure of the severity of the entanglement constraints and is proportional to the degree of chain interpenetration, and ζ takes into account the possibly nonaffine nature of the transformation of the domains of constraint with increasing strain.[7,9]

It should be noted however that the derivation of eq. 12 was based on the simplifying assumption that the fluctuations of the x components of end-to-end vectors in the real network $<(\Delta x)^2>$ relative to the fluctuactions in the phantom network scale in the same way as fluctuations of junctions $<(\Delta X)^2>$, i.e.[9]

$$\frac{<(\Delta x)^2>}{<(\Delta x)_{ph}^2>} = \frac{<(\Delta X)^2>}{<(\Delta X)_{ph}^2>} = 1 + B_x \qquad (15)$$

with similar expressions for the y and z components. Below we show that this assumption is approximate and derive more exact expression for the molecular deformation tensor in the constrained-junction and the constrained-chain theory.

The elastic free energy of the network according to the constrained-junction theory is

$$\Delta A_{el} = \Delta A_{ph} + \Delta A_c \qquad (16)$$

with the phantom part of the elastic free energy given by

$$\Delta A_{ph} = \frac{1}{2}\xi kT(\lambda_x^2 + \lambda_y^2 + \lambda_z^2 - 3) \qquad (17)$$

where $\xi = (1 - 2/\varphi)\nu$ is the cycle rank, and the contribution due to constraints being

$$\Delta A_c = \frac{1}{2}\mu kT \sum_{t=x,y,z} [B_t + D_t - \ln(B_t + 1) - \ln(D_t + 1)] \qquad (18)$$

230

Here μ is the number of junctions and the quantity D_t is related to B_t through

$$D_t = g_t B_t \tag{19}$$

where g_t is given by eq. 14. However, the elastic free energy of Gaussian chains is related to the end-to-end distribution function $W(r)$ by[1]

$$\Delta A_{el} = -kT \sum_v \ln \left[\frac{W(r_v)}{W^0(r_v)} \right] = \frac{3}{2} vkT \left(\frac{<r^2>}{<r^2>_0} - 1 \right) \tag{20}$$

where $W^0(r)$ is the distribution in the undeformed state.

The ratio

$$\Lambda_t^2 = \frac{<t^2>}{<t^2>_0} \qquad (t = x,y,z) \tag{21}$$

defines the molecular deformation tensor and therefore using eqs 16 - 18 and 20 - 21 we obtain

$$\Delta A_{el} = \frac{1}{2} vkT \sum_{t=x,y,z} (\Lambda_t^2 - 1) = \frac{1}{2} \mu kT \sum_{t=x,y,z} [B_t + D_t - \ln(B_t + 1) - \ln(D_t + 1) + \frac{\xi}{\mu}(\lambda_t^2 - 1)] \tag{22}$$

From eq 22, it follows that

$$\Lambda_t^2 = (1 - \frac{2}{\varphi})\lambda_t^2 + \frac{2}{\varphi}[1 + B_t + D_t - \ln(B_t + 1) - \ln(D_t + 1)] \tag{23}$$

Equation 23 gives the molecular deformation tensor for the constrained junction theory without the approximate assumption (eq. 15) of Erman and Flory.[9] It should be noted, however, that in addition to the molecular deformation tensor given by eq. 12, Erman and Flory[9] introduced the domain deformation tensor Θ^2 defined as

$$\Theta_t^2 = 1 + D_t \tag{24}$$

and the total effective microscopic tensor is a sum of Λ^2 and $(\Theta^2 - 1)$ scaled by an arbitrary factor b $(0 < b < 1)$

$$\Lambda_{t,eff}^2 = (1 - \frac{2}{\varphi})\lambda_t^2 + \frac{2}{\varphi}[1 + B_t + bD_t] \tag{25}$$

so that the only difference between our expression for the microscopic deformation tensor (eq. 23) and eq 25 is contained in logarithmic terms (for b = 1).

Recently Erman and Monnerie[8] extended Flory's idea[5] of constraints effecting fluctuations in real networks, relative to fluctuations in phantom networks. In this case, however, the constraints were viewed as affecting the fluctuations of the centers of masses of the chains, instead of the junctions. The resulting equations for B_t and D_t in the resulting constrained-chain theory[8] are very similar to those of Flory's constrained junction model[5,7] (eqs 13 and 19), and are given below

$$B_t = \frac{h^2\left(\frac{\kappa_G}{h}\lambda_t^2 - 1\right)}{\left(\lambda_t^2 + h\right)^2} \tag{26}$$

and

$$D_t = \frac{\lambda_t^2 B_t}{h} \tag{27}$$

where the parameter h is a function of the macroscopic deformation tensor λ

$$h(\lambda_t) = \kappa_G[1 + (\lambda_t^2 - 1)\Phi] \tag{28}$$

and the parameter κ_G is a measure of the strength of constraints effecting fluctuations of the chain, and is thus similar to the parameter κ in the constrained-junction theory.

If the fluctuations of the chain segments from their mean positions in the reference phantom network are independent of the applied strain (the classic result of the James and Guth theory), then the parameter Φ is given by the formula

$$\Phi = \left(1 - \frac{2}{\varphi}\right)^2 \left(\frac{1}{3} + \frac{2}{3n}\right) \tag{29}$$

where n is a number of segments per chain. Equations for B_t and $h(\lambda_t)$ printed in the original Erman-Monnerie paper[8] contain an error,[10] and thus differ slightly from our eqs 26 and 28.

The elastic free energy in the constrained-chain theory is given by eq. 16, where the phantom network contribution is described by eq.17 and the contribution due to constraints on chains is[8]

$$\Delta A_c = \frac{1}{2}\nu kT \sum_{t=x,y,z} [B_t + D_t - \ln(B_t + 1) - \ln(D_t + 1)] \tag{30}$$

with B_t and D_t defined by eqs 26 - 29. The comparison of eqs. 20 and 21 with the elastic free energy of the constrained chain leads to the following expression for the molecular deformation tensor

$$\Lambda_t^2 = \left(1 - \frac{2}{\varphi}\right)\lambda_t^2 + \frac{2}{\varphi} + B_t + D_t - \ln(B_t + 1) - \ln(D_t + 1) \tag{31}$$

232

The molecular deformation tensor for the constrained-chain theory was derived earlier by Galiatsatos.[11] This study unfortunately used the Erman-Monnerie equations in the earlier, uncorrected form, and assumed (by analogy with the constrained-junction theory) that the contribution to the molecular deformation tensor resulting from constraints is

$$\Lambda_{t,c}^{\ 2} = \frac{2}{\varphi}\, B_t$$

(32)

which is not correct. Specifically, the constrained part of the elastic free energy for the constrained junction model is proportional to the number of junctions

$$\mu = \frac{2}{\varphi}\, \nu$$

(33)

but for the constrained-chain model it is proportional to the number of chains ν. Therefore, the $2/\varphi$ factor in eq. 32 should be replaced by 1. In analogy with the Flory-Erman[9] theory, Galiatsatos[11] introduced the domain deformation tensor Θ^2 defined by eq. 24, so that the total effective molecular deformation tensor tensor (with the corrected B_t term) is

$$\Lambda_{t,eff}^{\ 2} = \left(1 - \frac{2}{\varphi}\right)\lambda_t^{\ 2} + \frac{2}{\varphi} + B_t + bD_t$$

(34)

where b is an arbitrary factor ($0 < b < 1$). The difference between our result (given by eq. 31) and this approximate result (eq. 34) is the logarithmic terms (for $b = 1$).

Knowledge of the molecular deformation tensor for real chains enables calculation of strain-induced crystallization in the Flory theory.[12] The assumption that $n - \eta$ segments of each of the chains melt gives an entropy change

$$\Delta S_a = \nu\,(n - \eta)\, s_f$$

(35)

where s_f is the entropy of fusion per segment. The melting of the $n - \eta$ segments leads to additional entropy change ΔS_b due to the change in the chain length distribution of the polymer:[2,12]

$$\Delta S_b = k \int \int \int \Omega\,(xyz) \ln W'(xyz')\, dxdydz - k \int \int \int \Omega'(xyz') \ln W'(xyz')\, dxdydz' =$$

(36)

$$= -\nu\, k\left[\frac{n\beta^2\eta^2 l^2}{(n-\eta)} - \frac{2\beta\eta l}{\pi^{1/2}}\, \frac{n}{(n-\eta)}\, \Lambda_z + \frac{n}{2\,(n-\eta)}\,(\Lambda_x^2 + \Lambda_y^2 + \Lambda_z^2) - \frac{3}{2}\right]$$

and the total configurational entropy of crystallization is

$$\Delta S = \Delta S_a + \Delta S_b$$

(37)

If one assumes that the change in chain length distribution is not accompanied by a change in internal energy, then the free energy of the system with respect to totally crystalline chains is[2,12]

$$A = vRT[n\theta(1-\gamma) + (\beta \ln)^2 \frac{(1-\gamma)^2}{\gamma} - \frac{2(\beta \ln)}{\pi^{1/2}} \frac{(1-\gamma)}{\gamma} \Lambda_z + \frac{1}{2\gamma}(\Lambda_x^2 + \Lambda_y^2 + \Lambda_z^2) - n\theta - \frac{3}{2}]$$

(38)

with

$$\gamma = (n - \eta)/n$$

(39)

and

$$\theta = (\frac{s_f}{k} - \frac{h_f}{kT}) = \frac{h_f}{k}(\frac{1}{T_m^o} - \frac{1}{T})$$

(40)

where h_f is a heat of fusion per segment, and $T_m^o = \frac{h_f}{s_f}$ is the incipient crystallization temperature of the undeformed polymer.

The conditions of longitudinal growth of crystallites are expressed by

$$(\frac{\partial A}{\partial \eta})_{\Lambda_x, \Lambda_y, \Lambda_z} = 0 \qquad \text{or} \qquad (\frac{\partial A}{\partial \gamma})_{\Lambda_x, \Lambda_y, \Lambda_z} = 0$$

(41)

which leads to the equation[2,12]

$$\theta \gamma^2 + n\beta^2 l^2 (1 - \gamma^2) = \Psi(\Lambda_x, \Lambda_y, \Lambda_z)$$

(42)

with

$$\Psi(\Lambda_x, \Lambda_y, \Lambda_z) = (\frac{6}{\pi n})^{1/2} \Lambda_z - (\Lambda_x^2 + \Lambda_y^2 + \Lambda_z^2)/2n$$

(43)

The physically realistic value of the degree of the crystallinity $\omega = 1 - \gamma$ is

$$\omega = 1 - \frac{[\frac{3}{2} - \Psi(\Lambda_x, \Lambda_y, \Lambda_z)]^{1/2}}{[\frac{3}{2} - \theta]^{1/2}}$$

(44)

The theory of Flory[2] is a useful approximation at relatively small values of the degree of the crystallinity ω. The crystallization temperature changes with elongation as

$$\frac{1}{T_m} = \frac{1}{T_m^o} - \frac{k}{h_f} \Psi(\Lambda_x, \Lambda_y, \Lambda_z)$$

(45)

where the function $\Psi(\Lambda_x,\Lambda_y,\Lambda_z)$ given in eq. 43. The retractive force for a stretched network at constant γ is[2]

$$f = (\partial A/\partial \alpha)_\gamma = \left[\frac{\partial A}{\partial \Lambda_x^2} \frac{\partial \Lambda_x^2}{\partial \alpha} + \frac{\partial A}{\partial \Lambda_y^2} \frac{\partial \Lambda_y^2}{\partial \alpha} + \frac{\partial A}{\partial \Lambda_z^2} \frac{\partial \Lambda_z^2}{\partial \alpha} \right]_\gamma \qquad (46)$$

where α is the linear extension ratio defined as the ratio of final length to the initial length and A is given by eq. 38. For uniaxial stretching in the z direction, the macroscopic deformation tensor λ is

$$\lambda_1 = \lambda_z = v_2^{-1/3} \alpha$$

$$\lambda_2 = \lambda_x = \lambda_y = v_2^{-1/3} \alpha^{-1/2} \qquad (47)$$

with v_2 being the volume fraction of polymer during the stress-strain measurements.

Using eq. 38 for the free energy we obtain the following expression for the retractive force

$$f = v\,RT \left\{ \left[-\sqrt{\frac{6n}{\pi}} \frac{(1-\gamma)}{\gamma} \Lambda_1^{-1} + \frac{1}{\gamma} \right] H(\lambda_1) v_2^{-2/3} \alpha - \frac{1}{\gamma} H(\lambda_2) v_2^{-2/3} \alpha^{-2} \right\} \qquad (48)$$

with the function $H(\lambda_t)$ defined as

$$H(\lambda_t) = \frac{\partial \Lambda_t^2}{\partial \lambda_t^2} \qquad (49)$$

For the constrained-junction model (eq. 23) the function $H(\lambda_t)$ becomes

$$H(\lambda_t) = \left(1 - \frac{2}{\varphi}\right) + \frac{2}{\varphi} \left[\frac{B_t \dot{B}_t}{1 + B_t} + \frac{D_t \dot{D}_t}{1 + D_t} \right] \qquad (50)$$

with B_t and D_t given by equations 13, 14 and 19, and values of $\dot{B}$ and $\dot{g}$ defined by

$$\dot{B}_t = B_t \left\{ [2\lambda_t(\lambda_t - 1)]^{-1} + (1 - 2\zeta\lambda_t)[2\lambda_t(1 + \lambda_t - \zeta\lambda_t^2)]^{-1} - 2\dot{g}_t(1 + g_t)^{-1} \right\} \qquad (51)$$

$$\dot{D}_t = \dot{B}_t g_t + B_t \dot{g}_t \qquad (52)$$

with

$$\dot{g}_t = \kappa^{-1} - \zeta(1 - 3\lambda_t/2) \qquad (53)$$

For the constrained-chain model, eq. 31 indicates

$$H(\alpha_t) = (1 - \frac{2}{\varphi}) + [\frac{B_t \dot{B}_t}{1 + B_t} + \frac{D_t \dot{D}_t}{1 + D_t}]$$

(54)

with B_t and D_t given by eqs. 26 and 27, $h(\lambda_t)$ by eq.28, and $\dot{B}_t$ and $\dot{D}_t$ defined as

$$\dot{B}_t = B_t (\frac{\kappa_G}{h} \lambda_t^2 - 1)[\frac{2}{(\alpha_t^2 + h)} - \frac{1}{(\alpha_t^2 - \frac{h}{\kappa_G})}]$$

(55)

and

$$\dot{D}_t = B_t (\frac{1}{h} - \frac{\lambda_t^2 \kappa_G \Phi}{h^2}) + \frac{\lambda_t^2}{h} \dot{B}_t$$

(56)

The theory also enables the calculation of the reduced stress (modulus) defined by

$$[f^*] = \frac{f^* v_2^{1/3}}{(\alpha - \alpha^{-2})}$$

(57)

where f^* is the force per unit undeformed area.

ILLUSTRATIVE NUMERICAL CALCULATIONS AND DISCUSSION

Illustrative calculations have been carried out to compare the constrained-junction model and the constrained-chain model in the Flory theory of strain-induced crystallization. The calculations were performed for five values of the parameter κ or κ_G (2, 5, 10, 20, and ∞). The number of segments n, the heat of fusion H_f per mol of repeat units, the incipient crystallization temperature T_m, and the temperature T were assumed to have the following values: 150, 9.195 kJ/mol, 272 K, and 278 K, respectively.[12] Values of n were arbitrarily chosen, but the values of H_f and T_m correspond to those for cis-1,4-polybutadiene.[13,14] The ζ parameter in the constrained-junction model was set to zero. Also compared are the effect of the approximation given by eq. 12 for the molecular deformation tensor in the constrained-junction theory with the results based on the molecular deformation tensor derived in this paper (eq. 23). Detailed analysis of the strain-induced crystallization calculated using eq. 12 and its comparison with experimental data were given in a previous study.[12]

Figure 1a shows values of the degree of crystallinity ω as a function of the deformation α, for the constrained-junction model with the molecular deformation tensor given by eq. 23. Figure 1b shows the same plots for the constrained-junction model, using the Erman-Flory equation for the molecular deformation tensor (eq. 12). Figure 1c shows the degree of crystallinity ω vs. elongation α for the Erman-Monnerie constrained-chain model. The molecular deformation tensor for the constrained-chain model was assumed to be given by eq. 31. In order to study the effect of constraints on the strain-induced crystallization, the parameter κ (and κ_G) was varied. In the limiting case $\kappa = \infty$, corresponding to affine deformation, the plot in Fig. 1b corresponds to the classical result of Flory,[2] since for $\kappa = \infty$ the deformation is affine.[5] The small difference between Fig. 1a and 1b is due to the logarithmic term ln(B + 1) which in the affine limit becomes lnΛ. For the constrained-chain model in the limit κ_G = ∞, the elastic free energy does not reach the affine limit,[5] as easily seen from eqs. 26 - 30, and the degree of crystallinity is lower than for the constrained-junction model. In the limit κ_G = ∞ the parameter D goes to 0, as in the constrained-junction model, but the parameter B shows the behavior

$$B \rightarrow \frac{\lambda^2}{[1 + (\lambda^2 - 1)\psi]} - 1$$

(58)

instead of $B \rightarrow \lambda^2 - 1$, as for the constrained-junction case.

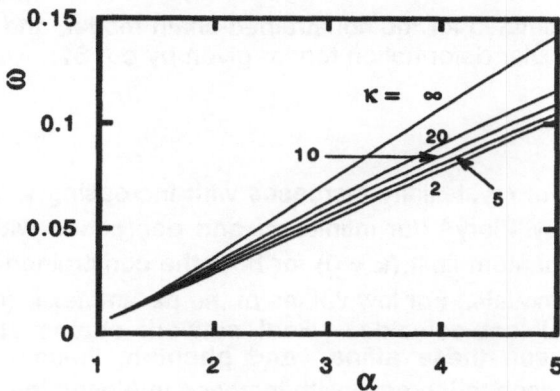

Figure 1a.   Degree of crystallinity ω vs. elongation α calculated from the constrained-junction model, and the molecular deformation tensor given by eq. 23.

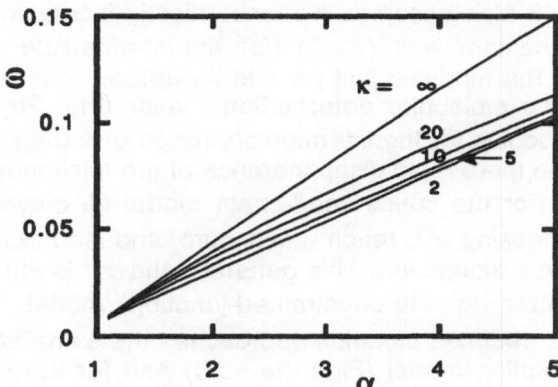

Figure 1b.   Degree of crystallinity ω vs. elongation α calculated for the constrained-junction model, and the molecular deformation tensor given by eq. 12.

237

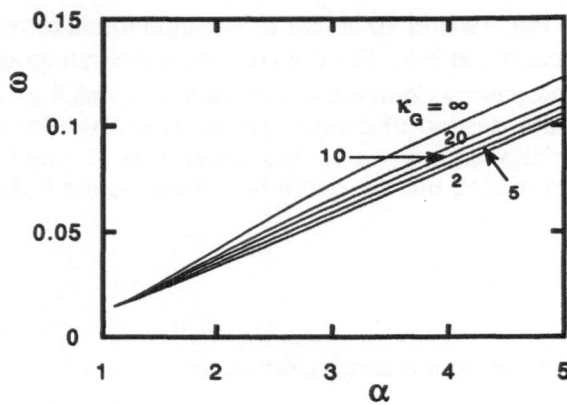

Figure 1c. Degree of crystallinity ω vs. elongation α calculated for the constrained-chain model, and the molecular deformation tensor given by eq. 31.

The degree of crystallinity increases with increasing κ, approaching the values predicted by Flory[2] (for infinite κ) and decreases with decreasing κ, approaching the phantom limit (κ = 0) for both the constrained-junction and the constrained-chain models. For low values of the parameter κ (or κ_G) the curves in Figs. 1a - c almost coincide. Real network chains show a behavior intermediate between these affine and phantom limits. The degree of crystallinity is also seen to increase with increase in elongation α, as expected.

Figures 2a - c show the behavior of the reduced nominal stress (modulus) $[f^*]/\nu RT$ as a function of the inverse elongation α^{-1}. All parameters are the same as in Figures 1a - c. There is an interesting difference in the behavior between the constrained-junction model and the constrained-chain model. For the constrained-junction model in the affine limit (κ = ∞), the modulus is a monotonically increasing function of the elongation, as seen in Figs. 2a and 2b. It also becomes a monotonically increasing function of the elongation in the phantom limit (κ = 0). For the intermediate values of κ, there is a downturn in the modulus just prior to its upturn both in Figs 1a and 1b. Equation 12 for the molecular deformation tensor (Fig. 2b) leads to bigger downturn in the modulus for the intermediate range of κ than does equation 23 (Fig. 2a), and to the more rapid disappearance of the minimum of the curves for low values of κ. For the constrained-chain model all curves in Fig. 2c first decrease with decreasing α^{-1}, reach a minimum, and start to increase for small values of the inverse elongation. This general behavior is shown even in the case $\kappa_G = \infty$, contrary to the constrained-junction model. For very small values of α^{-1}, the reduced modulus approaches the same affine limit both for the constrained-junction model (Figs. 2a - 2b) and for the constrained-chain model (Fig. 2c).

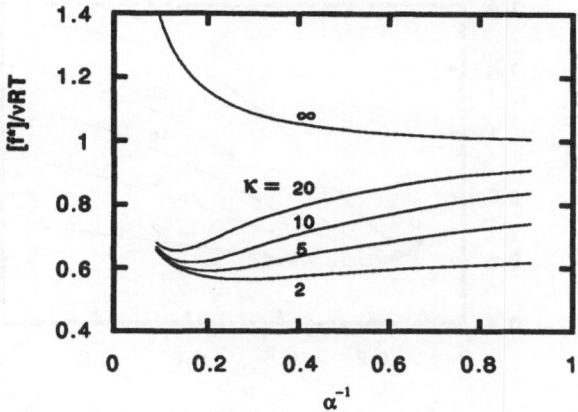

Figure 2a.   The reduced nominal stress $[f^*]/\nu kT$ vs. α^{-1} calculated for the constrained-junction model and the molecular deformation tensor given by eq. 23.

An upturn in the modulus has been observed experimentally in several studies.[1,15] Such an increase is predicted both by Flory's theory[2] and by our non-affine model. In our theory, it occurs at higher elongations as a consequence of the fact that the microscopic deformation Λ_z in our model is less than the macroscopic deformation λ_z. Such an upturn in modulus is often credited to limited chain extensibility and to the non-Gaussian behavior of real

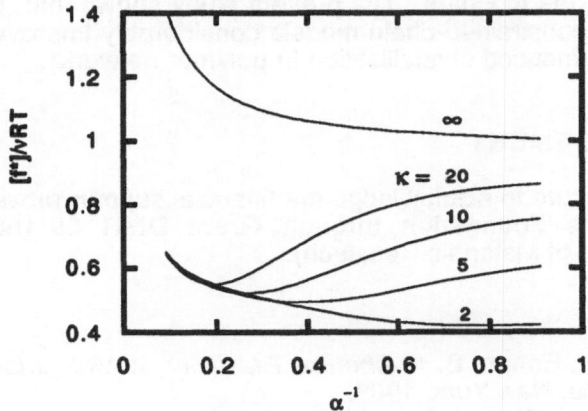

Figure 2b.   The reduced nominal stress $[f^*]/\nu kT$ vs. α^{-1} calculated for the constrained-junction model, and the molecular deformation tensor given by eq. 12.

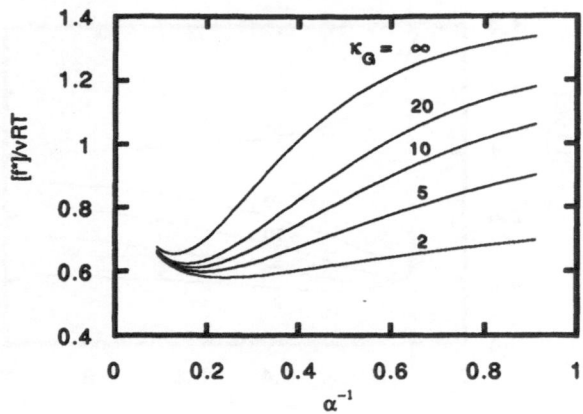

Figure 2c. The reduced nominal stress $[f^*]/\nu kT$ vs. α^{-1} calculated for the constrained-chain model and the molecular deformation tensor given by eq. 31.

chains.[1] All calculations in our theory are for Gaussian chains, and the calculated upturn in the modulus is due to strain-induced crystallization, especially since the minima of curves in Figs. 2a - c are observed for elongation ratios smaller than the maximum elongation ratio for freely jointed chains: α_{max} = $\sqrt{n}$ (i.e., $\alpha_{max} \approx 12.25$ for n = 150).

The stress-strain isotherms shown in Figs 2a - c are in satisfactory qualitative agreement with experimental data. In particular, the transition from the affine to the phantom limit of deformation is accompanied by a reduction in the modulus followed by an upturn, as is frequently observed experimentally. The classical Flory theory based on the affine deformation of the network gives only an upturn in the modulus. The present study shows that the constrained-junction and the constrained-chain models considerably improve the theoretical analysis of strain-induced crystallization in polymer networks.

ACKNOWLEDGEMENT

It is a pleasure to acknowledge the financial support provided JEM by the National Science Foundation through Grant DMR 89-18002 (Polymers Program, Division of Materials Research).

REFERENCES

(1)   Mark, J. E.; Erman, B. *Rubberlike Elasticity. A Molecular Primer*, Wiley-Interscience, New York, 1988.
(2)   Flory, P. J. *J. Chem. Phys.* **1947**,*15*, 397.
(3)   James, H. M.; Guth, E. *J. Chem. Phys.* **1947**, *15*, 669 .
(4)   Flory, P. J. *Proc. Roy. Soc. London, Ser. A,* **1976**, *351*, 351.
(5)   Flory, P. J. *J. Chem. Phys.* **1977**,*66*, 5720.
(6)   Erman, B.; Flory, P. J. *J. Chem. Phys.* **1978**,*68*, 5363
(7)   Flory, P. J.; Erman, B. *Macromolecules* **1982**,*15*, 800.
(8)   Erman, B.; Monnerie, L. *Macromolecules* **1989**,*22*, 3342.

(9)   Erman, B.; Flory, P. J. *Macromolecules* **1983**,*16*, 1601.
(10)   Erratum in preparation.
(11)   Galiatsatos, V. *Macromolecules* **1990**,*23*, 3817.
(12)   Sharaf, M. A.; Kloczkowski, A.; Mark, J. E. *Comput. Polym.Sci,* submitted.
(13)   Mark, J. E. *Polym. Eng. Sci.* **1979**,*19*, 409.
(14)   Sharaf, M. A. *Ph.D. Thesis in Chemistry*, **1979**, The University of Michigan.
(15)   Smith, Jr., K. J.; Greene, A.; Ciferri, A. *Kolloid Z - Z. Polym.* **1964**, *194*, 49.

KINETICS OF GELATION MONITORED THROUGH DYNAMICS

Donald F. Hodgson, Qun Yu, and Eric J. Amis

Department of Chemistry, University of Southern California

Los Angeles, California 90089-0482

Introduction

During the past decade significant advances have been made in developing an understanding of the underlying physics controlling the gelation process in a variety of systems. Of particular interest has been the desire to explore gelation in terms of critical phenomena using critical point exponents for diverging static quantities, such as the weight average molecular weight and cluster sizes. This has encompassed tests of scaling predictions from the Flory-Stockmayer (FS) mean field theory and also percolation models at the gelation threshold.[1-3] Of more recent interest has been the desire to examine the gelation critical point using dynamics.[4-8] Predictions and experimental values for the divergence exponents of the zero shear rate viscosity and modulus near the gel point have been reported for a number of systems.[9-11] However problems exist comparing theory and experiment because of the inability to perform a true zero shear rate experiment particularly given the divergence of the longest relaxation time in the gelling solution.

This has led to consideration of the frequency dependence of the components of the complex shear modulus $G^*(\omega) = G'(\omega) + iG''(\omega)$, where $G'(\omega)$ and $G''(\omega)$ are the storage and loss shear moduli respectively. At the gel point they are predicted[5,7,8] to scale with frequency as $G' \sim G'' \sim \omega^\Delta$, with the prediction of $\Delta = 0.67$ at the gel point using concepts of fractal dynamics. Percolation calculations based on the analogy of the gelation phenomenon to resistor-superconductor networks gives $\Delta = 0.72 \pm 0.2$.[12-14] Experiments to measure this frequency exponent have been conducted on a large number of chemical gels, networks of chemically crosslinked polyfunctional monomers, oligimers, or functionalized polymers. The observed scaling exponent at the gel point is $\Delta = 0.70 \pm 0.05$ for epoxy resins,[15] 0.69 ± 0.02 for polyurethanes,[16] 0.69 ± 0.02 for polyesters,[8] and 0.72 ± 0.04 for tetraethoxy silane.[17,18] Winter et al. found $\Delta = 0.5$ for end crosslinked PDMS chains and $0.5 < \Delta < 0.7$ for polyurethanes with varied stoichiometry.[19-21] Results for physical gels, i.e. gels that form by weak attractive forces such as hydrogen bonding, van der Waals attractions, or specific electrostatic interactions, have been much less numerous. One system, the crystallization and gelation of PVC plastisols, revealed critical dynamics with $\Delta = 0.8$.[22]

Synthesis, Characterization, and Theory of Polymeric Networks and Gels
Edited by S.M. Aharoni, Plenum Press, New York, 1992

243

The subject of this study is the measurement of the dynamics during the gelation of aqueous gelatin solutions, a physical gel. Gelatin is produced by the acid or base catalyzed denaturation and degradation of collagen fibrils.[23] Collagen fibrils are a supramolecular polypeptide structure consisting of associated triple helices, which in turn are composed of alpha chains rich in glycine.[23,24] Typically gelatin is obtained commercially as the single alpha chain form. Following dissolution in water at an elevated temperature, cooled solutions form optically clear gels. It has been shown that this gelation proceeds by physical crosslinking produced by the partial reformation of the triple helix regions between the polypeptide macromolecules, with the eventual crosslinking into a macroscopic network.[23,25,26] In this paper we will describe the kinetics of gelatin gel formation using dynamic viscoelastic measurements. As described above, the critical scaling exponent Δ has been measured for a number of chemical gels, however there have been few studies of the changes in Δ with a systematic variation of gelation conditions. Here we will present results for gels formed under a wide variety of conditions. In the final section of this paper use of dynamic light scattering to characterize the gelation process will also be discussed.

Experimental

The Rousselot gelatin samples used in this study were a gift from the Eastman Kodak Co. For these studies, gelatin solutions were gravimetrically prepared at the appropriate concentration by dissolving gelatin in distilled, deionized water at 45 °C. A small amount of sodium azide was added as an anti-bacterial agent. For the experimental measurements described below, solutions were rapidly cooled to a quench temperature ranging from 22 to 30 °C from the dissolution temperature of 45 °C. The time dependent evolution of the dynamic light scattering and viscoelasticity were determined immediately following equilibration at the quench temperature.

The pH of gelatin solutions were determined using a combination glass–calomel pH electrode (Fisher #13-620-286) with a calibrated pH meter (Orion model SA520). For the gelatin sample the isoelectric point was determined via titration to be at pH = 4.5. Solutions prepared by dissolving gelatin as obtained, in distilled water had a pH of 5.8. The solution pH was modified by addition of small aliquots of concentrated NaOH or HCl solutions. Ionic strengths were varied by dissolution of the gelatin in solvents prepared with a given concentration of NaCl. Deionization was performed using Dowex cation and anion exchange resins prewashed with aqueous HCl and NaOH and rinsed with deionized water.

Measurement of the dynamic viscoelastic properties of the gelatin solutions during the gelation process was performed using the multiple lumped resonator (MLR) instrument constructed in our laboratory.[27,28] The MLR determines G' and G" over a frequency range of 104 to 9600 Hz. The basic element of the MLR is a mechanical resonator, machined from a single titanium rod, suspended in a precision bore Pyrex tube which serves as the sample cell for the solution under investigation. The resonator consists of five identical fat segments (lumps) connected and held by six thin segments of varying diameters (springs). The thin segments behave as torsional springs when the resonator is clamped by the top and bottom. A sinusoidal current passing through a pair of coils drives a magnet in the resonator causing small torsional oscillations (0.001° maximum angular displacement). Changes in the positions and bandwidths of the five natural resonance frequencies for the device are determined as a function of reaction time or concentration of the solution surrounding the resonator. G' and G" are determined from analysis of the resonance shifts and bandwidth broadening. The time necessary to collect data for each set of five frequencies was about 0.5% of the total duration of the reaction to gelation. Because of the small amplitude of the resonator motion in the MLR, the VE measurements on these gelling system are performed essentially without perturbation of the cluster growth and gel structure.

Dynamic light scattering measurements were performed on quenched gelatin solutions using an Ar ion laser (Spectra Physics 2020-3) operating at 514.5 nm with 300-600 mW. A commercial light scattering goniometer (Brookhaven Instruments, BI-200SM) was used with its original integrated optics allowing measurement of scattering at angles between 10° and 150°. A refractive index matching bath of filtered toluene surrounds the scattering cell and its temperature was controlled to ±0.05 °C. Photon autocorrelations were performed with a multi-tau autocorrelator (ALV Instruments 5000). The correlator, counter, and goniometer stepping motor were controlled by a microcomputer (IBM AT type) for automated data acquisition.

Gelatin solutions prepared for dynamic light scattering measurements were hot filtered (45 °C) through 0.2 mm PVDF acrodisc filters (Gelman) into dust-free tubes with teflon lined caps. Following filtration the solutions were allowed to equilibrate at 45 °C for 30 minutes prior to quenching in the goniometer. Correlation functions were measured immediately following temperature equilibration. Since the measurements presented here are kinetic in nature, measurements were performed at fixed time intervals from the initial quench through gelation.

Results and Discussion

Measurements of $G'(\omega)$ and $G''(\omega)$ as a function of time following quenching to 25 °C are shown for a 1.75% w/w gelatin solution in Figures 1 and 2, respectively. The storage moduli show a gradual smooth increase with time from a non-zero initial value representative of G' for the free gelatin chains in solution. The loss moduli also show a smooth increase, however the initial values are much higher and the increase is less dramatic. As expected, the magnitude of both $G'(\omega)$ and $G''(\omega)$ increase as the measurement frequency increases. An interesting feature of these plots is the apparent lack of a transition or inflection in the time evolution even though macroscopic tests, such as inverting a test tube containing an identical sample incubated at the same temperature, indicate that the solution has indeed gelled. This is surprising considering that during this time period the solution has been converted from a low viscosity liquid into an elastic solid. The smooth growth of G' and G" is qualitatively in agreement with measurements performed at much lower frequency by Djabourov et al.[29]

More information about the sol-gel transformation is obtained in a plot of the data as the corresponding loss tangent, $\tan \delta = G''/G'$, for each measurement frequency as a function of gelation time. Shown in Figure 3 is a plot of $\tan \delta$ for the same system shown in Figures 1 and 2. A steady decrease in $\tan \delta$ is observed, with the decrease most severe for the lowest measurement frequency. This results in values of $\tan \delta$ becoming frequency independent at a particular reaction time, 2.30×10^4 s for the data shown in Figure 3. It has been demonstrated in many systems undergoing chemical or physical gelation that this behavior is indicative of the gel point. Since the following kinetic analysis of the gelation process relies on our ability to locate the exact time when the incipient gel is formed, we should

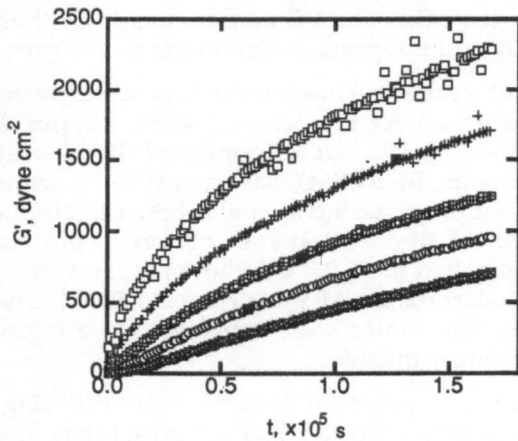

Figure 1. Change of dynamic storage moduli at five frequencies (⊠ 272, ○ 865, ⊞ 2017, + 4676, □ 9400 Hz) with time following a temperature quench to 25 °C of a 1.75% w/w gelatin solution.

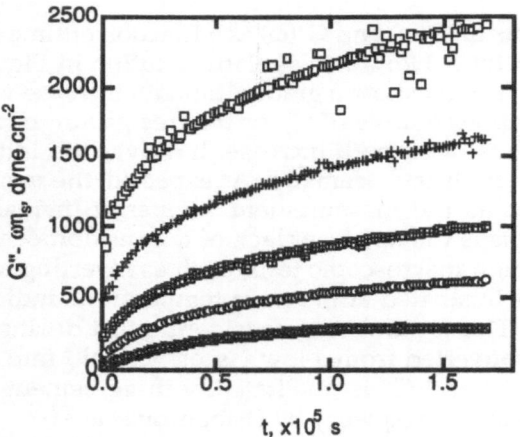

Figure 2. Change in dynamic loss moduli at five frequencies (⊠ 272, ○ 865, ⊞ 2017, + 4676, □ 9400 Hz) with time following a temperature quench to 25 °C of a 1.75% w/w gelatin solution.

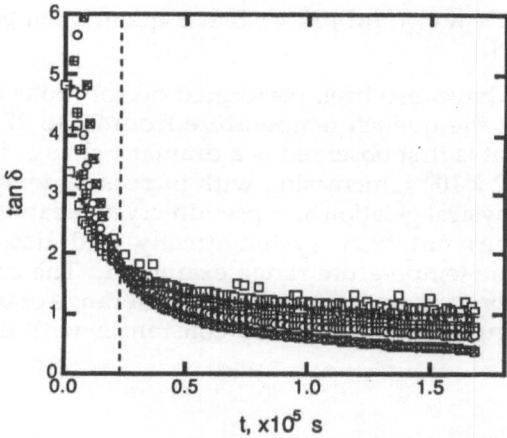

Figure 3. Viscoelastic loss tangent as a function of time for gelatin solution shown in Figures 1 and 2. The dashed vertical line indicates the gel point as determined by the frequency independent crossing of tan δ.

note that the identification of the gel point is taken as the extent of reaction where a frequency independent loss tangent (reflecting G'(ω) and G"(ω) scaling with the same frequency exponent) is observed. While the evidence collected on many gelling systems indicate that this is so, the gel point has previously been identified as the point where G' = G". In Figure 1 we have indicated the G' = G" points by filled squares. These crossing points are clearly frequency dependent and occur earliest at lowest frequencies. Since the gel point occurs at a single time this criterion of gelation is difficult to apply.

Examination of the frequency dependence of the dynamic moduli during the gelation process is found by performing linear fits of log G'(ω) and log G"(ω) versus log ω to find the dynamic scaling exponent Δ, where $G(\omega) \sim \omega^{\Delta}$. At the gel point this power law relation is strictly valid for all frequencies between a maximum frequency characteristic of the relaxation time of the largest cluster in solution and a minimum frequency characteristic of the shortest segmental relaxation time. However for our somewhat limited frequency range an apparent power law always holds. Values of Δ obtained for the storage and loss moduli shown in Figures 1 and 2 are plotted as a function of reaction time in Figure 4. The Δ for both G'(ω) and G"(ω) decrease monotonically and smoothly, with a single point in time where $G'(\omega) \sim G''(\omega) \sim \omega^{\Delta}$. This time corresponds to the observation of a frequency independent loss tangent shown in Figure 3. (The frequency independence of tan δ and the scaling of the storage and loss moduli with the same frequency exponent are after all equivalent statements.) The two figures are useful however because they allow a self-consistent evaluation of the gel times and the numerical values of Δ can be checked using $\Delta = 2\delta/\pi$. For the data shown in

Figure 4 with c = 1.75% w/w and pH = 5.8 at a quench temperature of 25 °C we obtain a $\Delta = 0.68 \pm 0.03$.

Measurements have also been performed on solutions with c = 1.75% w/w and pH = 5.8 varying the quench temperature from 22 to 27 °C with the results listed in Table 1. What is first observed is a dramatic change in the gel times, t_{gel}, from 5.70×10^3 to 1.42×10^5 s, increasing with increasing temperature, consistent with a treatment of physical gelation as a pseudo-crystallization. The other notable feature is that Δ does not vary systematically and has an average value $\Delta = 0.67 \pm 0.08$ over the temperature range examined. The concentration dependence of the gelation process was determined over a range of 0.90 to 10.0% w/w of gelatin. Over this range we obtain nearly constant Δ with experimental uncer-

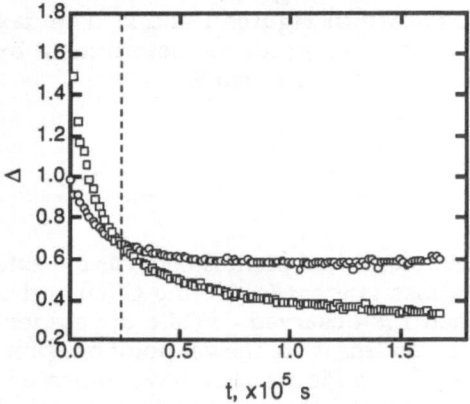

Figure 4. Changes of viscoelastic exponents Δ' for the storage ($\square$) and Δ'' for the loss ($\bigcirc$) shear moduli during the course of the gelatin gelation shown in Figures 1 and 2 as obtained from linear fits of log $G(\omega)$ versus log ω. The dashed vertical line at the gel point is at the position on the time axis determined in Figure 3.

tainty determining the spread in the values as shown in Table 1. Despite the constancy of the dynamic scaling exponent, a dramatic effect of concentration on the kinetics of the gelation is readily apparent. For a fixed quench temperature of 26 °C changing the concentration from 0.90 to 5.10% w/w resulted in gel times varying from 1.00×10^3 to 7.20×10^5 s. Using the values shown in Table 1 a linear plot of log t_{gel} versus log c can be constructed indicating the power law relationship $t_{gel} \sim c^{3.8\pm0.2}$. This is a very strong dependence indeed.

Since the solutions employed for the concentration and temperature dependences were dissolved in the form obtained from Kodak, with a solution pH of 5.8, it was anticipated that by changing the pH a polyelectrolyte effect would cause gels to be formed with expanded structure which would show different dynamics

Table 1. Temperature and concentration dependence of t_{gel} and Δ.

c, % w/w	T, °C	t_{gel}, s	tan δ	Δ
1.75	22.0	5,700	1.2 ± 0.3	0.59 ± 0.04
1.75	23.0	7,000	1.5 ± 0.3	0.70 ± 0.04
1.75	23.5	11,100	1.5 ± 0.3	0.64 ± 0.05
1.75	25.0	23,000	1.8 ± 0.1	0.68 ± 0.03
1.75	26.0	41,000	1.9 ± 0.1	0.73 ± 0.04
1.75	27.0	142,000	2.2 ± 0.1	0.67 ± 0.02
5.10	26.0	1,000	1.5 ± 0.5	0.64 ± 0.05
2.87	26.0	11,200	1.8 ± 0.1	0.73 ± 0.04
1.32	26.0	132,000	2.0 ± 0.1	0.72 ± 0.02
0.90	26.0	720,000	1.8 ± 0.1	0.72 ± 0.02
5.00	29.0	72,600	1.9 ± 0.2	0.75 ± 0.04
7.50	28.0	2,000	1.8 ± 0.3	0.71 ± 0.04
7.50	29.0	48,000	1.6 ± 0.1	0.67 ± 0.01
7.50	30.0	88,000	1.5 ± 0.2	0.63 ± 0.04
10.00	30.0	72,000	1.6 ± 0.3	0.67 ± 0.03

governing the gelation. As is shown in Table 2 no dramatic change in t_{gel} times was observed for a variation in pH from 4.0 to 9.0. At pH higher than 11 the gelatin solutions failed to gel regardless of the time allowed. Additionally, the dynamics appeared unchanged, with $\Delta = 0.70 \pm 0.03$ a constant over the investigated pH range. The relative pH insensitivity of t_{gel} appears puzzling, since at pH > 7.0 a gelatin solution is far from its isoelectric point, resulting in a large number of titrated acid groups and a large negative charge. To test whether this pH indifference was due to a screening of the polyelectrolyte effect of charged groups by an excess of salt remaining in the gelatin samples as obtained, the heated gelatin solutions were stirred with a mixed bed of cation and anion exchange resins. The variation of pH was repeated for these deionized solutions and the values obtained for t_{gel} and Δ are given in Table 2 and are plotted Figure 5 together with those for the "as obtained" samples. A much stronger dependence of t_{gel} on pH is apparent, with gel times increasing from 4.00×10^3 at the isoelectric point to 4.50×10^4 at pH = 10.0. Despite this change in the gel times, once again the dynamic scaling exponent for frequency, $\Delta = 0.72 \pm 0.07$ is independent of gelation conditions. This is also the case for variations in ionic strength as shown in Table 3. Upon increasing the ionic strength from zero added salt to 1.0 M in NaCl, an increase in the gel times is observed with a constant Δ of 0.67 ± 0.03.

It is clear from the results of the dynamic mechanical measurements described above that the exponent Δ has a value of 0.70 ± 0.08 independent of gelation conditions despite widely varying gel times. This "universal" constant is consistent with most of the cases previously reported for chemical gels.[8,15-21] However, there is no clear answer regarding whether this data lends support for the fractal dynamics models or the percolation models since the experimental uncertainty in these measurements prevents distinguishing between 0.67 and 0.72

Table 2. pH dependence of t_{gel} and Δ for deionized gelatin solutions (upper table) and gelatin as obtained (lower table).

c, % w/w	T, °C	pH	t_{gel}, s	tan δ	Δ
1.75	23.0	10.0	45,000	1.7 ± 0.2	0.68 ± 0.02
1.75	23.0	8.0	23,000	1.7 ± 0.2	0.68 ± 0.02
1.75	23.0	7.0	14,500	2.6 ± 0.2	0.76 ± 0.04
1.75	23.0	6.0	7,000	1.7 ± 0.2	0.67 ± 0.03
1.75	23.0	4.8	4,000	2.1 ± 0.3	0.79 ± 0.04
1.75	23.0	4.0	8,000	1.8 ± 0.2	0.68 ± 0.03
1.75	23.0	5.0	7,800	1.6 ± 0.4	0.70 ± 0.04
1.75	23.0	5.8	7,000	2.0 ± 0.4	0.71 ± 0.04
1.75	23.0	7.0	7,700	2.5 ± 0.3	0.70 ± 0.03
1.75	23.0	8.0	8,100	1.9 ± 0.1	0.70 ± 0.02
1.75	23.0	9.0	19,000	2.1 ± 0.1	0.73 ± 0.01

Table 3. Ionic strength dependence of t_{gel} and Δ for gelatin solutions.

c, % w/w	T, °C	NaCl, M	t_{gel}, s	tan δ	Δ
1.75	23.0	0.00	7,000	1.5 ± 0.3	0.70 ± 0.04
1.75	23.0	0.25	11,000	1.4 ± 0.2	0.66 ± 0.04
1.75	23.0	0.50	11,000	2.1 ± 0.4	0.66 ± 0.02
1.75	23.0	0.75	30,000	1.4 ± 0.1	0.66 ± 0.02
1.75	23.0	1.00	34,000	1.7 ± 0.1	0.69 ± 0.02

exponents. It seems unrealistic that any viscoelastic measurement can provide definitive proof of either model.

We now shift focus to a demonstration that the observed power law distribution of relaxation times at the gel point is not limited to dynamic mechanical measurements. In Figure 6 we show the time evolution of the field autocorrelation functions, $g^{(1)}(\tau)$, following a quench to 23 °C of a 1.50% w/w gelatin solution. At early times in the gelation process a bimodal correlation function is observed, with a sharp exponential decay at short correlation times and a much broader decay occurring at longer relation times. As the gelation proceeds the early relaxation process remains unchanged while the long time relaxation process continually shifts to longer times. At 1.17×10^4 s a power law is observed in the field autocorrelation function. As with the dynamic viscoelasticity, this power law relaxation can be identified with the gel point. Following the gel point the total magnitude of $g^{(1)}(\tau)$ decreases slightly and the long time relaxation shifts to shorter times.

Several studies have described the conformational and diffusive properties of gelatin solutions using light scattering techniques.[28-34] The primary focus of these studies has been to understand the bimodal relaxation spectrum observed by

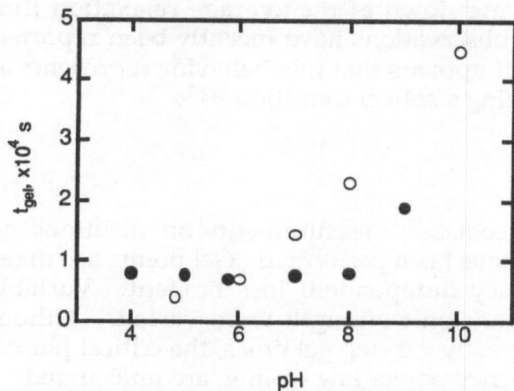

Figure 5. Gelation time as a function of solution pH for gelatin used as obtained (●) and deionized prior to pH adjustment (○).

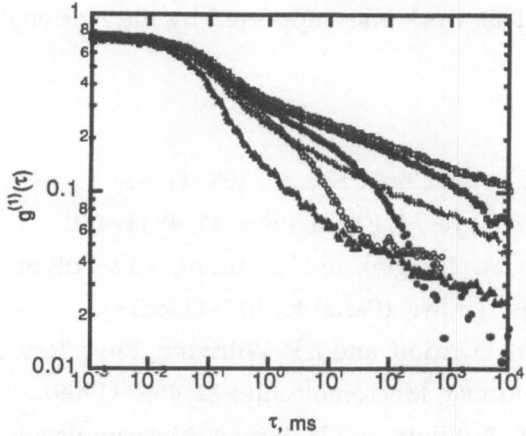

Figure 6. Field autocorrelation function with time (○ 2700, ● 7200, ⊞ 9900, □ 11700, + 18000, ▲ 27000 s) following a temperature quench to 23 °C of the 1.5% w/w gelatin solution.

dynamic light scattering at temperatures above the equilibrium melting temperature of the gel. Ren et al. have identified three regions of relaxation times in gelling gelatin solutions: an exponential decay at early times, a power law at intermediate times, and a stretched exponential at long times, with the characteristic time separating these regimes diverging at t_{gel} is approached.[34] This follows an earlier report of a power-law time decay of the autocorrelation function at the gel point, with a critical slowing down of the average relaxation time for gelling silicate systems.[35] Similar observations have recently been reported for physical gels of polysaccharides so it appears that this behavior represents a general observation for systems undergoing a sol-gel transition.[36,37]

Conclusions

Dynamic viscoelastic measurements on solutions of gelatin during the process of gelation have been performed. Gel points are determined by the observation of a frequency independent loss tangent. Variables of concentration, temperature, pH, and ionic strength were varied. Although these conditions produce gels with greatly varying gel times, the critical point dynamics, as characterized by the frequency power law scaling, are unchanged. An average value for the dynamic scaling exponent of the storage and loss shear moduli with frequency is $\Delta = 0.68 \pm 0.08$ which is between the values predicted by percolation models and fractal dynamics. A power law relation is also observed in the electric field autocorrelation function obtained by dynamic light scattering.

Acknowledgements

We thank Dr. Elizabeth Patton of Eastman Kodak Co. for provision of the gelatin samples. This work was supported by the National Science Foundation, DMR–8715567.

References

1. P. J. Flory, J. Am. Chem. Soc. 63, 3083 (1941).

2. W. H. Stockmayer, J. Chem. Phys. 11, 45 (1943).

3. D. Stauffer, A. Coniglio, and M. Adam, Adv. Polym. Sci. 44, 103 (1982).

4. M. E. Cates, J. Phys. (Paris) 46, 1059 (1985).

5. J. E. Martin, D. Adolf, and J. P. Wilcoxon, Phys. Rev. A 39, 1325 (1989).

6. M. Muthukumar, Macromolecules 22, 4658 (1989).

7. W. Hess, T. A. Vilgis, and H. Winter, Macromolecules 21, 2356 (1988).

8. M. Rubinstein, R. H. Colby, and J. R. Gillmor, in Space-Time Organization in Macromolecular Fluids, F. Tanaka, M Doi, and T. Ohta eds, Springer-Verlag, Berlin (1989).

9. M. Adam, M. Delsanti, and D. Durand, Macromolecules 18, 2285 (1985).

10. J. E. Martin and J. P. Wilcoxon, Phys. Rev. Lett. 61, 373 (1988).

11. B. Gauthier-Manuel, E. Guyon, S. Roux, S. Gits, and F. Lefaucheux, J. Phys. (Paris) 48, 869 (1987).

12. P.-G. de Gennes, Scaling Concepts in Polymer Physics; Cornell University Press: Ithaca, NY, 1979.

13. P.-G. de Gennes, C. R. Acad. Sci. Ser. B 286, 131 (1978).

14. J. G. Zabolinski, Phys. Rev. B 30, 4077 (1984).

15. J. E. Martin and J. P. Wilcoxon, Macromolecules 23, 3700 (1990).

16. D. Durand, M. Delsanti, M. Adam, and J. M. Luck, Europhys. Lett. 3, 297 (1987).

17. D. F. Hodgson and E. J. Amis, Phys. Rev. A. 61, 2620 (1990).

18. D. F. Hodgson and E. J. Amis, Macromolecules 23, 2512 (1990).

19. H. Winter and F. Chambon, J. Rheol. 30, 367 (1986).

20. F. Chambon and H. Winter, Polym. Bull. 13, 499 (1985).

21. H. H. Winter, P. Morganelli, and F. Chambon, Macromolecules 21, 532 (1988).

22. K. te Nijenhuis and H. H. Winter, P. Morganelli, and F. Chambon, Macromolecules 22, 411 (1989).

23. A. Veis, Macromeolecular Chemistry of Gelatin, Academic Press, New York, 1964.

24. D.R. Eyre, Science, 207, 1315 (1980).

25. M. Djabourov, Contemp. Phys. 29, 273 (1988).

26. M. Djabourov, J. Leblond, and P. Papon, J. Phys. (Paris) 49, 319 (1988).

27. M. Djabourov, J. Leblond, and P. Papon, J. Phys. (Paris) 49, 331 (1988).

28. H. Boedtker and P. Doty, J. Chem. Phys. 58, 968 (1958).

29. E. J. Amis, P. A. Janmey, J. D. Ferry, and H. Yu, Polym. Bull. 6, 13 (1981).

30. E. J. Amis, P. A. Janmey, J. D. Ferry, H. Yu, Macromolecules 16, 441 (1983).

31. T. Chang and H. Yu, Macromolecules 17, 115 (1984).

32. I. Pezron, M. Djabourov, and J. Leblond, Polymer 32, 3201 (1991).

33. T. Herning, M. Djabourov, J. Leblond, and G. Takerkart, Polymer 32, 3211 (1991).

34. S. Z. Ren, W. F. Shi, W. B. Zhang, and C. M. Sorensen, Phys. Rev. A 45, 2416 (1992).

35. J. E. Martin and J. P. Wilcoxon, Phys. Rev. Lett. 61, 373 (1988).

36. P. Lang and W. Burchard, Macromolecules 24, 814 (1991).

37. T. Coviello and W. Burchard, Macromolecules 25, 1011 (1992).

LIGHT SCATTERING STUDIES OF THE STRUCTURE OF RIGID POLYAMIDE GELS

M. E. McDonnell, K. Zero and S. M. Aharoni

Research and Technology
Allied-Signal Inc.
Morristown, NJ 07962

INTRODUCTION

Permanent gels are crosslinked networks swollen by a solvent that pervades the entire volume of the network. Macroscopically, the concentration of the polymer in the gel is uniform, but on the molecular level substantial fluctuations in polymer concentration are present. One characteristic of essentially all gels subjected previously to experimental measurements, but which is not implicit in the definition of a gel, is that the crosslink points are separated by randomly distributed segments. Generally, this random nature is the consequence of connecting the crosslink points with flexible segments. The mathematical consequences of assuming a Gaussian distribution of segments connecting the crosslinked points leads to the well established theory of rubber elasticity and the direct proportionality between the shear modulus and the number density of crosslinks[1,2]. It is easy to forget at times whether a property observed in a gel is truly a consequence of the crosslinks as opposed to a result of flexibility of the interconnecting segments. Recently, Aharoni and coworkers[3,4] have reported procedures to synthesize polyamide networks with rigid crosslinks and stiff intervening segments. These systems permit the study of gels which are essentially free of flexible chains connecting the crosslinks. In these systems, the crosslinks are rigid and the segments stiff and short, which make a Gaussian segment distribution inappropriate.

Much work has been done to characterize flexible gels in terms of a single polymer correlation length. This length first appeared in the self-consistent field theory of Edwards[5], but became widely used in the scaling approach of de Gennes[6]. The correlation length measures the distance over which segments remain spatially correlated in an analogous fashion to the Debye-Hückel screening length which measures the length over which charge effects remain important. Experimentally, the correlation length, ξ, of random coiled polymers in a semi-dilute solution is usually extracted from the diffusion coefficient, D, measured by dynamic light scattering:

$$\xi - k_B T/6\pi\eta D \tag{1}$$

where k_B is the Boltzmann constant; T, the absolute temperature; and η, the solvent viscosity. A single correlation length has also been generally successful for characterizing the properties of flexible gels. In contrast, this paper shows that two correlation parameters are necessary to describe

Synthesis, Characterization, and Theory of Polymeric Networks and Gels
Edited by S.M. Aharoni, Plenum Press, New York, 1992

gels with rigid crosslinks and stiff interconnecting segments. The interpretation of gels in terms of a single correlation length extracted from eq.
1 has been so widely applied that in this section we examine its origin and
implications. First, we consider the physical interpretation of dynamic
light scattering from gel networks, then apply scaling concepts of random
chains to obtain a basic correlation length for the gel, and finally review
the success of the model in characterizing flexible gels. In the discussion
section that follows, we show that the gels of interest here are characterized by two distinct correlation lengths.

A theoretical understanding for the applicability of dynamic light
scattering to gels was developed by Tanaka et al.[7] and has recently been
reviewed by Tanaka[8] and Patterson[9]. Newton's second law of motion is
applied to an element of gel by summing the applied forces: (1) those
resulting from the gel outside of the elements and (2) the drag friction
produced in the element by the gel liquid. Since the acceleration term is
negligible in the macroscopically static configuration of dynamic light
scattering, the net force is zero. Expressing the stress tensor of the gel
in terms of the displacement vector $\mathbf{x}$, leads to an equation of motion solved
by the diffusion equation

$$\mathbf{x} = D\nabla^2\mathbf{x}$$

where the diffusion coefficient D is given as

$$D = M_{os}/f = [K+(4/3)G]/f \tag{2}$$

for polarized scattering and M_{os}, K and G are, respectively, the longitudinal, bulk and shear moduli of the gel respectively and f is the frictional
coefficient per unit volume. The average intensity of the light scattered
by the gel in the low angle limit is also a measure of the longitudinal
modulus:

$$I_s \propto I_0 k^4 n^2 (dn/dc)^2 c^2 L k_B T/R^2 M_{os} \tag{3}$$

where I_0 is the incident intensity with scattered wave vector magnitude k;
n, the index of refraction of the solution of concentration c; and L, the
illuminated length monitored by the detector a distance R away.

The consequence of eq. 2 is that the fluctuations in light scattered by
a gel describe a diffusion process just as the fluctuations in a dilute
solution monitor the transitional diffusion of solute particles. The
diffusion coefficient for the gel is dependent on the elastic and the bulk
frictional properties of the matrix rather than the temperature and
frictional coefficient of a diffusing particle in the dilute solution. This
analysis places no restrictions on the conformation of the segment between
crosslinks.

These basic relations can be expressed in terms of other variables.
For example, the bulk and shear moduli can be related through the Poisson
ratio σ, the ratio of the transverse to longitudinal strain in the gel
during elongational stress:

$$G = K (1-2\sigma)/(1+2\sigma) \tag{4}$$

If the gel has a constant volume upon deformation, $\sigma = 1/2$ and $G = 0$ so eq.
2 becomes

$$D = K/f \tag{5}$$

Even if $\sigma \neq 1/2$, as long as the Poisson ratio is constant, eq. 5 is correct to within a multiplicative factor. The bulk modulus can alternatively be expressed in terms of the osmotic pressure Π by

$$K = c(d\Pi/dc) \tag{6}$$

The consequence of random chain configuration affects both the modulus and the friction term of eq. 5. These effects are seen most readily by scaling theory[6]. The size of an isolated random coil of linear segments with total mass M scales with the exponent ν:

$$R \propto M^\nu \tag{7}$$

where $\nu \approx 0.6$ describes the coil in a good solvent. The segments of adjacent chains begin to overlap as the concentration is raised to

$$c* = M/R^3 \propto M^{1-3\nu}$$

The condition that the correlation length be molecular weight independent above c* requires the correlation length

$$\xi \propto c^{-\nu/(3\nu-1)} \tag{8}$$

A semi-dilute solution can be considered as a group of cells of volume ξ^3 which on average contain a single chain of size ξ. Using Stokes law to describe the frictional coefficient of that chain portion, the frictional coefficient per volume is

$$f = 6\pi\eta\xi/\xi^3 = 6\pi\eta/\xi^2 \tag{9}$$

The equation should, perhaps, be written with a proportionality since the Stokes radius is proportional, rather than equal, to the correlation length. The constant of proportionality, however, should be approximately one.

Similarly, the scaling form of osmotic pressure follows from the requirements that it be inversely proportional to molecular weight in the dilute solution regime but molecular weight independent above c*:

$$\Pi/T \propto c^{3\nu/(3\nu-1)} \propto \xi^{-3} \tag{10}$$

Combining eqs. 5-10 gives

$$D = k_BT/6\pi\eta\xi \propto c^{\nu/(3\nu-1)} \tag{11}$$

As stated earlier, measurements of D have routinely been used to determine the correlation length. Since this derivation does not invoke the cross-linking of the segments it is equally applicable to random distributions of segments in gels and non-crosslinked semi-dilute solutions.

The dynamic light scattering studies of flexible gels have generally followed the predictions of eq. 11. For a gel in a good solvent when $\nu = 0.6$ the diffusion coefficient should vary as the 0.75 power of concentration. Reported values include $0.77 \pm .06$[10] and 0.63 ± 0.01[11] for polyacrylamide in water and 0.72[12] for polyvinylalcohol in water. Graphs imply comparable exponents for polystyrene[13] and polydimethylsiloxane[14] gels in toluene. In a theta solvent where there is no excluded volume, the correlation length varies as the 0.5 rather than the ν power of molecular weight. As a result, eq. 11 indicates that the diffusion coefficient is proportional to the concentration. Experimentally, the power 0.96 ± 0.06[15] was obtained for polyacrylamide in a water-methanol mixture. Davidson, et al.[16], found the concentration dependence of the diffusion coefficient to go from 1.19 to 0.46 for polystyrene in cyclohexane as the temperature was raised to change the solvent from a theta to a marginal solvent. These concentration dependencies for marginal conditions were noted to be consistent with scaling predictions of Schaeffer et al.[17] Recently Fang et al.[18] reported an unexpected concentration dependence of 0.22 for polyacrylamide gel diffusion coefficients. They suggest deviation from theory may be the consequence of the friction factor depending on a viscosity other than that of the bulk solvent. This explanation, however, would imply that other investigators should also not find agreement with scaling theory.

An alternate check of the applicability of the random configuration scaling model comes from the total scattered intensity. Combining eqs. 3, 5-8, and 10 gives

$$M_{os} \propto c^2/I_s \propto c^{3\nu/(3\nu-1)} \propto \xi^{-3} \tag{12}$$

The first proportionality holds for any solution; while the last two forms test the scaling model for random chains. This relation has been tested less often but has been shown to describe a variety of gels. In good solvents, the longitudinal modulus is predicted to vary as the $3\nu/(3\nu-1) \approx$ 2.250 power of concentration. Reported values include 2.35 ± 0.06[19], 2.18 ± 0.06[20], 2.32 ± 0.03[11] and 2.25[12] for polyacrylimide gels in water. At theta conditions, substitution of 1/2 for ν leads to a prediction of variation of longitudinal modulus as the cube of polymer concentration. The value 2.96 ± 0.12[15] was measured for polyacrylamide gel in a water-methanol mixture producing theta behavior. Both the static and dynamic light scattering measurements suggest that random gels are reasonably interpreted with a single correlation length. The random nature of the chain is introduced to evaluate both the longitudinal modulus and the frictional coefficient per volume. Specifically, the random nature of the chain is manifest in a concentration exponent such as ν. There may be a class of networks where the concentration dependencies are different from those yet observed, but a single scaling length still unifies the observations. In this case, eqs. 11 and 12 would require

$$M_{os} \propto \xi^{-3} \propto D^3 \tag{13}$$

This form will be shown below to be inadequate to characterize networks with rigid cross links and stiff interconnecting segments.

EXPERIMENTAL

Materials

Rigid polyamide gels were synthesized by the Yamazaki procedure as previously described[3]. Trifunctional and tetrafunctional crosslinks were formed with N, N', N"-tris(p-carboxyphenyl)-1,3,5-benzenetriamide and bis(3,5-dicarboxyphenyl)terephthalamide, respectively. Hexafunctional branch points were prepared by first reacting under Schotten-Baumann conditions 1,3,5-benzene tricarboxylic acid chloride with 5-amino isophthalic acid and then incorporating the extra functional moieties into the polymeric network during the one-step polymerization procedure. The rigid segments were made of poly(benzanilide-terephthalamide) that resulted from the condensation of 4,4-diaminobenzanilide and terephthalic acid. By judicious selection of difunctional amines, difunctional acids and cross-linking multifunctional acid, the average lengths of the stiff chain segments between crosslinks are controlled. Gels were polymerized in dimethylacetamide (DMAc) at a concentration of c_o. Slabs of gel were immersed in large excess volumes of DMAc. The solvent was changed several times over a ten week period to leach out portions of the network that were not connected to the gel. Although the final rinse removed negligible

Table I. Characteristics of Rigid Gels

Sample	c_o (%)	c (%)	l (Å)	F	$G \times 10^{-6}$ (dyne/cm)
88 G	5.0	~2.0[a]	52	3	_[b]
88 F	7.5	2.82	52	3	0.296
88 C	10.0	5.84	38.5	3	2.27
88 A	10.0	5.48	52	3	1.81
88 B	10.0	5.11	71	3	1.38
88 D	10.0	7.67	52	4	3.54
88 E	10.0	8.54	52	6	4.35

[a] Estimated
[b] Too soft to be measured

material from the gel, the amount of fragments confined in the gel but not chemically crosslinked to it is unknown. The final concentration c of each leached gel was measured. The characteristics of the seven gels used in this study are summarized in Table I. Comparison of samples 88 G, F, and A shows the effect of increasing gel concentration while segment length and functionality remain constant. Gels 88 C, A, and B demonstrate the consequence of increasing segment length with constant functionality and polymerization concentration. Samples 88 A, D, and E indicate the result of increasing branch point functionality while polymerization concentration and segment length remain unchanged.

Light Scattering

Light scattering measurements were made in a Langley-Ford LSA-II photon correlation spectrometer. The light source is a 50 mW helium neon laser. The sample is contained in a temperature controlled 1 cm path length fluorometer cell. Light scattered at specific angles is directed via optical fibers to an EMI 9863A/350 photomultiplier tube. The output signal is directed to a Langley-Ford 1096 digital correlator. Samples of swollen gel were cut to form pieces less than 1 cm in size. These were put in the spectrophotometer cell with DMAc so that the scattering volume was within the gel. The measured correlation function was not strongly dependent on the scattering volume selected from a specific gel. Intensity and autocorrelation function measurements were made at angles of 90, 64.7, and 26.1° from the transmitted beam. Interpretation of the results are discussed in the Data section.

Tensile Modulus

Shear modulus measurements were made with a Humboldt universal penetrometer. Procedures previously described were followed[3]. The results are tabulated in Table I.

DATA

The unnormalized intensity autocorrelation function

$$C(\tau) = <I(t)I(t+\tau)>$$

is measured by the digital correlator[21]. Angular brackets indicate time averages. The parameter of interest is the normalized scattered electric field intensity correlation

$$\phi(\tau) = <E_s^*(t)E_s(t+\tau)>/<|E_s|^2>$$

Table II. Light Scattering Values of Rigid Gels

Sample	I_s'	$D_z \times 10^7$ (cm^2/s)	$D_1 \times 10^7$ (cm^2/s)	A_1	$D_2 \times 10^7$ (cm^2/s)	A_2
88 G	5.71	3.18	4.06	0.614	0.670	0.386
88 F	4.43	14.7	14.50	0.807	1.01	0.193
88 C	1.563	59.6	203	0.481	22.4	0.519
88 A	1.000	41.5	86.1	0.613	10.7	0.387
88 B	0.549	40.5	90.6	0.646	8.17	0.354
88 D	0.1628	162.7	240	0.709	21.9	0.291
88 E	0.0500	256.7	480	0.940	20.2	0.060

Most solution measurements are performed in a homodyne configuration where light is scattered only by concentration fluctuations so

$$\phi(\tau) \propto [C(\tau) - B]^{1/2}$$

where B is the long time value of $C(\tau)$, the baseline value. A homodyne interpretation of these data leads to unreasonable values of the longitudinal moduli estimated from the total intensity scattering in the $\tau \to 0$ limit of eq. 3. The refractive index increment, $dn/dc = 0.24$ cm^3/g, for solutions with a low crosslink density gives an appropriate estimation for these gelled solutions. The total scattered intensity gives a value of M_{os} slightly less than the measured shear modulus. This is clearly incorrect, since we expect M_{os} to be much larger. These gels do not lose solvent when removed from solvent and gently pressed, so the volume appears to be constant, and the Poisson ratio must be about 0.5. Thus, eq. 4 indicates K >> G and so $M_{os} = K+(4/3)G >> G$. The conclusion from the inconsistency of calculated M_{os} and measured G is that most of the light scattered from the gels is elastically scattered from stationary inhomogeneities entrapped in the gel, rather than by concentration fluctuations; consequently, the samples do not produce homodyne scattering. The intense local stationary scatterers assure that the data is collected in the heterodyne limit where for a real electric field correlation function[22]

$$C(\tau) \propto I_{LO}^2 + 2gI_{LO}I_s\phi(\tau)$$

Here subscripts LO and s designate the light scattered by the stationary local oscillators and the concentration fluctuations respectively and g is a factor close to 1 when scattered light is collected over one coherence area. The proportionality constant is dependent on the efficiency of the photomultiplier. Thus

$$\phi(\tau) \propto C(\tau) - B$$

and

$$I_s \propto (C(0) - B)/B^{1/2}$$

where C(0) is the extrapolation of the C value to the $\tau \to 0$ limit, rather than the actual measurement at $\tau = 0$. Value of I_s relative to sample 88 A measured at scattering angle 64.7° are given in Table II. A prime is used throughout this paper to indicate that physical measurements are normalized by gel 88A.

Initially, the normalized correlation spectra were fit by the method of cumulants[21]. The spectra give a remarkably broad range of time constants. An f-test[23] shows that two exponential components with variable intensities give a statistically better fit than three cumulants. The two time constants are separated by about a factor of ten. Fig. 1 shows the two exponential fit to a typical sample. Physically the cumulant model is usually

used to describe a distribution of a single relaxation process while the
two-exponential deconvolution represents two distinct physical processes.
Additional fit parameters could describe a distribution of each of these
distinct times, but the quality of the data does not warrant it. The
cumulant fit procedure has the advantage of less coupling between successive
terms so that higher terms have a relatively small effect on the lowest
order ones which are of most interest. The general conclusions derived from
the data, however, are similar with either the first cumulant or the time
scale of the faster of the two exponentials. The two-exponential deconvolution is reported since it gives better overall fits to the data. Attempts
to describe the data in terms of partial heterodyning[24] did not significantly improve the data fit.

The diffusion coefficient is extracted from the time constant t_i of
each expontial decay with

$$D_i = 1/t_i q^2$$

where q is the magnitude of the change in the wave vectors of the scattered
and incident light:

$$q = |k_s - k_i| = (4\pi n/\lambda)\sin(\theta/2)$$

Within the accuracy of the data, most gels show the q^2 dependence predicted
by the Tanaka theory[7]. The exception is the lowest concentration or least
well formed network, 88 G. The second lowest concentrated sample shows
slight dependence, particularly in its slow mode. Typical angular dependences are shown in Fig. 2. Table II summarizes the diffusion coefficients of the different samples. Values for 88 G and F result from

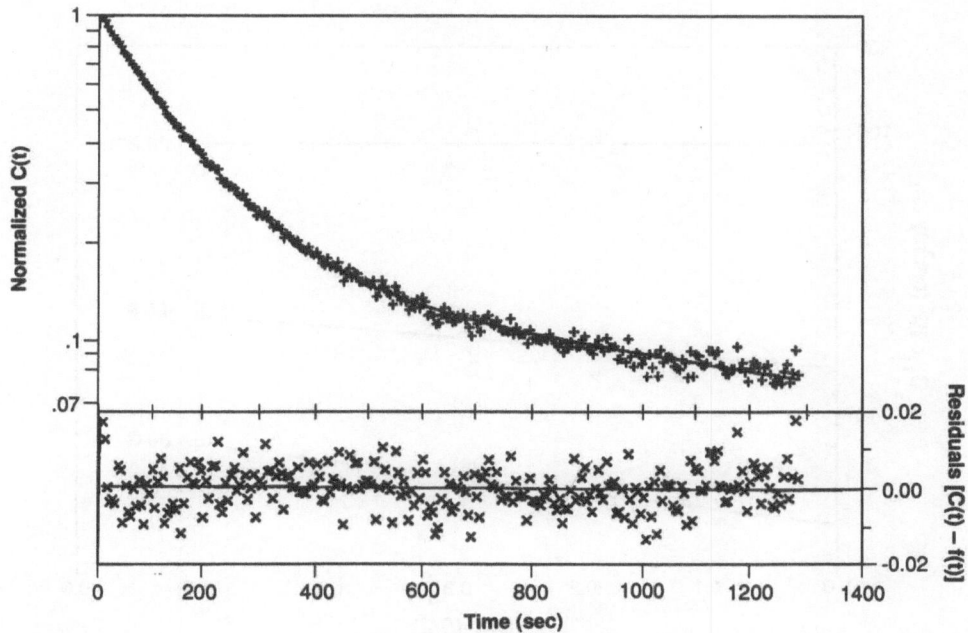

Fig. 1   The scattered intensity correlation function of gel 88 F measured
at an angle 26.1° from the incident beam. The solid line shows the
fit of the data to two exponentials. At the bottom, the residuals
of the fit are plotted.

extrapolation to zero scattering angles. The others are averages of the angle independent values. The small backflow corrections to convert the data from the solvent to a stationary laboratory reference frame were not made[25].

DISCUSSION

In this section, a systematic method is developed to extract two correlation lengths to describe these or any other gels. In the case of flexible gels, these two lengths turn out to be equivalent. First, however, we demonstrate that the rigid gels are not described by a single correlation length. If this were the case, eq. 13 demands that the longitudinal modulus scales inversely to the third power of the diffusion coefficient. This prediction is tested using both (1) the diffusion coefficient from the cumulant fit compared to the longitudinal modulus derived from the total scattered light intensity reported in Table II and (2) the fast diffusion coefficient dependence upon the longitudinal modulus derived from the component of the scattered light attributed to this mode. These longitudinal moduli are designated as $M_{os,t}'$ and $M_{os,1}'$ where subscripts t and 1 correspond to the value calculated from the total and the first (fast) component intensity while the prime indicates the parameter is a relative, rather than an absolute, value. The intensity used to calculate the longitudinal modulus should be the value obtained in the limit of zero scattering angle. The intensity measurements at scattering angle 64.7° are used since these values appear to have no systematic variation from those measured at 90°. Intensities taken at this angle should be appropriate to describe scattering from

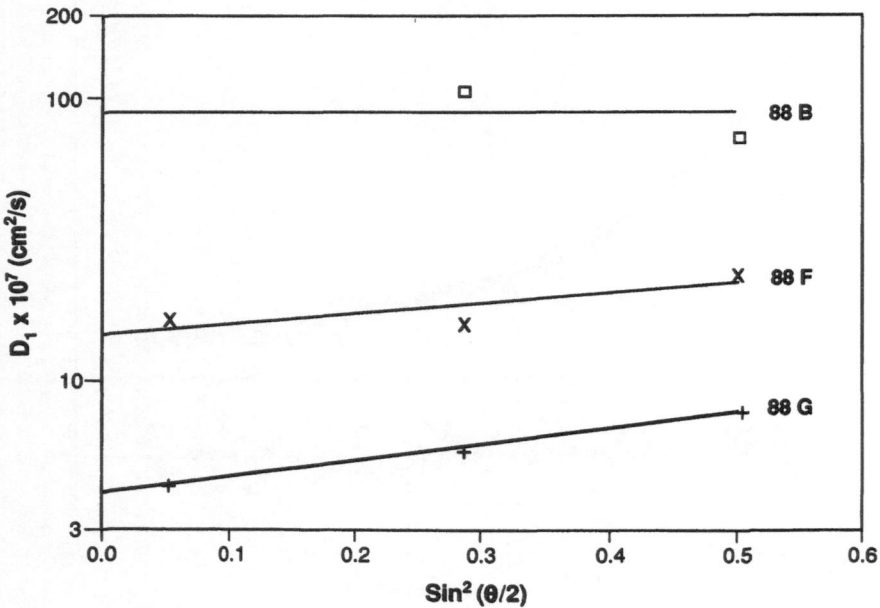

Fig. 2   The dependence of the fast diffusion mode on the square of the scattering angle. The straight lines show linear extrapolations of the data for low concentration gels, 88 G and F and average values for higher concentration gel 88 B.

structures smaller than a size $q^{-1} \approx 600\text{Å}$. Since $kT/6\pi\eta D$ is typically an order of magnitude smaller, extrapolation appears to be unnecessary.

The test of a single correlation length for each gel is shown in Fig. 3. Note that the longitudinal moduli data cover three orders of magnitude while the diffusion data cover two orders. The straight lines represent best least square fits of all the data except that for gel 88 G which has the lowest concentration. Throughout this paper, data associated with this gel is plotted but not used in data fits because (1) the final concentration of the gel was estimated, rather than measured, and (2) the non-q^2 dependence of the diffusion coefficient suggests that the Tanaka model of eq. 2 to correlate diffusion and longitudinal modulus may not be applicable. No matter whether a single or two-mode interpretation model is used, the longitudinal modulus and the diffusion coefficient correlate well with a power law over several orders of magnitude. The longitudinal modulus, however, does not scale as the third power of the diffusion coefficient, but much closer to the second. The conclusion is that the D and M_{os} are clearly related, but not solely in terms of a random chain configuration correlation length that is inversely proportional to the diffusion coefficient.

Two distinct correlation lengths follow from the data. The first length follows from the frictional coefficient per unit volume

$$f - M_{os}/D$$

which can alternately be viewed as the ratio of the impressed force per unit volume to the relative velocity of the gel to the solvent. Einstein's formula for the friction coefficient for a sphere moving through a solvent is often applied to polymers, but for a gel the analysis of Poiseuille for the flow of a fluid through a pipe is perhaps more appropriate. When solvent of viscosity η flows through a circular pipe of diameter $\xi_\perp$, the frictional coefficient per unit volume is given as

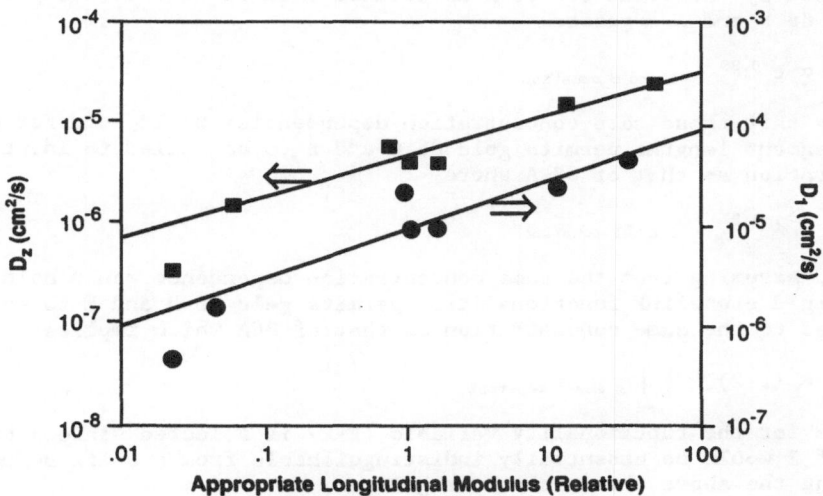

Fig. 3   The diffusion coefficient, either first cumulant average (■) or fast component (•) as a function of the longitudinal modulus attributable to the intensity of the diffusion mode.

Table III. Longitudinal Modulus, Correlation Lengths and Moduli Ratio
of Rigid Gels Relative to Sample 88A

Sample	$M_{os,1}'$	$\xi_{\perp}'$	$\xi_{\parallel}'$	$(G/M_{os,1}')$
88 G	0.0233	1.423	21.2	-
88 F	0.0454	1.926	5.94	3.60
88 C	0.926	1.596	0.424	1.354
88 A	1.000	1.000	1.000	1.000
88 B	1.503	0.837	0.950	0.507
88 D	10.40	0.517	0.359	0.1881
88 E	31.7	0.419	0.1794	0.0758

$$f = 32\eta/\xi_{\perp}^2 \qquad\qquad (14)$$

Comparison with the scaling law prediction for flexible coils, eq. 9, shows
that the two expressions are equivalent to within a factor of two. Thus

$$\xi_{\perp} = (32\eta/f)^{1/2} \; \alpha \; (D/M_{os})^{1/2} \qquad\qquad (15)$$

gives one length that we call the effective pore diameter or the perpendicular correlation length. The first name indicates that it has the same
frictional coefficient per unit volume of a pore of diameter $\xi_{\perp}$ while the
second name indicates that the basic length perpendicular to the velocity
direction is equivalent to the correlation length of a random coil.
Originally, de Gennes suggested that the correlation length be visualized as
a mesh size in an entangled solution. There is no net flow of solvent
through the gel, only uncorrelated random fluctuations throughout the
matrix. It follows that the perpendicular reference does not describe a
fixed axis within the gel, but only a convenient nomenclature designating
the origin of the correlation length.

The relative effective pore size

$$\xi_{\perp}' \; \propto \; (D_1/M_{os,1}')^{1/2}$$

normalized by the value for 88 A is tabulated in Table III. Comparing
samples 88 F and A suggests

$$\xi_{\perp}' \; \propto \; c^{-0.99} \; \Big|_{\text{1 and F constant}}$$

Assuming that these same concentration dependencies would hold for other
stiff segment lengths permits gels 88 C and B to be scaled to identical
concentration as that of 88 A where

$$\xi_{\perp}' \; \propto \; 1^{-1.27} \; \Big|_{\text{c and F constant}}$$

Finally, assuming that the same concentration dependence would hold for
other rigid crosslink functionalities permits gels 88 D and E to be
converted to the same concentration as that of 88A which implies

$$\xi_{\perp}' \; \propto \; (F-2)^{-0.31} \; \Big|_{\text{c and l constant}}$$

The form for the functionality variable (F-2) is selected since a branch
point of 2 would be essentially indistinguishable from a stiff segment.
Combining the above three relations gives

$$\xi_{\perp}' \; \propto \; c^{-0.99} \; 1^{-1.27}(F-2)^{-0.31} \qquad\qquad (16)$$

The values of $\xi_{\perp}$ normalized to gel 88 A are summarized in Table III. The dependence of the effective pore diameter on reduced-length scale is shown in Fig. 4. The uncertainty of these exponents is not known since the concentration dependence based on only two points is used to correct all the other data before extracting the other dependencies.

Physically, the functional dependencies of the perpendicular correlation length indicate that with increased concentration the effective pore size decreases, which certainly is reasonable. As the segment length is decreased at constant global concentration, the more compact structure in parts of space opens up larger pores in other regions. These larger pores produce the major effect for the frictional properties, which is consistent with the increase in $\xi_\perp'$ shown in eq. 16. The same argument of larger pores in other regions appearing with higher functionality at constant concentration suggests that $\xi_\perp$ must be related to F with a negative exponent. We conclude that eq. 16 is physically reasonable. It would be inappropriate to

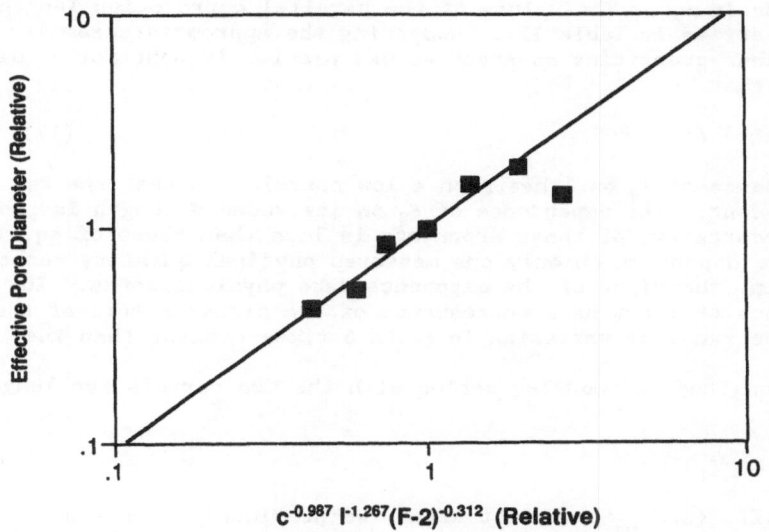

Fig. 4 The effective pore diameter $\xi_\perp$, as a function of its experimentally derived reduced variable $c^{-0.99} \, l^{-1.27} \, (F-2)^{-0.31}$.

suggest that the slow diffusion mode could be a manifestation of the longitudinal modulus via eq. 2 because this leads to illogical exponents for the effective pore size. Specifically, it would demand that the pore size increase with an increase in concentration. Since the inverse square dependence of longitudinal modulus on diffusion coefficient precludes the description of the data in a single correlation length, it follows that correlation of D_1 or M_{os}' with $\xi_\perp$ is not meaningful. An additional independent length must be introduced.

The second correlation length can be introduced from Einstein's recognition that a diffusion process is related to the friction factor $\zeta = k_B T / D$ where

$$\zeta = 6\pi\eta\xi_{\parallel}$$

scales as some length $\xi_{\parallel}$. To this point, discussion has been in terms of the friction factor per unit volume f rather than the friction factor of the scaling lengths. Since eq. 14 has already specified that the friction coefficient per volume is

$$f = 6\pi\eta/\xi_{\perp}^2$$

it follows the volume must be $\xi_{\parallel}\xi_{\perp}^2$. The length $\xi_{\parallel}$ shall be referred to as the effective pore length or the transverse correlation length. It is the quantity extracted from the diffusion coefficient that was initially identified as the correlation length of Gaussian random chains in eq. 1. As with the other correlation length, the parallel designation does not correspond to a permanent orientation within the gel.

Since $\xi_{\parallel}$ depends only on the dynamic light scattering results, it can be calculated in absolute, rather than relative, terms. The value of $\xi_{\parallel}$ for gel 88 A is 2.5Å, a length an order of magnitude less than the connecting segment length. Relative values of $\xi_{\parallel}'$ are reported to assist comparison with trends in $\xi_{\perp}'$. The values of the parallel correlation length of gel 88 A are summarized in Table III. Comparing the appropriate samples while keeping other quantities constant as was previously done for $\xi_{\perp}$ data, indicates that

$$\xi_{\parallel} \propto c^{-2.4} \ (F-2)^{-0.56} \tag{17}$$

The dependence of $\xi_{\parallel}$ on 1 has such a low correlation that the two appear to be independent. The dependence of $\xi_{\parallel}$ on its reduced length is shown in Fig. 5. The uncertainty of these exponents is less than those of eq. 16 for $\xi_{\perp}$ since $\xi_{\parallel}$ is dependent on only one measured physical quantity rather than two. Again, the signs of the exponents make physical sense. The observed independence at 1 may be a consequence of the planar nature of the cross-links. The range of variation in $\xi_{\parallel}$ is 6 times greater than that of $\xi_{\perp}$.

The longitudinal modulus scales with the two correlation lengths as follows

$$M_{os} \propto (\xi_{\perp}^2\xi_{\parallel})^{-1}$$

In Table III, $(G/M_{os,1}')$ is tabulated. We previously have argued that G/M would be a very small number since the Poisson ratio of these gels is close to 0.5. If most of the light scattered by the gels is scattered by stationary local oscillators, the quantity G/M_{os} will indeed be small, but the interesting observation from the normalized data in Table III is that the ratio is varying systematically with gel structure. Higher concentration, longer segment length, and higher functionality all produce gels with more constant volume and Poisson ratio closer to 0.5.

The physical significance of the slow diffusion mode deserves brief consideration. Koňák et al.[26] recently noted their correlation spectra of polymethylmethacrylate gels could be described by a fast component that could be attributed to the modulus of the gel and a slower broad distribution of relaxation times they attribute to diffusion of sol species in the gel. Certainly the present gels contain fragments entrapped in the gel which diffuse. The order of magnitude difference in the time scale of the two components might be more related to the time range assessable in a linear time scale correlation function than an accurate reflection of the actual physical process. The observation of a slow mode appears consistent with the sol diffusion hypothesis.

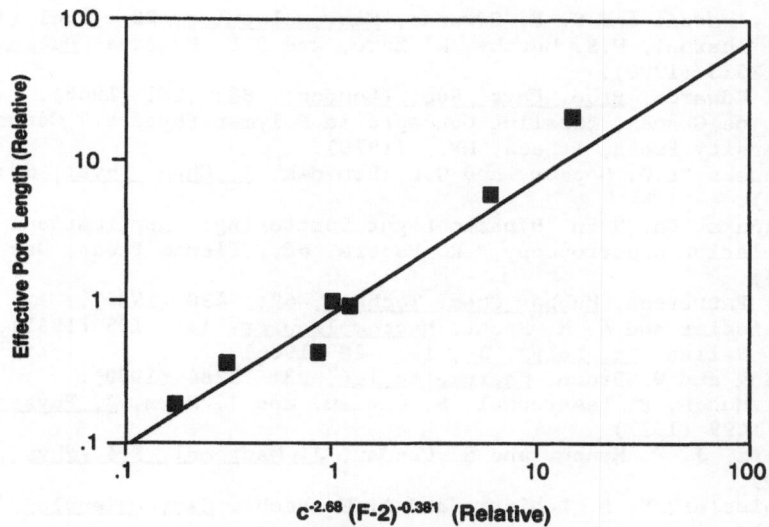

Fig. 5 The effective pore length, $\xi_\parallel$, as a function of its experimentally derived reduced variable $c^{-2.4}$ $(F-2)^{-0.56}$.

CONCLUSIONS

When a gel network loses its Gaussian segment distribution, it can no longer be described in terms of a single length scale. Two distinct length scales are obtained from dynamic and static light scattering data:

$$\xi_\parallel \propto T/\eta D \qquad \text{and}$$

$$\xi_\perp \propto (\eta D I_s/c^2 T)^{1/2} \propto (I_s/c^2 \xi_\parallel)^{1/2}$$

where I_s is the intensity of the light scattered by the gel of concentration c. For gels made of connected flexible chains, these two lengths become equivalent and correspond to the correlation length traditionally measured. For gels of concentration c made with rigid crosslinks of functionality F and stiff interconnecting segments with length l, we find

$$\xi_\parallel \propto c^{-2.4} (F-2)^{-0.56} \qquad \text{and}$$

$$\xi_\perp \propto c^{-0.99} l^{-1.27} (F-2)^{-0.31}$$

Although the signs of the exponents in the dependencies are physically reasonble, we are aware of no physical model that predicts such dependencies. We hope that the report of such observations will entice others to explain why it is so.

REFERENCES

1. P. J. Flory, Ch. 11, "Principle of Polymer Chemistry," Cornell University Press, Ithaca, NY. (1953).
2. L. R. G. Treloar, "The Physics of Rubber Elasticity," Clarendon Press, Oxford. (1975).

3. S. M. Aharoni and S. F. Edwards, <u>Macromolecules</u>, 22: 3361 (1989).
4. S. M. Aharoni, N.S. Murthy, K. Zero, and S.F. Edwards, <u>Macromolecules</u>, 23: 2533 (1990).
5. S. F. Edwards, <u>Proc. Phys. Soc. (London)</u>, 88: 265 (1966).
6. P.-G. de Gennes, "Scaling Concepts in Polymer Physics," Cornell University Press, Ithaca, NY. (1979).
7. T. Tanaka, L.O. Hocker, and G.B. Benedek, <u>J. Chem. Phys.</u>, 59: 5151 (1973).
8. T. Tanaka, Ch. 9 in "Dynamic Light Scattering: Applications of Photon Correlation Spectroscopy," R. Pecora, ed., Plenam Press, New York, (1985).
9. G. D. Patterson, <u>Rubber Chem. Technol</u>, 62: 498 (1989).
10. E. Geissler and A. M. Hecht, <u>Macromolecules</u>, 14: 185 (1981).
11. D. B. Sellen, <u>Br. Polym. J.</u>, 18: 28 (1986).
12. L. Fang and W. Brown, <u>Macromolecules</u>, 23: 3284 (1990).
13. J. P. Munch, P. Lemarechal, S. Candau, and J. Herz, <u>J. Phys. (Paris)</u>, 38: 1499 (1977).
14. J. Herz, J. P. Munch, and S. Candau, <u>J. Macromol. Sci.-Phys.</u>, B18: 267 (1980).
15. E. Geissler, H. B. Bohidar, and A. M. Hecht, <u>Macromolecules</u>, 18: 949 (1985).
16. N. S. Davidson, R. W. Richards, and A. Maconnachie, <u>Macromolecules</u>, 19: 434 (1986).
17. D. W. Schaefer, J. F. Joanny, and P. Pincus, <u>Macromolecules</u>, 13: 1280 (1980).
18. L. Fang, W. Brown, and C. Koňák, <u>Polymer</u>, 31: 1960 (1990).
19. A. M. Hecht and E. Geissler, <u>J. Phys. (Paris)</u>, 39: 631 (1978).
20. E. Geissler and A. M. Hecht, <u>Macromolecules</u>, 14: 185 (1981).
21. B. J. Berne and R. Pecora, "Dynamic Light Scattering," John Wiley & Sons, New York. (1976).
22. E. Jakeman, <u>J. Phys. A. Gen. Phys.</u>, 3: 201 (1970).
23. P. R. Bevington, Ch. 10, "Data Reduction and Error Analysis for the Physical Sciences," McGraw-Hill, New York. (1969).
24. D. B. Sellen, <u>J. Polym. Sci. B. Polym. Phys.</u>, 25: 699 (1987).
25. G. D. J. Phillies, <u>J. Chem. Phys.</u>, 60: 976 (1974).
26. Č. Koňák, J. Jakeš, W. Brown, and L. Fang, <u>Polymer</u>, 32: 1077 (1991).

INFLUENCE OF CURE SYSTEMS ON DIELECTRIC AND VISCOELASTIC RELAXATIONS IN CROSSLINKED CHLOROBUTYL RUBBER

Rodger N. Capps and Christopher S. Coughlin

Naval Research Laboratory
Underwater Sound Reference Detachment
P.O. Box 568337
Orlando, FL 32856

and

Linda L. Beumel

TRI/TESSCO
Orlando, FL 32806

INTRODUCTION

Dielectric spectroscopy has been used as a probe of molecular behavior in organic molecules for a number of years.[1],[2] In conjunction with dynamic mechanical spectroscopy, it has been used to correlate the dynamic mechanical and dielectric relaxation behavior of a number of polymer systems, as well as providing insight into the molecular structural factors responsible for multiple relaxation mechanisms in polymers.

In many instances where simultaneous measurements of dynamic mechanical and dielectric behavior were performed,[3-5] the polymers involved were comparatively simple systems. Although synthetic rubbers have been extensively characterized by dynamic mechanical measurements, fewer dielectric studies have been performed on these systems.[6-9] In a recent work[6], we showed that the dielectric behavior of filled chlorobutyl and butyl rubber was significantly influenced by carbon black particle size, type, and loading. Results obtained on unfilled systems indicated that the principal dielectric relaxation was a cooperative process involving molecular motion of chain segments coupled with thermally activated rotation of polar crosslinking molecules. It was also shown that the dielectric behavior followed the Kolrausch-Williams-Watts (KWW) "stretched exponential" function, which has recently received great interest as a sort of "universal relaxation" function, with applicability to a large number of physical systems.[10-13] In the present work, we have extended our study to compare the dielectric and viscoelastic behavior of chlorobutyl rubber (CIIR) cured only with zinc oxide, and with increasingly complex cure systems, as well as varying plasticizer loadings. For comparison purposes, we have also included the structurally related butyl rubber (IIR), which does not possess molecular dipoles due to chlorine atoms. The similarity between the observed

Synthesis, Characterization, and Theory of Polymeric Networks and Gels
Edited by S.M. Aharoni, Plenum Press, New York, 1992

viscoelastic behavior and dielectric behavior is discussed in terms of molecular motions.

EXPERIMENTAL

The polymer systems studied are summarized in Table 1. Compounded rubber samples were obtained from the Burke Rubber Company of San Jose, CA. Proper cure conditions were determined with a Monsanto oscillating disc rheometer as per ASTM D2084.

Experimental samples were compression molded at 155 degrees C. Dynamic mechanical and dielectric measurements were performed as previously described.[8] Tensile measurements were performed as per ASTM D412.

Table 1. Formulations for IIR and CIIR Rubbers

System	IIR	CIIR
Component (Parts per Hundred of Rubber (P.H.R.))		
Exxon Butyl 268	100.0	
Chlorobutyl HT1066		100.0
Zinc Oxide	5.0	5.0
Schenectady SP1055[*]	12.0	0-8.0
MBTS[**]		0-2.0
Stearic Acid	1.0	1.0
Diphenyl guanidine		0-0.5
Sunpar 120[***]	0-10.0	0-10.0

[*]   Brominated methylol phenol resin
[**] 2, 2'-Benzothiazyl disulfide
[***] Paraffinic oil

RESULTS

Dynamic Mechanical Behavior.

Fig. 1 shows the isochronal, temperature-dependent storage Young's modulus and loss tangent, obtained with a Polymer Laboratories Dynamic Mechanical Thermal Analyzer (DMTA), for three different cure systems in the chlorobutyl rubber. One primary transition is clearly evident in the loss tangent, with a secondary shoulder that has been classified by others[14] in the analogous bromobutyl as a β transition due to the rotation of pendant methyl groups. Similar behavior was observed for the butyl rubber. It can be seen that the increased crosslinking in the chlorobutyl rubber causes an increase in the storage modulus, a slight decrease in the peak magnitude of tan δ, and a shift upward in the temperature at which the maximum of tan δ occurs.

These trends are mirrored in the extended master frequency curves constructed by time-temperature superposition using measurements from the transfer function technique.[8] Fig. 2 shows master frequency curves for two different cure systems. All of the materials tested behaved as thermorheologically simple materials, and could be shifted equally well by either an Arrhenius shift function, or standard WLF form. Note that, although the tan δ scale is a linear one, the apparent superposition is good. The shoulder seen in the DMTA scans appears to manifest itself as a shoulder on the high frequency side of the tan δ curve in the master frequency curves, at approximately 100 kHz. This is interpreted as a

270

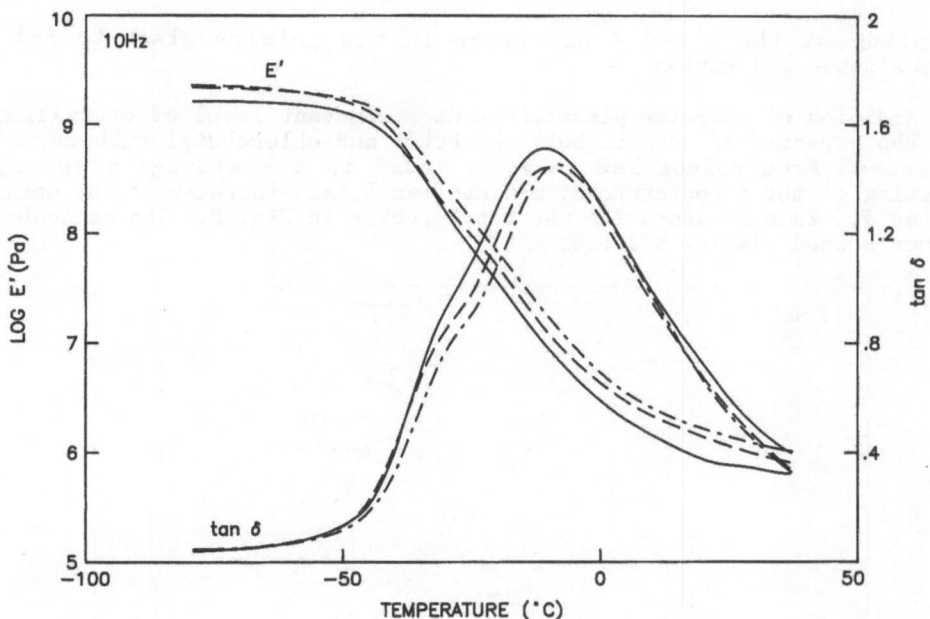

Fig. 1. Plot of storage Young's modulus and loss tangent for chlorobutyl
rubber as a function of temperature at 10 Hz: solid lines,
5 phr of ZnO; dashed lines, 5 phr ZnO and 6 phr SP1055; double
dashed lines, 5 phr ZnO, 6 phr SP1055, 2 phr MBTS, 0.5 phr DPG.

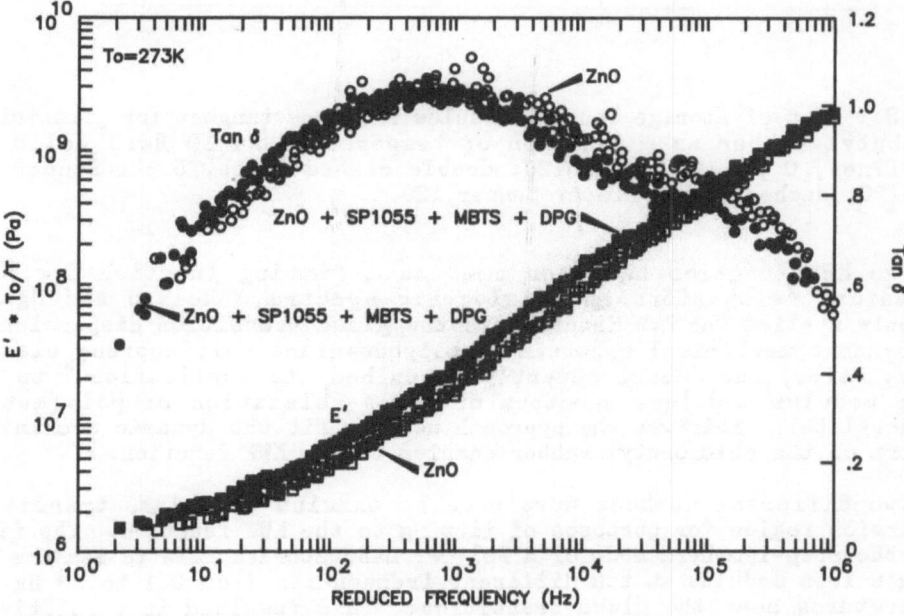

Fig. 2. Plot of storage Young's modulus and loss tangent as a function
of reduced frequency for chlorobutyl rubber with two different
cure systems at a reference temperature of 273.15 K: clear
squares and circles, 5 phr of ZnO; solid squares and circles,
5 phr ZnO, 6 phr SP1055, 2 phr MBTS, 0.5 phr DPG.

coupling of the α and β processes in the primary glass-to-rubber viscoelastic relaxation.

Addition of nonpolar plasticizer at a constant level of crosslinking had the expected effects in both the butyl and chlorobutyl rubbers. The increased free volume led to a decrease in the storage modulus, a lowering of the temperature of maximum tan δ, and increase in the maximum of tan δ. This is shown for the butyl rubber in Fig. 3. The chlorobutyl rubber showed similar effects.

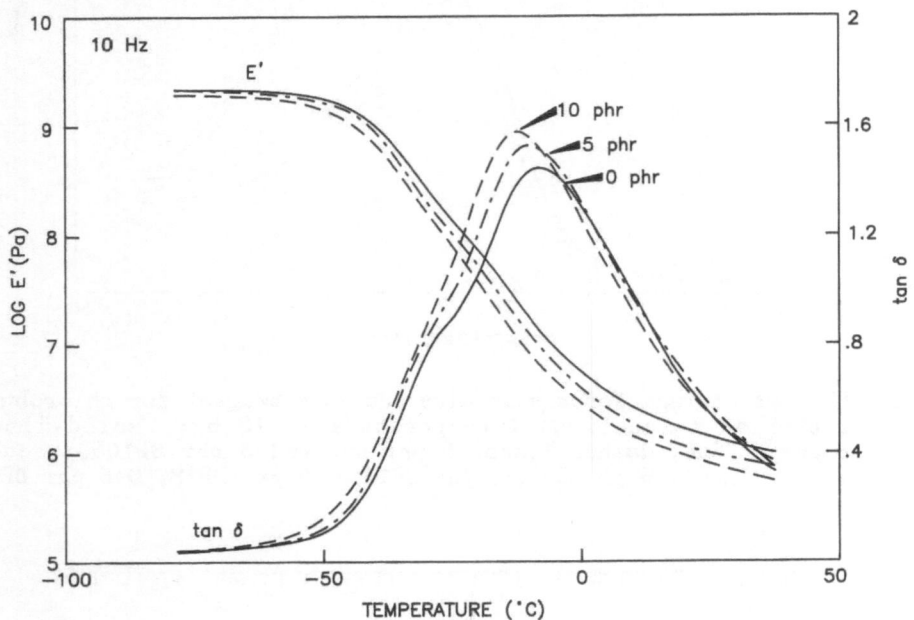

Fig. 3. Plot of storage Young's modulus and loss tangent for plasticized butyl rubber as a function of temperature at 10 Hz: solid lines, 0 phr of Sunpar 120; double dashed lines, 5 phr Sunpar 120; dashed lines, 10 phr Sunpar 120.

The KWW function has been used as a fitting function for both mechanical relaxations and dielectric spectra. Roland and Ngai[12] recently applied the KWW function to the glass transition dispersion of the dynamic mechanical spectrum of polybutadiene-polyisoprene blends. Muzea, Perez, and Johari recently described its application[13] to the shear modulus and loss spectrum of the β-relaxation of poly(methyl methacrylate). This was the approach used to fit the dynamic mechanical spectra of the chlorobutyl rubber samples to the KWW function.

Two different methods were used to examine the glass transition dispersion region for purposes of fitting to the KWW function. The first used the step-isotherm mode of a Polymer Laboratories DMTA to measure the Young's loss modulus at ten different frequencies from 0.1 to 30 Hz, at temperatures near the glass transition. This resulted in a relatively narrow isothermal frequency spectrum, and considerable scatter in KWW fits. The second approach used the transfer function technique and time temperature superposition to construct extended master frequency curves at a reference temperature of 243.15 K, so that the dispersion curve in the loss modulus covered approximately the same frequency range as the dielectric loss. These were then considered to represent isothermal frequency curves at the reference temperature. An example of such a curve is shown in Fig. 4. A third order polynomial was used to find the maximum in the loss modulus and frequency of maximum loss before fitting these to the KWW function.

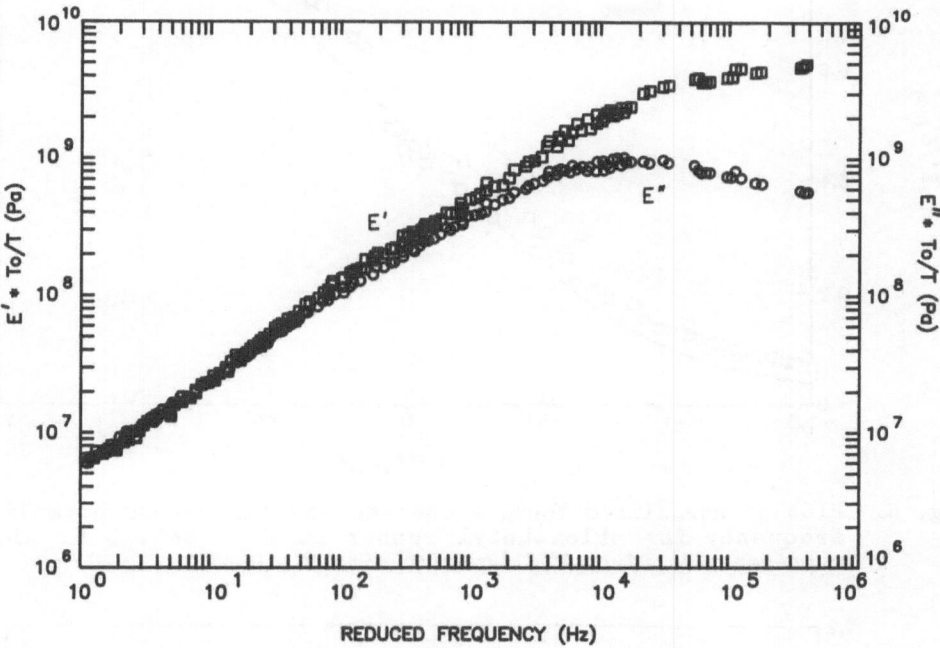

Fig. 4.   Reduced frequency curve of Young's storage and loss moduli for chlorobutyl rubber cured with 5 phr ZnO, 6 phr SP1055, MBTS, and DPG. T_o = 243.15 K.

Figs. 5 and 6 show the normalized dynamic mechanical spectra for one material, and the resultant fit to the KWW function using an exponent of 0.475. This is the chlorobutyl rubber with the most complex cure system, using ZnO, SP1055, MBTS, and DPG. In contrast, chlorobutyl rubbers cured with ZnO alone, and ZnO combined with SP1055, were fitted to the KWW function with an exponential factor of 0.375. This indicates a broader distribution of relaxation times in the dynamic mechanical behavior of the rubbers with the simpler cure systems.

Dielectric Behavior

The dielectric behavior of the chlorobutyl rubber was significantly affected by the crosslinking system used. Relative permitivitties were low for the systems, on the order of 2.7, and showed little frequency dependence. Fig. 7 shows the dielectric loss for four different cure systems, plotted as a function of frequency at 295.15 degrees K. In all cases, a secondary maximum is seen at approximately 22-25 kHz. The exact nature of this peak is not clear. It was also observed in the butyl rubbers, but not in other polymers with higher losses, such as peroxide cured acrylonitrile rubber, and sulfur cured natural rubber. It did not show any frequency shift as a function of temperature, indicating that it is probably an instrumental artifact, and not a separate transition.

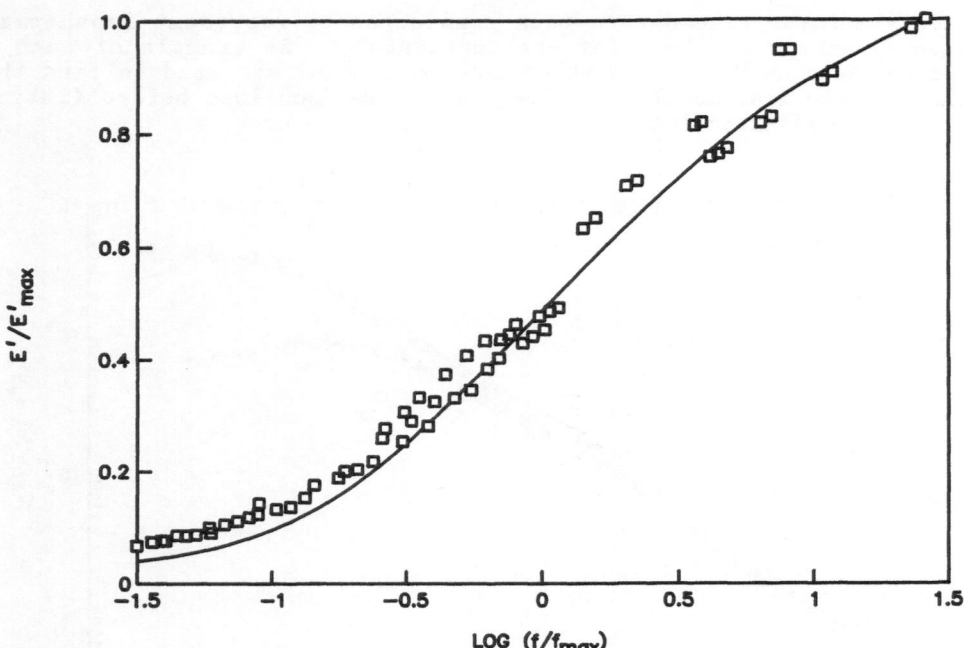

Fig. 5. Plot of normalized Young's storage modulus versus normalized
frequency for chlorobutyl rubber at T_o = 243.15 K. Line
represents fit from KWW function with β = 0.475.

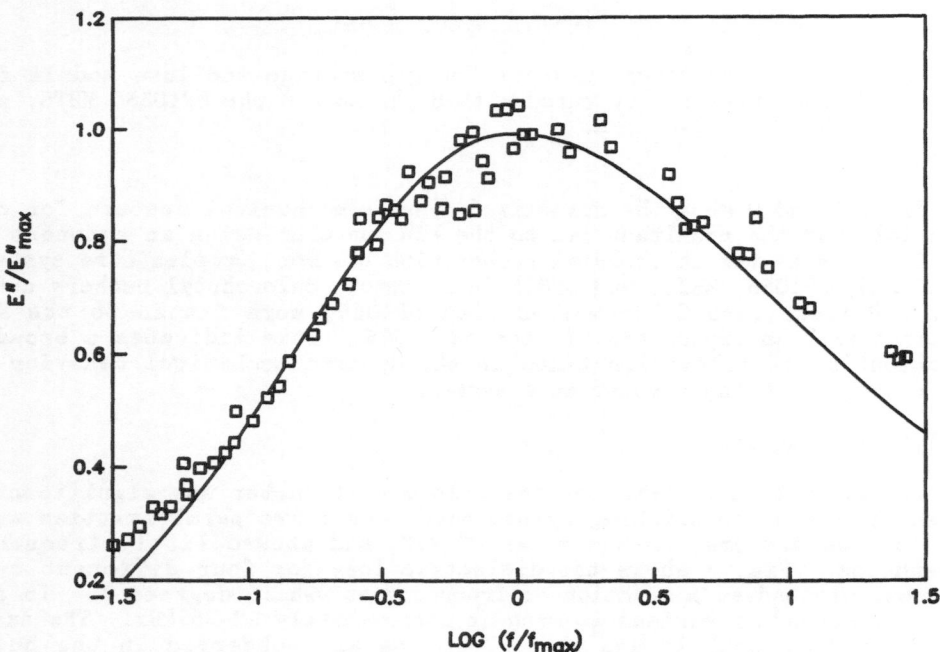

Fig. 6. Plot of normalized Young's loss modulus versus normalized
frequency for chlorobutyl rubber at T_o = 243.15 K. Line
represents fit from KWW function with β = 0.475.

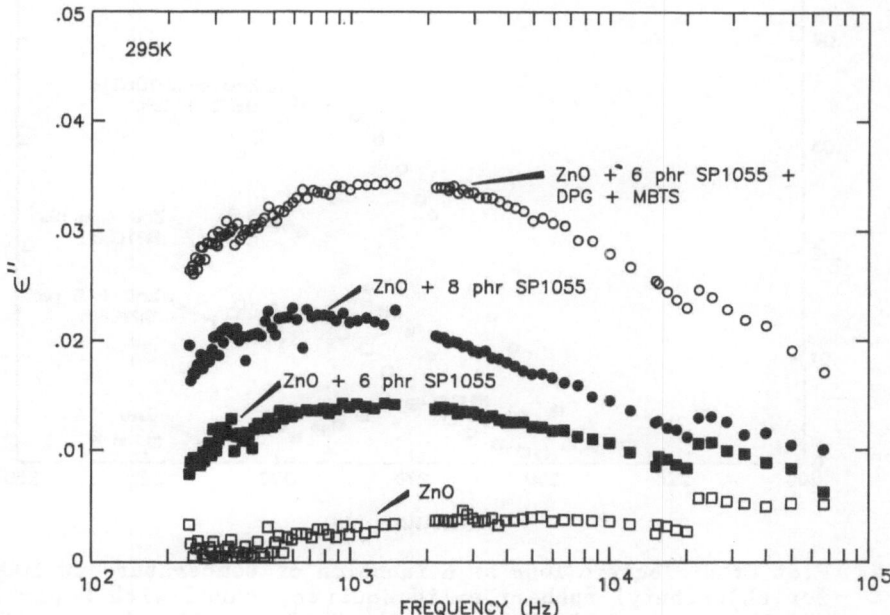

Fig. 7. Plot of dielectric loss as a function of frequency at 295.15 K for chlorobutyl rubber: clear squares, cured with 5 phr ZnO; solid squares, 5 phr ZnO and 6 phr SP1055; solid circles, 5 phr ZnO and 8 phr SP1055; clear circles, 5 phr ZnO, 6 phr SP1055, 2 phr MBTS, 0.5 phr DPG.

The effects of different cure systems upon dielectric relaxation behavior can be more clearly seen in Fig. 8, where the dielectric loss is plotted as a function of temperature at a reference frequency of 1000 Hz. The systems that use the brominated methylol phenol resin, SP1055, show relatively strong maxima in the range of room temperature, although some weaker frequency dependent dispersion is still present at lower temperatures. This is consistent with the primary mechanism of charge conduction in these materials being due to polar materials used in the crosslinking system.

Plots of log f_{max} versus 1/T were linear for the systems using brominated methylol phenol resin cures, indicative of a thermally activated process. An example of this is shown in Fig. 9, which depicts relaxation maps for a butyl and two different chlorobutyl rubber compounds. The range of log f_{max} versus 1/T values that could be plotted was limited by the fact that the peak of the dielectric loss tended to drop below the lowest frequency at which reliable measurements could be achieved as the temperature was lowered. It was found that the dielectric spectra taken at different temperatures could be superimposed reasonably well in a master curve of $\epsilon''/\epsilon''_{max}$ versus log (f/f_{max}), when the portion of the transition considered to be alpha in character was considered (Fig. 10).

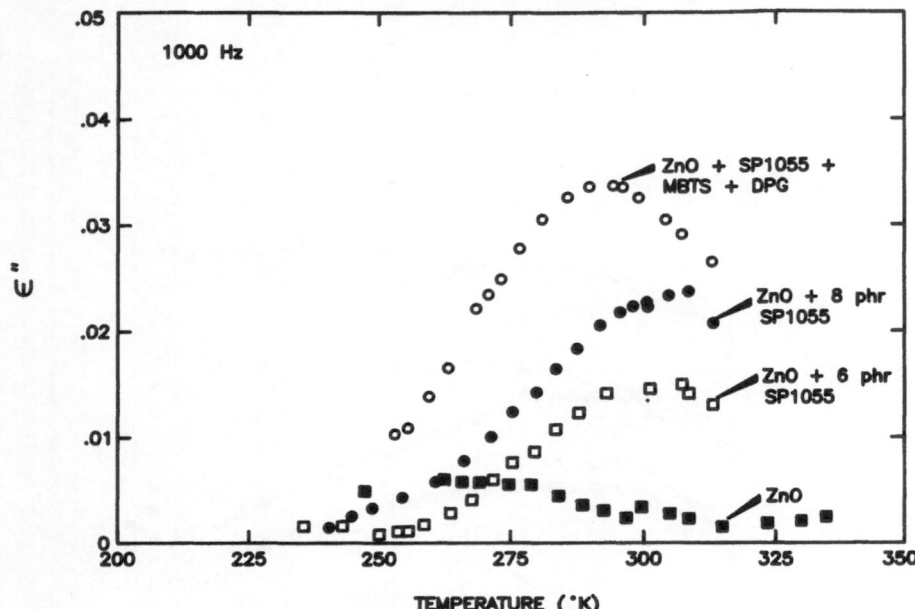

Fig. 8. Plot of dielectric loss as a function of temperature at 1000 Hz for chlorobutyl rubber: solid squares, cured with 5 phr ZnO; clear squares, 5 phr ZnO and 6 phr SP1055; solid circles, 5 phr ZnO and 8 phr SP1055; clear circles, 5 phr ZnO, 6 phr SP1055, 2 phr MBTS, 0.5 phr DPG.

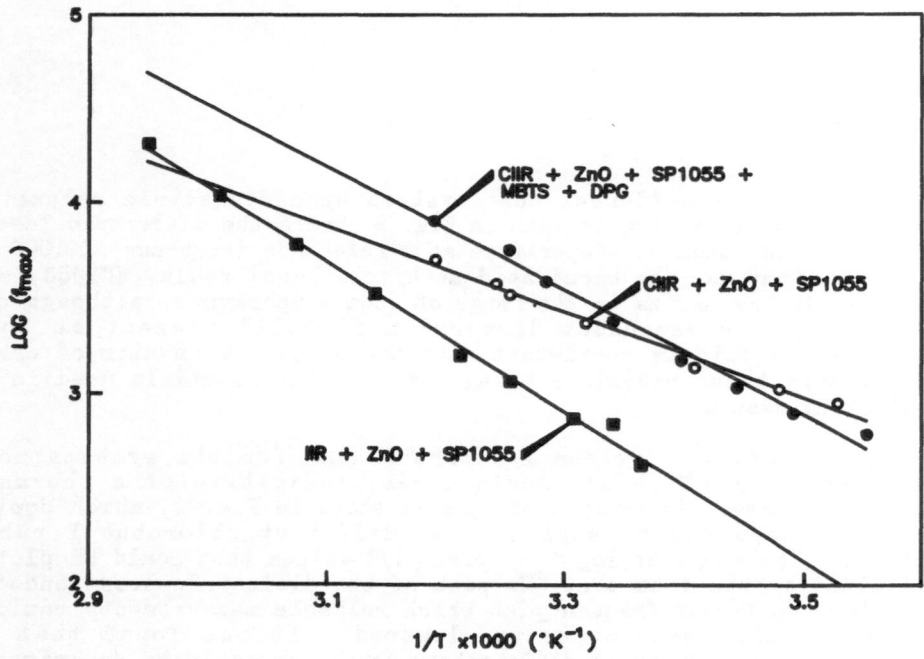

Fig. 9. Relaxation maps for butyl rubber and chlorobutyl rubber with two different cure systems.

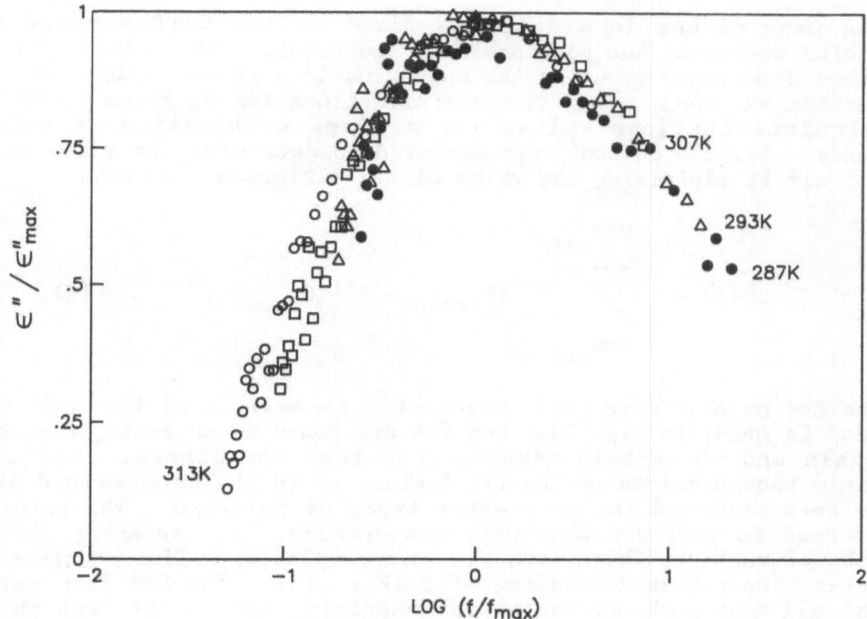

Fig. 10. Master curve of normalized dielectric loss versus f/f_{max} for chlorobutyl rubber cured with 5 phr ZnO and 6 phr SP1055 resin: open circle, 313K; open square, 307K; open triangle, 293K; and solid circle, 287K.

Addition of plasticizer showed an effect upon the dielectric relaxation, in both the butyl and chlorobutyl systems, that was analogous to the dynamic mechanical relaxation when plotted as a function of temperature at a fixed frequency. The maximum in the dielectric loss increased, and showed slight temperature shifts, as shown in Fig. 11.

The dielectric behavior of the systems could be fitted reasonably well using the KWW function. This was done using both the tables of Moynihan, Boesch, and Laberge[15], and a slight modification of the method of Weiss, Bendler, and Dishon (WBD)[16]. These authors define the dielectric loss function as

$$\epsilon'' = A \, z \, \mathcal{Q}_\beta(z) \tag{1}$$

where $z = \omega\tau$, β is the exponential factor (called a_β by WBD), τ is the relaxation time in the KWW function $\phi(t) = \exp -(t/\tau)^\beta$, $\omega = 2\pi f$, with

$$\mathcal{Q}_\beta(z) = \left(\frac{1}{\Pi}\right) \int_0^\infty e^{-u^\beta} \cos(zu) \, du, \tag{2}$$

and

$$A = \epsilon_0 - \epsilon_\infty \tag{3}$$

The peak of the loss data was first fitted with a third order polynomial to locate the peak position and height. The values of τ and A were then determined by using the approximations found in Ref. 16. This information was used, along with approximations for Q_β given in Ref. 17, to calculate the loss values for a given combination of beta and frequency. Brent's method[18] was employed to determine the value of beta iteratively by minimizing the value of the difference function

$$D(\beta) = \sum_{\omega=\omega_{min}}^{\omega=\omega_{max}} \left(\epsilon''(\omega)_{calc} - \epsilon''(\omega)_{data} \right)^2 \qquad (4).$$

The two methods gave good agreement. An example of the type of fit obtained is shown in Fig. 12. The fit was found to be better for certain materials and at certain temperatures than for others. All of the materials showed values of the fit factor, β, in the neighborhood of 0.4, as has been observed for many other types of polymers. The value of β also tended to vary somewhat with temperature. In comparing different chlorobutyl rubbers, those with the more complex crosslinking systems had a greater temperature dependence of β (Fig. 13). The KWW fits were not done at all temperatures for these materials, due to the fact that the loss tended to decrease in magnitude and shift to lower frequencies as the temperature was lowered. The entire transition region was therefore not covered at lower temperatures. The low temperature dielectric spectra of the zinc oxide cured rubber could be fitted to the KWW function, with a β value of 0.375.

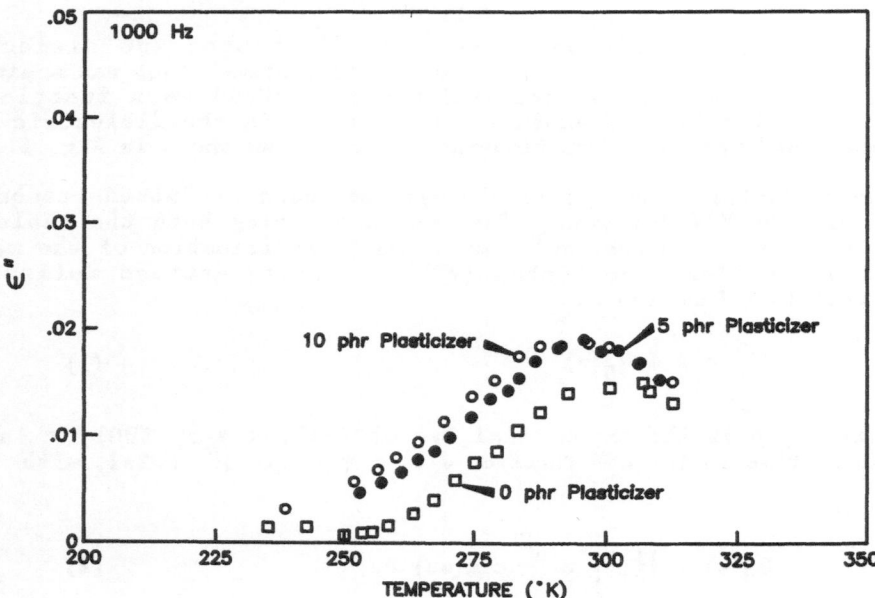

Fig. 11. Plot of dielectric loss as a function of temperature for chlorobutyl rubber with varying plasticizer loadings.

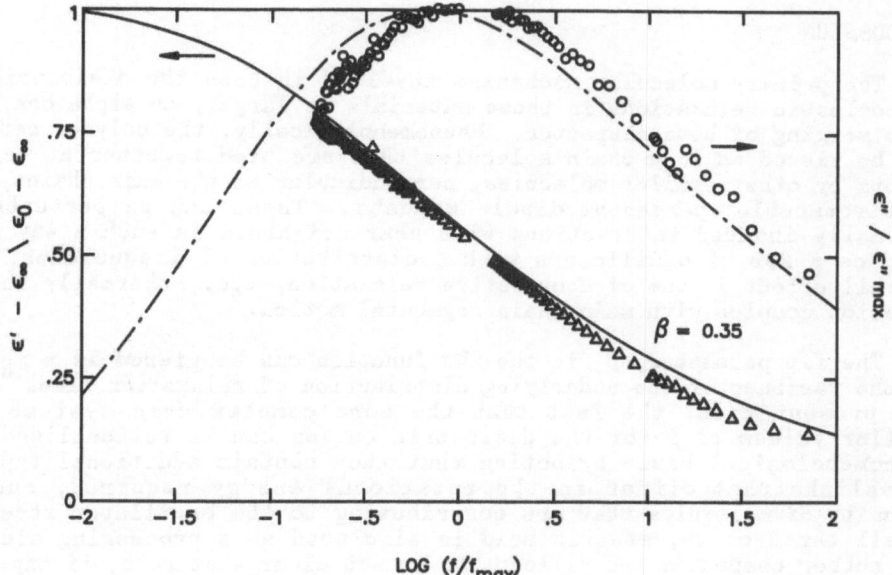

Fig. 12. Normalized dielectric spectrum for chlorobutyl rubber at
294.15 K. System cured with 5 phr ZnO, 6 phr SP1055, 2 phr
MBTS, 0.5 phr DPG. Lines represent splines through points
calculated from KWW function with $\beta = 0.35$.

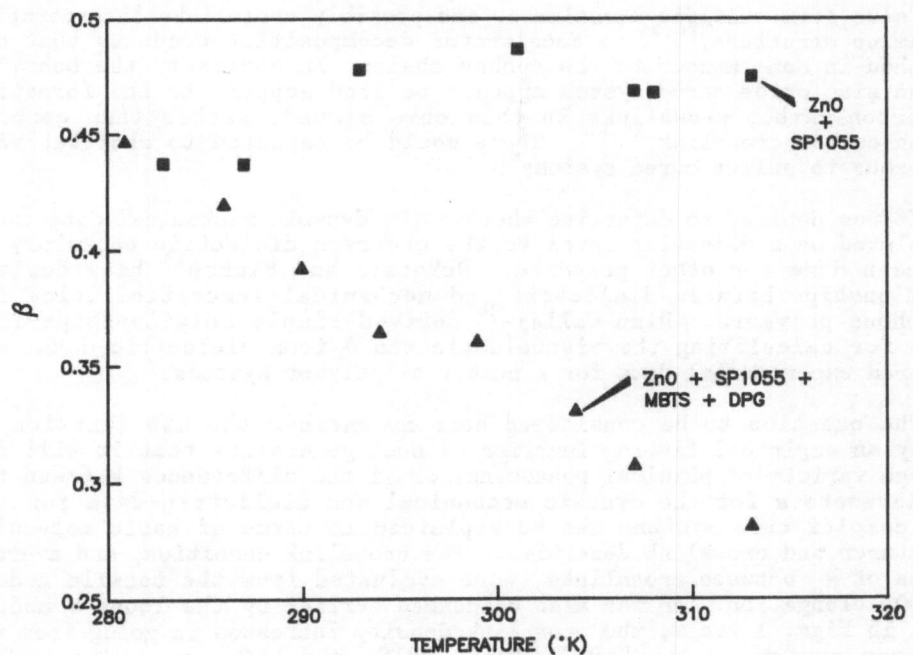

Fig. 13. Effect of temperature on value of exponential factor in KWW
function for chlorobutyl rubber with two different cure
systems.

279

DISCUSSION

The primary molecular mechanism involved in both the dielectric and viscoelastic relaxations in these materials is largely an alpha one, with some merging of beta character. Phenomenologically, the polymer networks can be viewed as long-chain molecules that are tied together at certain points by other smaller molecules, perpendicular to the main chains, that have rotatable, permanent dipole moments. These can be perturbed by thermally induced interactions with near neighbors in such a way as to produce a set of oscillators with a distribution of frequencies. The overall effect is one of cooperative relaxation, i.e., thermally induced rotation coupled with main chain segmental motion.

The fit parameter, β, in the KWW function can be viewed as a measure of the variance of the underlying distribution of relaxation times[19] In the present case, the fact that the more complex cure systems have smaller values of β for the dielectric curves can be rationalized on a phenomenological basis by noting that they contain additional types of crosslinks that differ in the rotational energy required, and the polarity of molecules that are contributing to the oscillator strength. In all three cases, stearic acid is also used as a processing aid when the rubber compounds are milled. It is not clear what role, if any, this plays in the dielectric behavior. Some small residual dipolar activity will also be present from butyl zimate and calcium stearate that are used as stabilizers in the butyl and chlorobutyl gums, respectively. The brominated methylol phenol resin-zinc oxide system will form carbon-carbon bonds, and bonds which incorporate the resin between polymer chains. The systems incorporating MBTS and DPG use these as cure accelerators. It is not clear from published literature whether they will also form sulfidic crosslinks, and possibly crosslinks incorporating the amine structure,[20,21] or accelerator decomposition products that are attached in some manner to the rubber chains. In contrast, the behavior of the zinc oxide-cured system appears to lend support to the formation of carbon-carbon crosslinks in this cure system, rather than carbon-oxygen-carbon crosslinks[22,23] These would be expected to show behavior analogous to sulfur cured systems[8].

It was desired to determine whether the dynamic mechanical data could be related on a molecular level to the observed dielectric behavior, as has been done for other polymers. DeMarzio and Bishop[24] have derived relationships between dielectric and mechanical susceptibilities for amorphous polymers. Diaz-Calleja[25] derived simple relationships from these for calculating the viscoelastic tan δ from dielectric data, and compared experimental data for a number of polymer systems.

The question to be considered here is whether the KWW function is simply an empirical fitting function of such generality that it will fit a large variety of physical phenomena, or if the differences between the fit parameters for the dynamic mechanical and dielectric data for the more complex cure systems can be explained in terms of basic molecular structures and crosslink densities. The crosslink densities, and average values of M_c between crosslinks, were evaluated from the tensile moduli at 100% elongation. As was also evidenced earlier by the Young's moduli shown in Figs. 1 and 2, the crosslink density increases in going from the ZnO cure system to the ZnO, SP1055, MBTS, and DPG cure. The number average molecular weight of segments between crosslinks decreases from approximately 22,00 to 14,000.

The dielectric relaxations are clearly cooperative ones. The dynamic mechanical relaxations are also cooperative, but appear to be influenced more by segmental motions of the main chains. The fit parameter for the dielectric and dynamic mechanical data are the same for the ZnO cured material, indicating a similar distribution of relaxation times and involvement of similar molecular motions for both processes. The base polymer itself is a random copolymer of polyisobutylene and trans-isoprene, with approximately 2 mole % unsaturation. In the case of the chlorobutyl, it also contains 1.1-1.3 weight % chlorine, existing in the form of approximately 20% tertiary allylic chlorides, and a mixture of isomeric secondary allylic chlorides in approximately equal concentrations[20]. These will differ in their reactivity, with the tertiary allylic chlorides being most reactive. These are the probable crosslinking sites for the ZnO cure. This system will have the most random distribution of crosslinking sites and largest average molecular weight of polymer between crosslink sites.

Increasing the complexity of the cure system increases both the crosslink density and the types of crosslinks formed. The average molecular weight of polymer between crosslinks will decrease, and the relaxation times for the polymer segments will become more uniform. Thus, the exponent in the KWW function will increase, indicating a more uniform distribution of relaxation times as the rubber becomes more highly crosslinked.

If this is a correct phenomenological explanation of the difference in the KWW fits to the dynamic mechanical behavior of chlorobutyl crosslinked with different curatives, then it is still not clear why the value of the exponent for the system cured with ZnO and SP1055 does not differ from that of the ZnO cured system. Part of the reason for this may be a temperature dependence of the KWW distribution for the rubbers cured with the three different cure systems. This can not be definitively established by the method used here, since shifting the master curves to another reference temperature will simply transpose them along the frequency axis, and have no effect on the KWW fit. All three cure systems were shifted to the same reference temperature, but the temperature dependence of their dynamic mechanical relaxations is different, as shown earlier in the Young's modulus and loss tangent curves of Figs. 1 and 2.

SUMMARY

The temperature and frequency-dependent Young's modulus and loss tangent have been examined as functions of cure systems and plasticizer loading in chemically crosslinked butyl and chlorobutyl rubbers. The crosslinked materials exhibited thermorheologically simple behavior. Dielectric permittivity and loss were measured as functions of frequency and temperature. The viscoelastic and dielectric behaviors were significantly influenced by the type of crosslinking system and by plasticizer loading. The Kolrausch-Williams-Watts "stretched exponential" function was found to be a reasonable fit to both the dielectric and dynamic mechanical spectra. The effects of molecular motions and cooperative behavior upon the two types of relaxations have been discussed.

ACKNOWLEDGEMENTS

This work was supported by the Office of Naval Research.

REFERENCES

1. P. Hedvig, <u>Dielectric Spectroscopy of Polymers</u>, Ch. 5 and references listed therein, Halsted Press, New York (1977).

2. N. G. McCrum, B. E. Read, and G. Williams, _Anelastic and Dielectric Effects in Polymeric Solids_, Ch. 10 and references listed therein, John Wiley & Sons, New York (1967).

3. B. E. Read, Polymer 30, 1439-1445 (1989).

4. P. Colomer-Vilanova, M. Montserrat-Ribas, M. A. Ribes-Greus, J. M. Meseguer-Duenas, J. L. Gomez-Ribelles, and R. Diaz-Calleja, Polym.-Plast. Technol. Eng. 28, 635-647 (1989).

5. E. R. Fitzgerald, Polymer Bull. 3, 129-134 (1980).

6. R. N. Capps and J. Burns, J. Non-Crystalline Solids 131, 877-882 (1991). See also Refs. 7-9 in this paper.

7. S. P. Kabin and G. P. Mikhailov, J. Tech. Phys. (USSR) 26, 493-497 (1956).

8. R. Bakule and A. Havranek, J. Polym. Sci. C53, 347-356 (1975).

9. D. Boese and F. Kremer, Macromolecules 23, 829-835 (1990).

10. _Relaxations in Complex Systems_, K. L. Ngai and G. B. Wright, Eds., Naval Research Laboratory, Washington, D. C. , Oct. 1984.

11. J. Non-Crystalline Solids, Vols. 131-133.

12. C. M. Roland and K. L. Ngai, Macromolecules 24, 5315-5319 (1991).

13. E. Mazeau, J. Perez, and G. P. Johari, Macromolecules 24, 4713-4723 (1991).

14. N. K. Dutta and D. K. Tripathy, Polym. Degr. and Stability 30, 231-256 (1990).

15. C. T. Moynihan, L. P. Boesch, and N. L. Laberge, Phys. Chem. Glasses 14, 122-125 (1973).

16. G. H. Weiss, J. T. Bendler, and M. Dishon, J. Chem. Phys. 83, 1424-1427 (1985).

17. M. Dishon, G. H. Weiss, and J. T. Bendler, J. Res. Nat. Bur. Stand. 90, 27-39 (1985).

18. W. H. Press, B. P. Flannery, S. A. Teukolsky, W. T. Vetterling, _Numerical Recipes: The Art of Scientific Computing_, Cambridge University Press, NY, 1986, pp. 283ff.

19. C. P. Lindsey and G. D. Patterson, J. Chem. Phys. 73, 3348-3357 (1980).

20. R. L. Zapp and P. Hous, "Butyl and Chlorobutyl Rubber", in _Rubber Technology_, 2nd. Edition, Maurice Morton, Ed., Van Nostrand Reinhold, New York, 1973.

21. L. Bateman, C. G. Moore, M. Porter, and B. Savile, "Chemistry of Vulcanization", in _The Chemistry and Physics of Rubber-like Substances_, L. Bateman, Ed., John Wiley & Sons, New York, 1963.

22. S. W. Schmitt, in _Vanderbilt Rubber Handbook_, R. O. Babbit, Ed., Norwalk, CT., 1978, p.144.

23. I. Juntz, R. L. Zapp, and R. J. Pancirov, Rubber Chem. Technol. 57, 813-825 (1984).

24. E. A. DiMarzio and M. Bishop, J. Chem. Phys. 60, 3802-3811 (1974).

25. R. Diaz-Calleja, Polymer 19, 235-236 (1978).

NETWORK FORMATION THEORIES AND THEIR APPLICATION
TO SYSTEMS OF INDUSTRIAL IMPORTANCE

Karel Dušek and Ján Šomvársky

Institute of Macromolecular Chemistry
Czechoslovak Academy of Sciences
162 06 Prague 6, Czechoslovakia

INTRODUCTION

The structure of polymer networks is closely related to the network formation (structure growth) process. A covalent polymer network is a giant macromolecule of dimensions commensurable with the macroscopic dimensions of a given object. Although it can be characterized by a number of average structural parameters, like sol fraction or crosslinking density, its internal structure can be very different[1] varying from that of a random loosely crosslinked network of vulcanized rubber to that of very dense networks of some thermosets or ceramers. Also, micronetworks (microgels) or other microscopic precursors (dendritic structures, starburst polymers, etc.) are formed first and then they grow into a macronetwork with the same or other types of structure growth processes. Vinyl–divinyl copolymerization can serve as an example.[2] Therefore, the understanding of structure growth is necessary for understanding the network structure, and the properties of polymer networks can be correlated with their structure. An important role in undestanding of the network structure via the network formation process is played by the network formation theories.

In the last decade, the network formation theories aimed at more complex network formation processes than are, for example, random polycondensation or random vulcanization of primary chains. The development was often iniciated by needs of industry to understand and be able to control a given technological process or to develop a new one. Sometimes, the needs of industry have stimulated a theoretical research the results of which have had much more general validity.

The following topical problems of complex network formation can serve as examples:

 (a) reactions with complex reaction mechanisms (like in some epoxy or isocyanate systems),

 (b) chainwise network build-up (initiated reactions),

 (c) diffusion-controlled network build-up.

THEORETICAL DEVELOPMENT

Correlations in Structure Growth

The factors determining the growth of network structures are as follows:

Synthesis, Characterization, and Theory of Polymeric Networks and Gels
Edited by S.M. Aharoni, Plenum Press, New York, 1992

(a) *connectivity correlations* given by *short-range "chemical"* effects — rules of making (or breaking) bonds between constituent units determined by reaction paths (mechanism) and their kinetics. The short-range effects and history of the cross-linking process may result in so-caled *long-range connectivity* correlations by which the structure may "remember" the sequential orders of units and bonds in structure how they have developed in time,

(b) *long-range spatial correlations* given either by the possibility of reaction of two functional groups bound to the same molecule (cyclization), or unequal accessibility for the reaction of groups on molecules of different size and symmetry (excluded volume effect), or by diffusion control of the crosslinking reaction.

Whereas terms the short-range chemical and spatial correlations are more or less clear, the term long-range connectivity correlations deserves a comment.

These correlations are to be understood as such that are not related to any interactions in space and yet have an effect on structure (sequential order, degree-of-polymerization distributions) and related properties compared with the structure generated by random combination of building units of a given size. Historically, these correlations were called stochastic[3,4] when it was found that structures generated by statistical theories from units differ from those generated by kinetic theories (see below). This was because special kinetics and mechanisms caused a kind of long-range correlations in the structure generated by a method considering only short-range effects.

The long-range connectivity correlations in the structure can arise from

(1) certain kinetics and mechanisms which, in case of linear polymerizations, yield degree-of-polymerization distributions differing from the most probable or pseudo most probable ones. The term pseudo means that the first terms of the distribution are non-random but the rest is of the most-probable type.[4,5] The kinetics and mechanisms are controlled by short-range correlations (e.g. substitution effect, or initiated reactions) but they result in long-range connectivity correlations.

(2) history of network formation determined by a set of variables of the process which controls the development of sequences in time. Continuous monomers feed regimes or staging of a given reaction can serve as examples. Staging means that the final product is obtained by mixing portions of monomers in several stages. The final structure depends on the parameters of the individual stages and is different from that which results from mixing of monomers in one step even for the same conversion of functional groups resulting in the same numbers of bonds of different types.

An example corresponding to an initiated (rapid initiation) living polymerization of a bifunctional monomer in which the initiator has been added in three portions is shown in Fig. 1.[6] The staging of initiator addition results in a trimodal degree-of-polymerization distribution which depends on the monomer conversions at which the initiator portions have been added.

(3) higher-shell substitution effects really existing in the system (non-equilibrium or equilibrium), if the structure is generated by the statistical method that works with a distribution of units calculated under the assumption of a lower-shell substitution effect. The x-th substitution effect means that the reactivity of a group of a given unit depends on the states of $(x - 1)$-th neighbors; for the first-shell effect, it depends only on the state of this unit.

Network Formation Theories

The presently existing network formation theories can be grouped in three categories:

(1) *Statistical theories* by which the structures are generated from smaller units (monomer units or other structural fragments) by their random combination respecting the type of the really existing bonds between these units. The fragments can be either smaller or

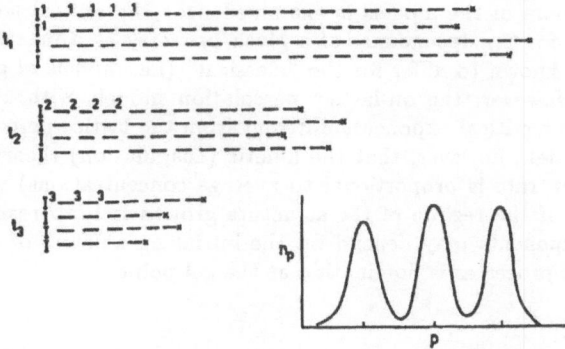

Figure 1. An initiated polymerization of a bifunctional monomer. The initiator was added in three portions at times t_1, t_2, and t_3; the bonds formed in the respective stages are designated 1, 2, 3; n_p is the fraction of p-mers.

larger (superspecies) than monomer units. This generation occurs at any extent of reaction, so that any correlation with the history (a kind of preference of combining certain units, which is translated along the structure) is lost. The method is mean-field, time and space correlation can be taken into account only by some (perturbation) approximations.

(2) *Kinetic (coagulation) theories* in which the existing structure of reacting species is kept intact (not split into units). The structure growth is described by a (infinite) set of differential equations for the time evolution of concentrations of structures. The rate constants depend on the number and reactivities of the functional groups in the reacting species. Physical effects on the growth rates of species can be incorporated. The approach is mean-field, connectivity correlations are taken into account exactly. Space correlations can be dealt with as an approximation, but the rate constant of pair of species can be made dependent on their structure which is not possible within the approximations to the statistical theories.

The kinetic differential equations can also include diffusion terms and deal with diffusion controlled spatial fluctuations. Stochastic differential equations describe diffusion controlled processes in which the initial conditions fluctuate about a mean value. Description of these processes is dealt with in a monograph.[7]

(3) *Simulation in space* is performed in finite systems by computer. The most frequent simulations are based on lattice percolations either random[8] or kinetic (initiated).[9] Off-lattice simulation methods have also been used.[10] The simulations in space are non-mean-field and can take into account connectivity and space correlations. The classical percolation methods suffer from the rigidity of the lattice corresponding to a complete diffusion control of the network forming reaction (only spatially neighboring groups can react and their positions are fixed), or the mobility varies in an uncontrollable way.[10] Recently, an off-lattice percolation simulation of diffusion controlled A_f homopolymerization has been published by Gupta et al.[11] The reaction probability for two groups is a step function of the distance between their positions which are a priori fixed in space.

For simulation of aggregation, there exist various models of diffusion controlled processes (ballistic, diffusion-limited aggregation, etc., models).[12]

The network formation simulations have not yet advanced so far to be able to generate network structures as a result of minimization of free energy of interacting network chains.

Often, the behaviors of the models is examined near the critical treshold (gel point). The critical exponents for the dependence of a given property as a function of the distance from the gel point are known to differ for the "classical" (i.e. models of group (1)) and the percolation models. However, the off-lattice percolation models with a variable diffusion control give values of the critical exponents different from the lattice percolation ones.[10,11,13] It should not be forgotten, however, that the kinetic (coagulation) theory being essentially mean-field (the reaction rate is proportional to average concentrations) yields non-classical critical exponents.[14,15] If the regime of the structure growth is both reaction and diffusion limited, the critical exponents may depend on the initial parameters of the system. They determine which of the processes is dominating at the gel point.

Statistical vs. Kinetic Theories

The network formation theories differ not only in their rigor of treatment of the long-range connectivity correlations but also in the easiness in calculating various structural parameters when applied to real systems. The theories of group (1) are easy; their great advantage for structure-property correlations is the possibility to characterize details of the infinite gel structure (elastically active network chains and pendant chains and their distributions, elastically active crosslinks, lengths of various sequences having property, etc.). Such chacterizatization is made possible by a simple calculation (using recursive relations) of a probability which determines whether a bond has finite or infinite continuation along the bond sequences (cf. extinction probability).

The solution by the kinetic (coagulation) theory is less simple but manageable. Only in the simplest cases, the infinite set of differential equations of chemical kinetics has an analytical solution. Most frequently, the coagulation equations describing the rates of formation of species with the product kernel $K_{ij} \sim ij$ can be transformed into a single differential equation for the generating function of distribution. From this differential equation, differential equations for the time evolution of the moments of the distributions (of degrees of polymerization, molecular weights, composition, etc.) can be derived. The values of the respective moments as a function of time are calculated numerically.[16,17] This approach has also been used for modeling free-radical copolymerizations (cf. e.g. Refs.[18-20] or other complex initiated reactions[21]). In more complex cases, when a single equation for the generating function of the distribution cannot be derived, one can analyze the equations with respect to asymptotic solutions in the given time region or at the singular point (gel point) which is, however, not sufficient for obtaining the necessary structural information in the whole conversion range. In that case, a Monte-Carlo simulation of the process as described by the differential equations for a finite system can be used. For finite systems, many realizations of distributions at a given time are possible and the kinetic differential equations are transformed into the stochastic ones. The principles of the stochastic kinetics were reviewed by McQuarrie.[22] Several methods of solution are possible, and one of them is the above mentioned Monte-Carlo simulation.

The Monte-Carlo simulation of structure growth during an A_3 step polyaddition with a substitution effect was developed by Mikeš and Dušek.[23] The random number generator selects a given pair of molecules for reaction; the probability of finding of a molecule is weighted by its "reaction ability" which is given by the numbers and reactivities of the unreacted functional groups. With increasing size of the system given by the initial number of monomer units, the difference between the largest and the second-largest molecule rapidly increases when a certain conversion is surpassed. This difference diverges when the system size approaches infinity and the region where the difference becomes marked is tranformed into a point which is equal to the gel point. The mass fraction of the largest molecule becomes equal to the gel fraction.

A similar simulation technique was recently used for stochastic simulation of polymerization and coagulation processes by Breuer et al.[24] kinetically controlled growth processes.

Another "stochastic" element can be introduced into the kinetic differential equations by letting the density of reacting group fluctuate in space which gives rise to microscopically different reacting rates. The local changes in concentrations of reactants determined the gradients and diffusion of reactants.[25] This fluctuation-diffusion control may have an important effect on kinetics and critical behavior of the system and one can observe a transition from the mean field to the fluctuation dominated behavior.

We are not going to review the methods of simulation of polymer networks growth in space where the lattice and off-lattice percolation methods are still not the most suitable ones for description of the formation – structure relations. In the next section, the possibility of simulation of the effect of long-range time and space correlations within the framework of the statistical and kinetic theories will be briefly discussed.

Approximations of the effects of long-range correlations

Statistical theories – connectivity correlations. Approximations of long-range *connectivity correlations*: There exist two groups of methods applicable to statistical theories which can approximate these long-range correlations:

(a) making the conditional probability of forming (by their reconstruction) a given bond dependent on the types and states of differently distant neighbors,

(b) approximating the time sequence of bond formation by splitting the conversion interval into a number of smaller increments and distinguishing between bonds formed in the respective conversions increments.

The statistical theories are rigorous for description of equilibrium controlled systems where the time sequence of bonds cannot be distinguished. Within the equilibrium control, there may exist substitution effects which determine how the units are arranged in clusters, i.e. the structure and structure growth can be determined by the state of neighbors in different distances (n-th shell substitution effect). The long-range correlations arising from short range-effects in kinetically controlled systems (substitution effects even first-shell, initiation, etc.), appearing as "infinite-shell" when considered from the viewpoint of the statistical theory (cf. e.g. initiated structure growth,[24] can be approximated by treating the system as if an n-th shell s.e. existed. This means that the probability of formation of a bond between two (monomer) units is conditional depending on the types and states of neighbors up to the n-th generation. Consequently, the overlapping units in the $n-1$ shells neighboring the reacting units must be identical.

Practically, it means that the branched polymers and the network must be generated from larger structural fragments. The distribution of these fragments is generated using the conventional chemical kinetics corresponding, in the linear case to triads, tetrads, etc. An example of generation of a linear copolymer is shown in Fig. 2.

The higher is the functionality of the monomer unit, the more complex are these fragments. It is expected that the introduction of just the second-shell s.e. can represent a good approximation. In fact, a stochastically "impure" fragment method,[26] consisting in random combination of larger fragments (without any overlap), may offer a very reasonable approximation. In a number of kinetically controlled systems with short-range correlations due to the s.e., initiated mechanism, etc., these effects are not translated along the sequences to infinity but are localized in clusters of units. This happens if this translation is interrupted by a random bond, i.e. such the formation of which does not depend on the states of any unit in the system and is given only by the overall conversion of that group.

Then, it is sufficient to generate rigorously (using the kinetic method) the distribution of these clusters (superspecies — a term introduced by Miller[27]) and to recombine them ran-

$$AABABBABA \begin{cases} [A] \\ [B] \end{cases}$$

$$\begin{matrix} \text{(A)A} \\ \text{(A)B} \end{matrix} \quad \text{dyads}$$

$$\begin{matrix} \text{(BA)A} \\ \text{(BA)B} \end{matrix} \quad \text{triads}$$

$$\begin{matrix} \text{(ABA)A} \\ \text{(ABA)B} \end{matrix} \quad \text{tetrads}$$

Figure 2. Build-up of a linear copolymer from polyads using conditional probabilities of addition of A and B units.

domly. The technique consists in cutting the connections between the groups of independent reactivity and labelling the points of cut (1st step). In the 2nd step, clusters are generated from this distribution which bear the labelled point of cut and in the 3rd step the points of cut are coupled into bonds for which operation the techniques of the statistical methods are used[3,4,28] (Fig. 3). Also the approach to vulcanization of primary chain distributions used for the first time by Dobson and Gordon[29] and later extended by Miller and Macosko[30] within the framework of the recursive method belongs to this category. In fact, the steps 1 and 3 (without 2) are inherent to the classical Flory-Stockmayer treatments and to all derived theories, but the first treatment via the formation of superspecies was perhaps the cyclization effect treatment in free radical copolymerization by Dušek and Ilavský.[31]

The other approximation method (b) consists in approximating the right time sequence of bonds. The conversion interval $\langle 0, \alpha \rangle$ is split into smaller intervals and the bonds formed in different intervals are distinguished. Within the framework of the statistical theories it means that the halfbonds issuing from units and to be recombined into full bonds are labelled, with respect to the conversion interval in which they have reacted, and only bonds with the same label are recombined.

The labelling is schematically shown in Fig. 4 for a trifunctional monomer when the conversion is split into three intervals marked 1, 2, and 3. It is seen that the number of

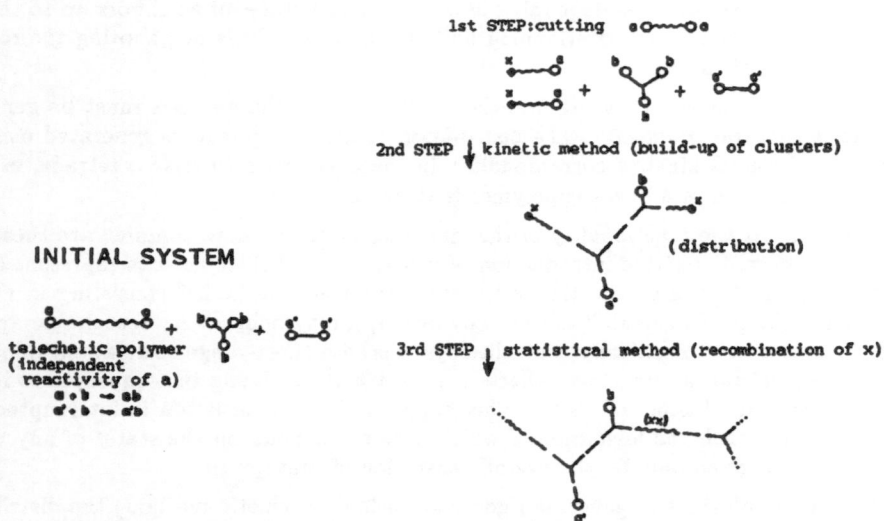

Figure 3. Combination of kinetic and statistical methods Ref.[28]

one-stage

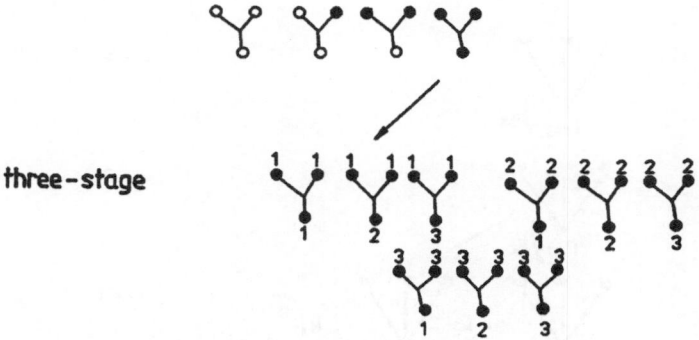

three-stage

Figure 4. Labelling of reacted functional groups with respect
to the stage in which the bonds have been formed.

states of the monomer units increases considerably. The probabilities (fractions) of states are
obtained by solution of the respective differential equations of chemical kinetics.

It is clear that the approximation approaches the correct solution by the kinetic theory
if the conversion intervals are made infinitesimally small.

This procedure was first developed for multistage processes[32] and later on developed by
Sarmoria and Miller[33] and Dotson[34] as an approximative method for long-range connectivity
correlations for stepwise and chainwise network build-up, respectively.

Statistical Theories – Spatial Correlations. For approximation of *spatial correlations*
within the framework of the statistical theories the following methods are used:

(a) distinguishing between inter- and intramolecular bonds, i.e. reacted functional groups
engaged in the respective bonds (spanning-tree approximation),

(b) considering the formation of only the smallest ring,

(c) combination of methods (a) and (b), i.e. calculation of the population of the smallest
ring and using the spanning-tree approximation to the larger ones.

A general method for a branching process with ring formation was suggested by
Kuchanov.[35] The structure is build up from fragments containing cyclic structures and connected with the neighboring clusters only by a single bond. The problem is to define the
structure of such cluster and to calculate its fraction. For real systems much deeper approximations are used.

The method (a) is simple in application because it makes use of simple formalism of the
TBP with first shell substitution effect. It is called spanning tree-approximation[36] because all
existing rings are cut and the half-bonds obtained are transformed into ring closing functional
groups (reacted but not taking part in branching).

Because the reacted functional groups can be engaged either in the inter- or intramolecular bonds, the number of reaction states of monomer units increases as can be seen for the
example of a trifunctional monomer (Fig. 5). The calculation of the probabilities (fractions)
of the states is, however, more complicated. It is necessary to calculate the probability that a
functional group on a unit reacts intramolecularly relative to the probability of an intermolecular reaction. The latter is proportional to the density of unreacted groups in the system.

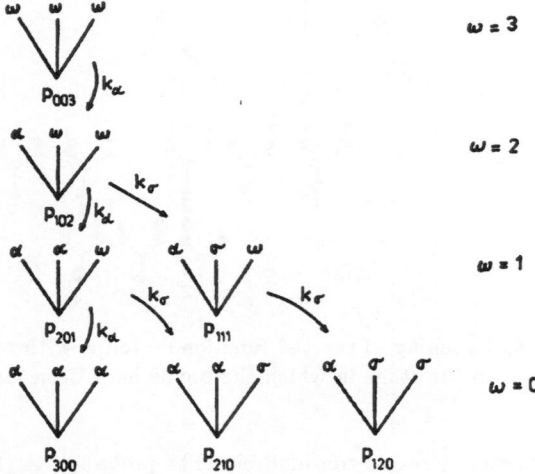

Figure 5. Distribution of units of a trifunctional monomer in different reaction states (ω, α, σ – unreacted, intermolecularly reacted, and intramolecularly reacted functional groups, respectively).

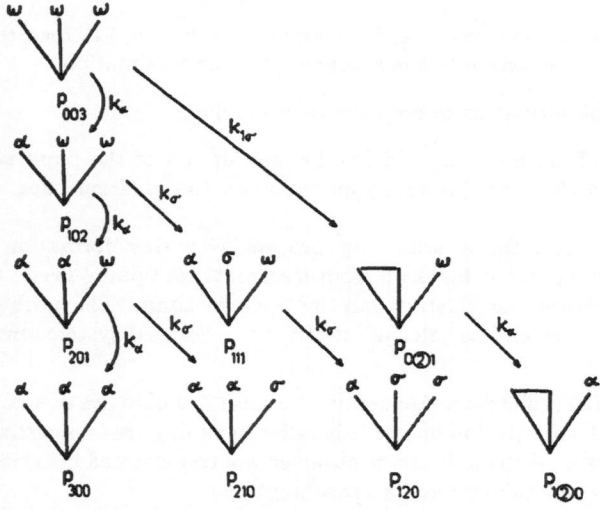

Figure 6. Distribution of Fig. 5 by inclusion units with smallest ring.

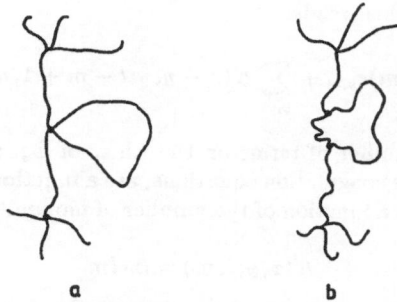

Figure 7. Ring structures in a network.

For calculation of the probability of cyclization, all existing unreacted groups in the molecule can be reaction partners. The cyclization probability is determined by the distance along the connecting paths in an (assumed) tree-like structure. If the conformations of the conecting paths are described by the random-flight model (Gaussian chains), the ring-closure probability is proportional to the −3/2 power of the number of units of connecting sequence. Therefore, the calculation of the cyclization probability involves summations over all possible topological distances of the groups.

The time dependence of the population of reaction states (shown e.g. in Fig. 5) is described by a set of differential equations containing the cyclization terms. In some systems, the population of the smallest ring may represent the largest fraction of ring structure (e.g. over 50%). Then, it is better to calculate the fraction of the smallest rings exactly and to include this ring structure among the building units. The effect of larger cycles can possibly be dealt with using the spanning-tree approximation. The modification of the distribution of building units is seen in Fig. 6. The smallest ring method was developed by Stepto et al. (cf. e.g. Refs.[37,38]) in extensive studies of formation of networks from telechelic polymers, and some large cyclic fragments were considered by Sarmoria et al.[39]

Beyond the gel point, the treatment and interpretation of the effects of ring formation is even more difficult because ring formation in the network always occurs. In fact, the elasticity of the networks can be related to the cycle rank — the number of cuts to be done in order to convert a graph with cycles into a spanning tree. However, some of these rings (the smaller ones) can be elastically inactive (Fig. 7a) because they are not deformed in equilibrium as a result of an external strain.

An approximate calculation of the fraction of these elastically inactive cycles is also possible.[40,41]

In general, the elastic response of a network in which the elastically active network chains have the same end-to-end vectors but are of different length (cf. e.g. the structure of Fig. 7b) is complicated due to different relative extensions of these chains.

Kinetic (Coagulation) Theories – Spatial Correlations. In the majority of cases, the kinetic theories treat irreversible growth processes which can be described by formation of the molecule $A_{x,l}$ composed of x units and bearing l unreacted functional groups

$$A_{x-y,l-m+1} + A_{y,m+1} \longrightarrow A_{x,l} \tag{1}$$

Generally, x and l are vectors because the molecule can be composed of more types of units and can bear functional groups of different reactivity and/or type.

The corresponding differential equation expressing the change of concentration of the species $A_{x,l}$, $c_{x,l}$, with time then reads:

$$\frac{dc_{x,l}}{dt} = -c_{x,l}\sum_{y,m} K(x,y;l,m)c_{y,m} + \sum_{y,m} K(x-y,y;l-m+1,m+1)c_{x-y,l-m+1}c_{y,m+1} \quad (2)$$

If x and l are vectors the number of terms on the r.h.s. of Eq. (2) is larger than two. Rate constants, called kernel of the coagulation equations, are a function of the number of unreacted functional groups and can be a function of the number of monomer units. For the simple mass action law

$$K(x,y;l,m) = k \cdot lm \quad (3)$$

where k is a constant corresponding to the rate constant for reaction of two groups. Since for tree-like structures, l is a linear function of x

$$lm \sim xy \quad (4)$$

and the kernel (3) can be written down as

$$K(x,y) \sim xy \quad (5)$$

which is the classical "product" kernel. It was already found by Stockmayer[42] that Eq. (2) with the kernel (3) or (5) and satisfying Eq. (4) and corresponding to single monomer units with groups of the same and independent reactivities gives before the gel point a solution identical with that obtained by the statistical theory. This equivalence was later proved to be valid also beyond the gel point.[43] However, the equivalence does not hold any more, if the number of terms on the r.h.s. of Eq. (2) is larger than two (e.g. groups of different and dependent reactivity). The produkt kernel (3) corresponds rather to a step reaction in which groups on any species can react with their partners. For initiated reactions, for instance, only monomers can reacted with the activated groups, so that $m = 1$ and $K(x,y;l,1) \sim l$; for linear polymerization initiated with a monofunctional initiator also $l = 1$ and $K(x,1;1,1) =$ constant.

Furthermore, the classical kinetic approach can be extended by inclusion of degradation described by the scheme

$$A_{x,l} \longrightarrow A_{x-y,l-m} + A_{y,m} \quad (6)$$

or, if the functional groups are regenerated

$$A_{x,l} \longrightarrow A_{x-y,l-m+1} + A_{y,m+1} \quad (7)$$

Degradation is a monomolecular reaction and the form of the kernel determines the mechanism (random scission, preferred zipp degradation, bond activation, etc. – cf. e.g.[44,45]).

If the degradation is random, and possibly coupled with crosslinking, statistical theory can be used.[46]

The kinetic theory can succesfully approximate the effects arising from spatial correlations. The important difference, compared to approximations offered by the statistical theories, is that they can make the magnitude of the effect size and structure dependent.

The chemical effect of spatial correlations is cyclization which can be described as

$$A_{x,l} \longrightarrow A_{x,l-2} \quad (8)$$

and its effect by the corresponding differential equation

$$\left(\frac{dc_{x,l}}{dt}\right)_c = -K_c(x,l)c_{x,l} + K_c(x,l+2)c_{x,l+2} \quad (9)$$

The possible forms of the kernel will be discussed below.

292

The excluded volume effects due to thermodynamic exclusion or steric hindrances affect both the structure growth and cyclization rates. Steric hindrances are likely to affect the probability of collisions of functional groups. They will become more important in large clusters where groups at and near the surface have a greater chance to react than those in the cluster interior. This effect can be roughly approximated by an exponent form of the product kernel

$$K(x, y; l, m) \sim x^\nu y^\mu \tag{10}$$

where the exponents are lower than 1. The behavior of structure evolution described by Smoluchowski equations with kernels of the form (10) was widely examined (cf. e.g. Refs.[14,15] and a recent review[47]), in particular with respect of the critical behavior.

If $1 \leq \nu + \mu$, the system will gel. The values of the exponents determine the relative rates of the reactions of large – large, large – small, or small – small clusters.

The physical analogy of the decreasing exponent is the shell or surface effect when only a part of the groups near the surface can take part in the reaction, and is dependent on the fractal dimension of the objects. However, the exponent is expected to vary (decrease) with increasing compactness of the cluster being, for instance, close to 1 for small clusters and comming down up to 2/3 in the limit of very large compact clusters. One of the possible empirical dependences of the exponent s in the kernel

$$K(x, y; l, m) \sim l^{s_1} m^{s_2} \tag{11}$$

is

$$s_1 = \frac{2}{3} + \frac{1}{3} c \frac{l}{x} \tag{12}$$

where c is a constant such that $cl/x = 1$ for $x = 1$ and s_2 has the same functional form.

Also the cyclization kernel is expected to depend on structure. It can be assumed that the probability of the ring closure is a function of l:

$$K_c(x, l) \sim l^r \tag{13}$$

The number of ring closures is proportional to l, the probability of finding an unreacted group in the cluster, and to the probability that one of the partners occurs in a close vicinity. Therefore, r will increase initially and become constant for larger clusters or may even decrease to zero when l/x becomes small. Again, a possible empirical dependence of the form

$$r = c_1 x^{1/2} - c_2 x/l \tag{14}$$

can be used for investigation of the effect on cluster growth.

Various types of kernels can be used to model the effect of the diffusion control. In fact, the original Smoluchowski equations described the rate of coagulation of diffusing clusters; the rate is proportional to the probability of their collision (proportional to the sum of their projections, R, and to the frequency of their collisons, proportional to the sum of their diffusion coefficients, D. Thus the corresponding kernel is

$$K(x; l) \sim (D_x + D_l)(R_x + R_l) \sim (x^{-1/3} + l^{-1/3})(x^{1/3} + l^{1/3}) \tag{15}$$

It has repeatedly been stated that the Smoluchowski equation cannot take into account differences in symmetry of the reacting clusters. However, even this limitation can be somewhat relaxed because the symmetry is a function of the distribution of units in different reaction states (differing in the number of reacted functional groups). For a tree-like structure, a cluster composed predominantly of units with two reacted groups is much more asymmetric than a cluster composed of equivalent amounts of units with three and one reacted groups. This information can be stored throughout the process of simulation and corresponding weightings can be applied.

The possibility to describe various processes characteristic of structures evolution by varying functional forms of the kernels make the Smoluchowski equations attractive. The Monte-Carlo solution of the corresponding differential equations is a suitable method for finding formation-structure relationships.

APPLICATION TO COMPLEX SYSTEMS OF INDUSTRIAL IMPORTANCE

In this part, application of the theories to systems of industrial importance will be illustrated by several examples. Without referring to the pertinent literature sources, one can list the most important application fields:

- network formation from various telechelic polymers involving various end group chemistries: urethane and urea formation, silanol condensation, hydrosilylation, cyclotrimerization of isocyanates, cyanates and acetylenic compounds, esterification. Polydimethylsiloxane and polyurethane networks have been mostly studied as model networks;

- curing of epoxy resins: various curing mechanisms, effect of group reactivities on network formation, special network formation mechanisms for resins based on N,N-diglycidylamines;

- chain copolymerization of polyvinyl or polyunsaturated compounds;

- multistage processes in which the monomers react in several steps and are eventually crosslinked.

The complexness of these systems may be felt on the level of kinetics and mechanisms of the elementary crosslinking reactions (input information for the theory), or on the level of the theory (strong long-range correlations even for a simple mechanism), or both.

Crosslinking of polyurethanes by a variety of reactions when isocyanate groups are in excess belongs to the first group: complex mechanism and kinetics, simple (statistical theory). Scheme 1 shows various reaction paths in the system composed initially of isocyanate and hydroxyl groups. An urethane group reacts with isocyanate group to give an allophanate group (a branch point). Traces of water react with isocyanate via amine to urea, and biuret (branch point) groups are formed from urea. Isocyanate groups can also trimerize to yield isocyanurate rings (branch point). As a result of side reactions, the originally linear system obtained by reaction of diisocyanate with diol can gel. In case of a diisocyanate with independent reactivities of isocyanate groups, the building units can be represented by halves of the diisocyanate and various structural fragments formed thereof: urethane, urea, allophanate, biuret, isocyanurate. They are listed in Table 1.[48]

The branched and crosslinked structures are generated by proper combination of half-bonds: II with II, and IH with HI. In the absence of cyclotrimerization, the theory predicts two gel points for a diol – diisocyanate systems as a result of side reactions: the first one is a sol – gel transition that occurs at a slight excess of isocyanate when long polyurethane chains are crosslinked by a few allophanate and biuret groups. The gel – sol transition is predicted to occur at a high excess of isocyanate where many branch points have been formed but the branches are terminated by unreacted isocyanate groups. Such behavior has been found experimentally (Fig. 8) and also a quantitative agreement was good for reaction at lower temperatures (Fig. 9).

The differences between theoretical and experimental values of the sol fraction observed for reaction at higher temperatures (120 – 140°C)[48] (Fig. 9) can be interpreted by a weak interference of other, isocyanate deactivating, reactions not accounted for in Scheme 1.

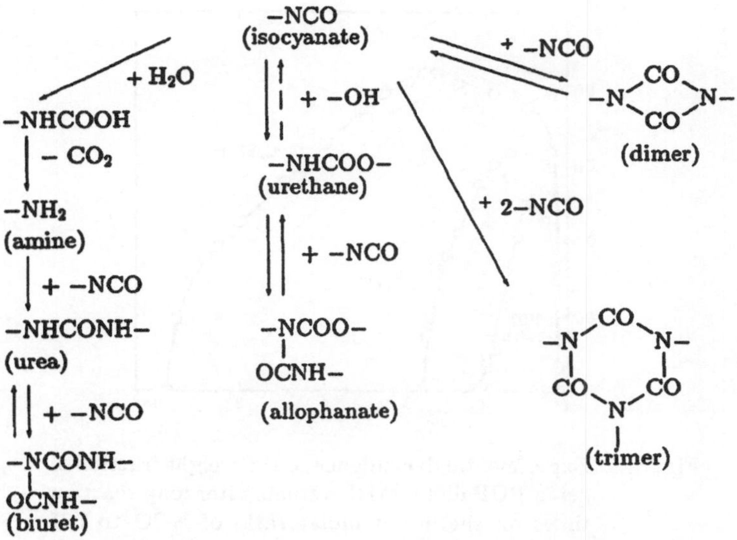

Scheme 1 content:

$$-\text{NHCOOH} \xleftarrow{+ H_2O} \quad \begin{array}{c} -\text{NCO} \\ \text{(isocyanate)} \end{array} \xrightarrow{+ -\text{NCO}} \begin{array}{c} -\text{N} \overset{\text{CO}}{\underset{\text{CO}}{\diamond}} \text{N} - \\ \text{(dimer)} \end{array}$$

```
                              -NCO
                          (isocyanate)
           + H₂O              ↑↓
  -NHCOOH                     ↑│ + -OH              + -NCO
     │                        ↑↓              -N  CO  N-
     │ - CO₂                  -NHCOO-               CO
     ↓                        (urethane)          (dimer)
  -NH₂
  (amine)                      ↑↓ + -NCO           + 2 -NCO
     │
     │ + -NCO                 -NCOO-
     ↓                        │
  -NHCONH-                    OCNH-
  (urea)                      (allophanate)                CO
                                                    -N      N-
     ↑↓ + -NCO                                         CO    CO
     │                                                     N
     ↓                                                     │
  -NCONH-                                             (trimer)
  │
  OCNH-
  (biuret)
```

Scheme 1. Main reactions of the isocyanate group during
polyurethane formation.

Table 1. Fragments used for build-up of polyurethane structures.

Unit	Formula and types of bonds
isocyanate	$\xleftarrow{\text{II}}$ NCO
urethane	$\xleftarrow{\text{II}}$ NHCO $\xrightarrow{\text{IH}}$ (O)
urea	$\xleftarrow{\text{II}}$ NHCONH $\xrightarrow{\text{II}}$
biuret	$\xleftarrow{\text{II}}$ NHCO–N–CONHH $\xrightarrow{\text{II}}$
	$\downarrow$ II
isocyanurate	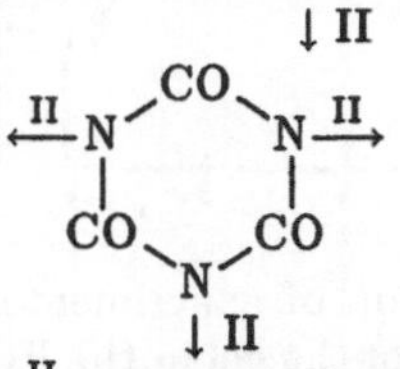 $\xleftarrow{\text{II}}$ N $\overset{\text{CO}}{\underset{\text{CO}}{\diamond}}$ N $\xrightarrow{\text{II}}$ CO–N–CO $\downarrow$ II
allophanate	$\xleftarrow{\text{II}}$ NHCO–N–CO $\xrightarrow{\text{IH}}$ (O) $\downarrow$ II
reacted hydroxyl	O $\xrightarrow{\text{HI}}$

Curing of epoxy resins is another example where the application of the branching theories (in most cases of statistical ones) has helped to understand the effects of changing reactivities, molar ratios, compositions, and functionalities of components on network build-up and network structure.[49] In the last years, the network formation theories have been applied to more complicated systems. For amine curing of polyepoxides based on N,N-diglycidylamines, strong cyclization is characteristic where the majority of rings is of a small size: six-, seven-, and eight-membered. Moreover, there exists a strong substitution effect between the glycidyl groups bound to the same nitrogen atom. Therefore, the mechanism of the network formation process is much more complicated than that of polyepoxides based on Bisphenol A types of polyepoxides.[50] In the theoretical treatment based on the theory of branching processes, larger fragments including the small cyclics had to be used.[51] However, one should be aware of the fact that long-range connectivity correlations may exist in these systems. Their affect is expected to be relatively weak, however.

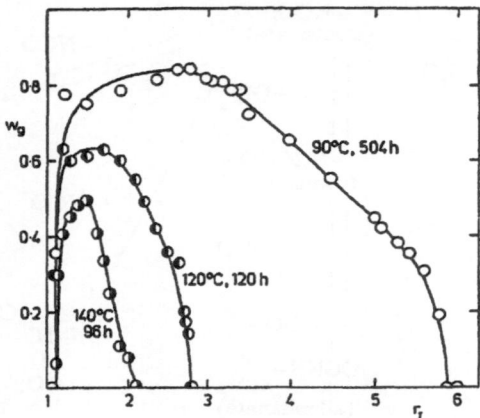

Figure 8. Experimental dependence of the weight fraction of gel in POP diol – MDI systems after long reaction times on the initial molar ratio of NCO to OH groups, r_I, at temperatures indicated. M_n of POP diol 1230 g/mol; data from Ref.[48]

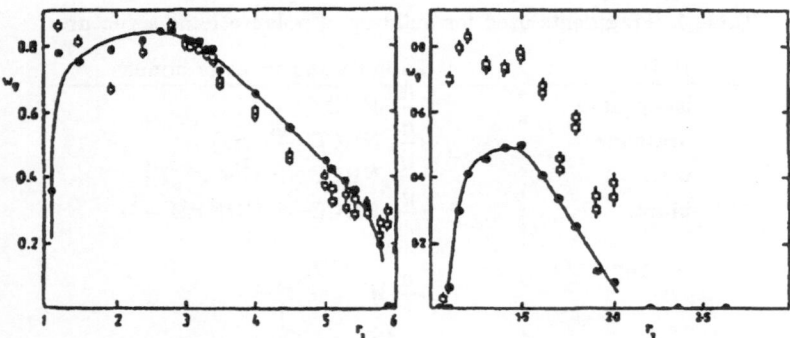

Figure 9. Comparison of experimental (•) and theoretical (o) weight fractions of the gel in the POP diol – MDI systems in dependence on r_I reaction at 90°C and 140°C.[48]

Theoretical treatment of polyetherification released by the reaction of epoxy group with an amino group is another example of a relatively complicated network formation mechanism. Generaly, it is an initiated chain growth process which has to be treated using the kinetic theory. However due to several competitive initiations of chain growth and strong chain transfer coupled with reinitiation make also the theoretical treatment somewhat complicated but possible.[21] The main problem is to find the values of the corresponding rate constants.

Multistage processes are very important in technology for a variety of practical reasons but it was not clear whether staging has an effect on the final structure of the network and how important this effect can be. Therefore, a theoretical procedure has been devised in which the distributions obtained at the end of stage one serve as input information for stage 2, etc. (Scheme 2).[52]

The process in Scheme 2 allows combination of the kinetic method in one stage and

```
                STAGE 1      monomers 1
                              │ reaction 1
                              ▼
                             PRODUCTS 1
                STAGE 2      PRODUCTS 1 +
                             monomers 2
                              │ reaction 2
                              ▼
                             PRODUCTS 2
                              ⋮
                STAGE M      PRODUCTS M − 1 +
                             monomers M
                              │ reaction M
                              ▼
                             PRODUCTS M
```

Scheme 2. Theoretical treatment of a multistage process.

the statistical method in the other one. However, the transition from one stage to the other involves conversions of weight distributions into number distribution — a difficult step which involves integration. It has come out later from analysis of a concrete system[53] that an analytical integration is always possible if the statistical method was used for network build-up in all stages. This finding stimulated a more general conclusion that a multistage process can be described by a statistical theory if the bonds (reacted functionalities) formed in different stages are distinguished.[32] This lead to the development of a method of appproximation of long-range connectivity correlation by an arbitrary splitting of the conversion interval into smaller ones[33] (see above).

The application to a concrete case is schematically shown in Fig. 10:[32] a bifunctional

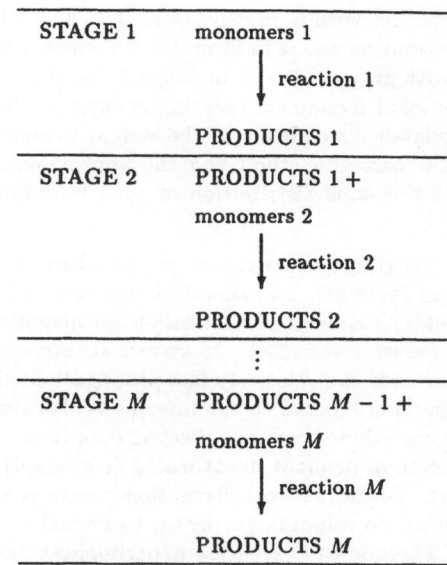

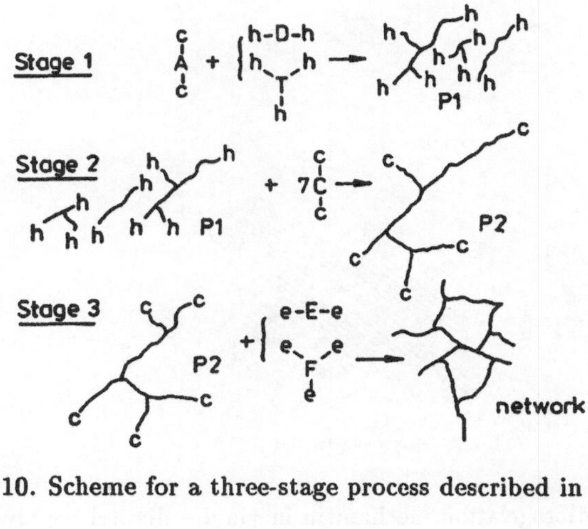

Figure 10. Scheme for a three-stage process described in the text.

monomer A (groups c) reacts first with a mixture of bi- (D) and trifunctional (T) monomers (groups h), the resulting distribution reacts in Stage 2 with another monomer (groups c) (some groups of A can still react with groups h), and in Stage 3 this distribution is crosslinked with a mixture of bi- and trifunctional monomers carrying groups e. Figure 6 of Ref.[32] shows the difference between the calculated dependence of the weight-average molecular weight of the product in Stage 3, using the correct method and the approximation disregarding the time sequences of bond formation the same distribution of units in different reaction stages.

Chain copolymerization of polyunsaturated monomers represents one of the most complicated network forming systems. The chemical mechanism is rather complicated and involves a number of elementary reactions like initiation, propagation, chain transfer (several variants), and termination (several variants). Moreover strong spatial correlations exist in these systems. The primary reason is a relatively fast propagation with respect to termination which results in strong cyclization relative to the intermolecular reaction in the beginning of polymerization. Also, a strong exluded volume effect is operative. As a result of cyclization and fast propagation, a fraction of pendant unsaturated (e.g. vinyl) groups becomes strongly shielded and unable to react. Often, network formation proceeds via microgelation and the gel point conversions are shifted to values larger by up to several orders of magnitude relative to an ideal ring-free case. There exists extensive experimental data on such mechanism of network formation[54,55] via a "monochain", "two-chain", ... Microgels are visualised in Fig. 11.

However, the structure effect is often at least partly diffusion controled (cf. Trommdorff effect occuring even in linear bulk polymerizations).

The complex chemical mechanism in network formation can succesfully be dealt with using kinetic theory (cf. e.g. Refs.[49,19]).

No good theory exists at the moment for theoretical description of the strong spatial

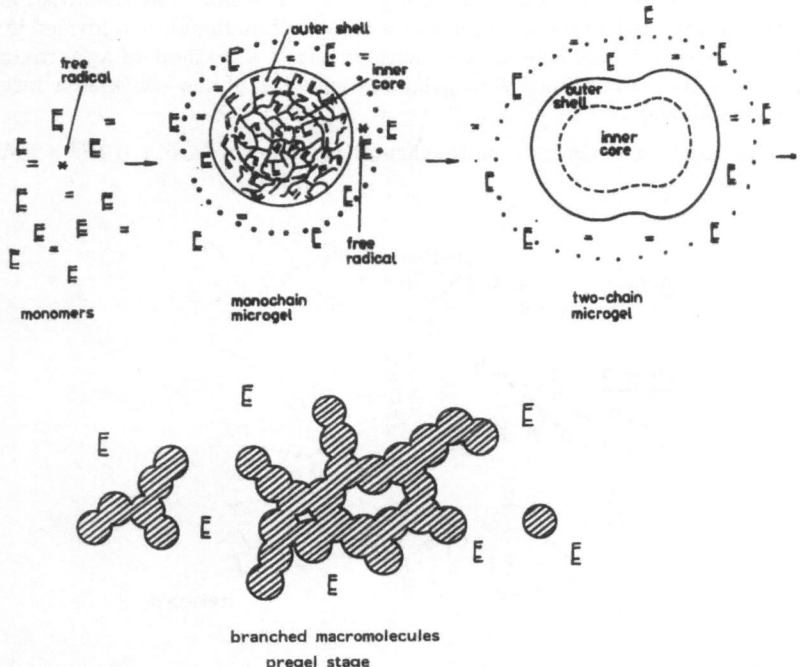

Figure 11. Microgelation mechanism in vinyl – divinyl copolymerization.

correlations although a rough picture of structural inhomogeneity is obtained by initiated (kinetic) percolation mentioned above. This topic is dealt with in more detail in this book by Dotson.

CONCLUSION

The goal in the area of network formation theories is to develop a simulation in space where the growing structures interact in space and both the complex kinetics of bond formation and the kinetics of conformational rearrangements can be taken into account. The development of simulations techniques covering a sufficiently large space will probably require some time. Meanwhile, the statistical and kinetic theories and their combinations equipped with approximations capable of simulating the effect of spatial and sometimes also connectivity correlations can serve well for the elucidation of the formation – structure relations even for chemically complex systems.

REFERENCES

1. K. Dušek, Polymer networks: A challenge to theorist and technologist, *J. Macromol. Sci. – Chem.* A28: 843 (1991).
2. K. Dušek, Network formation in chain crosslinking (co)polymerisation, *in:* "Advances in Polymerisation. 3.," R. N. Haward, ed., Applied Science Publishers, Barking (1982).
3. K. Dušek, Formation – structure relationship in polymer networks, *Brit. Polym. J.* 17: 185 (1985).
4. K. Dušek and J. Šomvársky, Build-of polymer networks by initiated polyreactions. I. Comparison of the kinetic and statistical approaches to living polymerization type build-up, *Polym. Bull.* 13: 313 (1985).
5. S. I. Kuchanov, "Application of the kinetic method in polymer chemistry" (in Russian), Nauka, Moscow 1978.
6. K. Dušek, Polymer networks. Formation and structure, *Rec. Trav. Chim. Pays-Bas* 110: 507 (1991).
7. N. G. van Kampen, "Stochastic processes in physics and chemistry," North Holland, Amsterdam (1981).
8. D. Stauffer, A. Coniglio, and M. Adam, Gelation and critical phenomena, *Adv. Polym. Sci.* 44: 103 (1981).
9. H. M. J. Boots, J. G. Kloosterboer, G. M. M. van de Hei, and R. B. Pandey, Inhomogeneity during the bulk polymerisation of divinyl compounds: differential scanning calorimetry experiments and percolation theory, *Br. Polym. J.* 17: 219 (1985).
10. Y. K. Leung and B. E. Eichinger, Computer simulations of end-linked elements. I. Trifunctional networks cured in bulk, *J. Chem. Phys.* 80: 3877, 3885 (1984).
11. A. M. Gupta, R. C. Hendrickson, and C. W. Macosko, Monte Carlo description of A_f homopolymerization: Diffusional effects, *J. Chem. Phys.* 95: 2097 (1991).
12. "Kinetics of aggregation and gelation," F. Family and D. P. Landau, eds., Elsevier (1984).
13. L. Y. Shy, Y. K. Leung, and B. E. Eichinger, Critical exponent for off-lattice gelation of polymer chains, *Macromolecules* 18: 983 (1985).
14. F. Leyvraz and H. R. Tschudi, Critical kinetics near gelation, *J. Phys. A: Math. Gen.* 15: 1951 (1982).
15. E. M. Hendrics, M. H. Ernst, and R. M. Ziff, Coagulation equations with gelation, *J. Stat. Phys.* 31: 519 (1983).
16. S. I. Kuchanov and E. S. Povolotskaya, Calculation of gel point for non-equilibrium polycondensation taking into account the substitution effect, *Vysokomol. Soedin.* A24: 2179 (1982).
17. H. Galina and A. Szustalewicz, A kinetic approach to network formation in an alternating stepwise copolymerizations, *Macromolecules* 23: 3833 (1990).

18. H. Tobita and A. E. Hamielec, A kinetic model for network formation in free radical polymerization, *Macromol. Chem., Macromol. Symp.* 20/21: 501 (1988).

19. H. Tobita and A. E. Hamielec, Modeling of network formation in free radical polymerization, *Macromolecules* 22: 3098 (1989).

20. A. G. Mikos, C. G. Tsakoudis, and N. A. Peppas, Kinetic modeling of copolymerization/cross-linking reactions, *Macromolecules* 19: 2174 (1986).

21. K. Dušek, J. Šomvársky, M. Ilavský, and L. Matějka, Gelation and network formation by polyetherification of polyepoxides initiated by hydroxyl groups: Theory, *Comput. Polym. Sci.* 1: 90 (1991).

22. D. A. McQuarrie, "Methuen's review series on applied probability. Volume 8. Stochastic approach to chemical kinetics," Methuen, London (1967).

23. J. Mikeš and K. Dušek, Simulation of the polymer network formation by the Monte-Carlo method, *Macromolecules* 15: 93 (1982).

24. H. P. Breuer, J. Honerkamp, and F. Petruccione, Stochastic simulation of polymerization reactions, *Comput. Polym. Sci.* 1: 233 (1991).

25. K. Kang and S. Redner, Fluctuation-dominated kinetics in diffusion-controlled reactions, *Phys. Rev. A* 32: 435 (1985).

26. R. J. J. Williams, C. C. Riccardi, and K. Dušek, Build-up of polymer networks by initiated polyreaction. 3. Analysis of the fragment approach to living polymerization type of build-up, *Polym. Bull.* 17: 515 (1987).

27. D. R. Miller and C. W. Macosko, Calculation of average network parameters using combined kinetic and Markovian analysis, *in:* "Biological and synthetic polymer networks," O. Kramer, ed., Elsevier 1988, p. 219.

28. K. Dušek and W. J. MacKnight, Cross-linking and structure of polymer networks, *in:* "Cross-linked polymers. Chemistry, properties, and applications," R. A. Dickie and R. S. Bauer, eds., *ACS Symp. Ser.* No. 367, American Chemical Society, Washington, D.C., 1988, p. 2.

29. G. R. Dobson and M. Gordon, Theory of branching processes and statistics of rubber elasticity, *J. Chem. Phys.* 43: 705 (1965).

30. D. R. Miller and C. W. Macosko, Network parameters for crosslinking of chains with length and size distribution, *J. Polym. Sci., Part. B: Polym. Phys.* 26: 1 (1988).

31. K. Dušek and M. Ilavský, Cyclization in crosslinking polymerization. I. Chain copolymerization of a bisunsaturated monomer (monodisperse case); II. Chain copolymerization of a bisunsaturated monomer (polydisperse case), *J. Polym. Sci., Symposium No. 53*, 13: 57, 75 (1975).

32. G. P. J. M. Ttiemersma–Thone, B. J. R. Scholtens, K. Dušek, and M. Gordon, Theories of network formation in multi-stage processes, *J. Polym. Sci., Part. B: Polym. Phys.* 20: 463 (1991).

33. C. Sarmoria and D. R. Miller, Models for the first shell substitution effect in stepwise polymerization, *Macromolecules* 24: 1833 (1991).

34. N. A. Dotson, Correlations on nonlinear free-radical polymerizations: Substitution effect, *Macromolecules* 25: 308 (1992).

35. S. I. Kuchanov, S. V. Korolyov, and M. G. Slinko, Configurational statistics of branched polycondensation type polymers, *Vysokomol. Soedin.* A26: 263 (1984).

36. M. Gordon and G. R. Scantlebury, The theory of branching processes and kinetically controlled ring – chain competition processes, *J. Polym. Sci., Part C* 16: 3933 (1968).

37. R. F. T. Stepto, Intramolecular reaction and gelation in condensation or random polycondensation, *in:* "Developments in Polymerisation. 3.," R. N. Haward, ed., Applied Science Publ., Barking 1982, p. 81.

38. R. F. T. Stepto, Intramolecular reaction: effect on network formation and properties, *in:* "Cross-linked polymers. Chemistry, properties, and applications," R. A. Dickie, S. S. Labana, and R. S. Bauer, eds., *ACS Smp. Ser.* No. 367, American Chemical Society, Washington, D.C., 1988, p. 28.

39. C. Sarmoria, E. Valles, and D. R. Miller, Ring – chain competition kinetic models for linear and nonlinear step reaction copolymerizations, *Makromol. Chem., Macromol. Symp.* 2: 69 (1986).

40. K. Dušek, M. Gordon, and S. B. Ross-Murphy, Graphlike state of matter. 10. Cyclization and concentration of elastically active network chains in polymer networks, *Macromolecules* 11: 236 (1978).

41. K. Dušek and V. Vojta, Concentration of elastically active network chains and cyclisation in networks obtained by alternating stepwise polyadition, *Br. Polym. J.* 9: 164 (1977).

42. W. H. Stockmayer, Theory of molecular size distribution and gel formation in branched chain polymers, *J. Chem. Phys.* 11: 45 (1943).

43. K. Dušek, Correspondence between the theory of branching processes and the kinetic theory for random crosslinking in the post-gel stage, *Polym. Bull.* 1: 523 (1979).

44. R. M. Ziff and E. D. McGrady, The kinetics of cluster framentation and depolymerisation, J. Phys. A: Math. Gen. 18: 3027 (1985).

45. R. M. Ziff and E. D. McGrady, Kinetics of polymer degradation, Macromolecules 19: 2513 (1986).

46. M. Demjanenko and K. Dušek, Statistics of degradation and crosslinking of polymer chains with the use of the theory of branching processes, Macromolecules 13: 571 (1980).

47. J. E. Martin and D. Adolf, The sol-ge transition in chemical gels, *Annu. Rev. Phys. Chem.* 42: 311 (1991).

48. K. Dušek, M. Špírková, and I. Havlíček, Network formation in polyurethanes due to side reactions, *Macromolecules* 23: 1774 (1990).

49. K. Dušek, Network formation in curing of epoxy resins, *Adv. Polym. Sci.* 81: 167 (1986).

50. L. Matějka and K. Dušek, The mechanism of curing of epoxides based on diglycidylamine with aromatic amine. I. Reaction between diglycidylaniline, *Macromolecules* 22: 2902, 2911 (1991).

51. L. Matějka and K. Dušek, Influence of the reaction mechanism on the network formation of amine cured N,N–diglycidylamine epoxy resins, *Polymer* 32: 3195 (1991).

52. K. Dušek, B. J. R. Scholtens, and G. P. J. M. Tiemersma–Thoone, Theoretical treatment of network formation by a multistage process, *Polym. Bull.* 17: 239 (1987).

53. G. P. J. M. Tiemersma–Thoone, B. J. R. Scholtens, and K. Dušek, A stochastic description of copolymerization and network formation in a three-stage process, *Proc. 1st Internat. Conf. Ind. Appl. Mathematics*, Amsterdam 1989, p. 295.

54. A. Matsumoto, H. Matsou, H. Ando, and M. Oiwa, Solvent effect in the copolymerization of methyl methacrylate with oligoglycol dimethacrylate, *Eur. Polym. J.* 25: 237 (1989).

55. A. Matsumoto, H. Ando, and M. Oiwa, Gelation in the copolymerization of methyl methacrylate with trimethylolpropane trimethacrylate, *Eur. Polym. J.* 25: 385 (1989).

OSMOTIC PROPERTIES OF SWOLLEN POLYMER NETWORKS

Ferenc Horkay,[a,b]
Erik Geissler,[c]
Anne-Marie Hecht,[c]

[a]Department of Colloid Science,
Loránd Eötvös University,
H-1117 Budapest, Pázmány s. 2,
Hungary

[b]Institute of Macromolecular Chemistry,
University of Freiburg,
Stefan Meier Str. 31.
D-7800 Freiburg, FRG

[c]Laboratoire do Spectrométrie Physique,
Université Joseph Fourier de Grenoble,
B.P. 87, F-38402 St. Martin D'Heres
Cedex, France

INTRODUCTION

In the past few years several studies have been performed on the swelling and elastic behavior of polymer gels[1-8]. This field of polymer physics has recently come of interest because of newly developed theoretical approaches and significant improvements in the experimental techniques used for investigating swollen network systems.

Swelling equilibrium measurements and elastic modulus measurements are relatively simple methods for characterizing the network. Scaling theory is a powerful tool to describe the behavior of polymer solutions and gels. In addition to the information obtained from macroscopic measurements, small angle neutron scattering (SANS) and small angle X-ray scattering (SAXS) provide information at the molecular scale[9-17]. It has been found that the cross-linking process is in general accompanied by redistribution of the polymer segments. Any changes in the thermodynamic interaction of the system are reflected both through the macroscopic and microscopic properties of the swollen network. Consequently a comparison between the results of macroscopic osmotic and elastic investigations and those deduced from scattering observations is expected to contribute to a more quantitative understanding of the basic physical properties of the swollen polymer networks.

In this work we summarize recent results of osmotic investigations performed on permanently cross-linked polymer gels. The macroscopic thermodynamic response of chemically different network systems is compared with thermodynamic properties obtained from scattering observations for the same gels.

THEORETICAL SECTION

Thermodynamic Considerations

A thermodynamic treatment of a swollen cross-linked polymer was first proposed by Flory and Rehner[18,19]. The theory is based on the principal assumption that the change in the free energy accompanied by swelling of a network can be separated into an elastic free energy (ΔF_{el}) and a mixing free energy (ΔF_{mix}) term. The latter term is identified with the free energy of mixing of the solvent molecules with the uncross-linked polymer of infinite molecular weight.

At equilibrium the chemical potentials of the solvent inside and outside the swollen network are equal:

$$RT\ln a_1 = \partial\Delta F_{el}/\partial n_1 + \partial\Delta F_{mix}/\partial n_1 \tag{1}$$

where a_1 is the solvent activity and n_1 is the number of moles of solvent in the gel.

The elastic free energy term is given by the theory of rubber elasticity. According to the phantom network model due to James and Guth[20,21]

$$\Delta F_{el} = (\zeta kT/2)(\lambda_x^2 + \lambda_y^2 + \lambda_z^2 - 3) \tag{2}$$

where ζ is the cycle rank of the network and λ_x, λ_y and λ_z are the principal deformation ratios. In the case of isotropic swelling ($\lambda_x = \lambda_y = \lambda_z$) eq.2 yields

$$\partial\Delta F_{el}/\partial n_1 = GV_1 \tag{3}$$

where G is the elastic modulus of the gel and V_1 is the molar volume of the diluent.

The expression for the mixing term is given in the Flory-Huggins theory by

$$\partial\Delta F_{mix}/\partial n_1 = RT[\ln(1-\varphi)+\varphi+\chi\varphi^2] = -\Pi_{mix}V_1 \tag{4}$$

where φ is the volume fraction of the polymer χ is the Flory-Huggins interaction parameter and Π_{mix} is the mixing pressure.

More recent scaling considerations[22] predict that the concentration dependence of Π_{mix} in the semidilute regime ($\varphi^* \ll \varphi \ll 1$) obeys a power law behaviour, i.e.

$$\Pi_{mix} = A\varphi^n \qquad\qquad (5)$$

where the exponent $n=9/4$ (good solvent condition) or $n=3$ (theta condition).

The swelling pressure of the gel ($\omega = -RT\ln a_1/V_1$) is obtained from eqs. 3 and 5

$$\omega = A\varphi^n - G \qquad\qquad (6)$$

Scattering Formalism

Scattering of radiation passing through a polymer system reveals the local distribution of the polymer segments. In a small angle neutron scattering (SANS) experiment the intensity of elastically scattered neutrons, $I(Q)$, is measured as a function of the transfer wave vector, $Q=(4\pi/\lambda)\sin(\theta/2)$, where λ is the incident wavelength and θ is the scattering angle.

The spectra from neutral polymer solutions in the region $Q\xi \leq 1$ (ξ is the polymer-polymer correlation length in the solution) usually adopt a Lorentzian form

$$I(Q) = a\,\frac{(\rho_p-\rho_s)^2 kT\varphi^2}{K_{os}}\,\frac{1}{1+Q^2\xi^2} \qquad\qquad (7)$$

where a is an apparatus constant, ρ_p and ρ_s are the scattering densities of the polymer and solvent respectively, and k is the Boltzmann constant. K_{os} is the osmotic compressional modulus, $K_{os}=\varphi(\partial\Pi/\partial\varphi)$, where Π is the osmotic pressure.

It is reasonable to assume that at small scale the polymer molecule behaves as a wormlike chain of radius r_c, i.e. at larger Q values the scattering function tends to $(1/Q)\exp(-Q^2 r_c^2/2)$. The form factor then becomes[16]

$$S(Q) = A(1+Qr_c/\sqrt{2})\exp(-Q^2 r_c^2/2)/(1+Q^2\xi^2) \qquad\qquad (8)$$

For gels, the scattering response is more complex than for polymer solutions. In the network the distribution of the cross-links is usually not uniform. The local polymer concentration in the densely cross-linked regions exceeds the average concentration. The distribution of the polymer in the swollen network can be considered as a static structure with a mean square amplitude $\langle\delta\varphi^2\rangle$ of static concentration fluctuations. Such permanent departures are expected to participate little in the dynamic fluctuations. The dynamic part, $\langle\Delta\varphi^2\rangle_{dyn}$, is assumed to be represented by a structure factor of the same form as that of the solution (see (eq. 8)). The total amplitude of the concentration fluctuations is the sum of the static and the dynamic contribution[15-17]

$$\langle\Delta\varphi^2\rangle = \langle\Delta\varphi^2\rangle_{dyn} + \langle\delta\varphi^2\rangle \qquad\qquad (9)$$

The structure factor of the static component is assumed to have the form $\exp(-Q^s\Xi^s)$, where Ξ is the mean size of the static non-uniformities, and s is a constant[16]. (The value s=2 corresponds to a Gaussian spatial distribution for the non-uniformities in the gel.) The total scattering intensity is given by

$$I(Q) = a\,(\rho_p - \rho_s)^2 \left[\frac{kT\varphi^2}{M_{os}}\,S(Q) + f(s)\,\exp(-Q^s\Xi^s) \right] \qquad (10)$$

where M_{os} is the osmotic modulus of the gel and $f(s)$ is a normalisation factor.

EXPERIMENTAL PART

General Considerations

The theoretical basis of the comparison between macroscopic and microscopic osmotic properties of swollen network systems is given by eqs 6 and 10.

According to eq.6 the thermodynamics of the gel is governed by the swelling pressure. The swelling pressure can be measured directly using a swelling pressure osmometer described by Pennings and Prins[23]. Similar information is obtained from osmotic deswelling when the diluent activity is lowered by dissolving a polymer in the surrounding media, or from vapor sorption measurements[1-3,24-26]. In the latter experiment the quantity of the solvent absorbed by the cross-linked polymer and the corresponding linear polymer is determined as a function of the vapor pressure of the solvent. All these methods allow model independent determination of the solvent activity difference between the cross-linked and uncross-linked systems. From the concentration dependence of the swelling pressure both the mixing pressure, Π_{mix}, and the volume elastic modulus of the gel, G, can be obtained. The modulus can also be determined from separate measurements (e.g. from stress-strain isotherms).

The dynamic part of the scattering spectrum of the gel is controlled by the osmotic modulus, M_{os}, describing the restoring force in osmotic plane waves

$$M_{os} = \varphi(\partial\omega/\partial\varphi) + (4/3)G \qquad (11)$$

In principle M_{os} in eq.10 is identical with the quantity obtainable from macroscopic swelling pressure and modulus measurements. Thus the scattering spectrum of a gel fitted in a wide range of wave vectors to eq.10 allows the thermodynamic parameters to be evaluated and hence the microscopic and the macroscopic thermodynamic response of the swollen network to be compared.

Model Systems

The validity of the proposed relationships has been tested on two entirely different types of network systems: end-linked

polydimethylsiloxane (PDMS), and randomly cross-linked poly-
(vinyl acetate) (PVAc).

The PDMS networks were obtained by end-linking of
hydroxyl terminated PDMS chains of molecular weight 40.000 with
the cross-linking agent ethyl triacetoxy silane[27]. After
completion of the reaction, the sol fraction was extracted by
octane. The swelling and mechanical properties of the PDMS gels
were studied in toluene at 25 °C (good solvent).

PVAc gels were prepared by acetylation of chemically
cross-linked poly(vinyl alcohol) networks using a method described elsewhere[28,29]. Cross-linking was performed in aqueous
poly(vinyl alcohol) solutions using glutaraldehyde as crosslinking agent. The poly(vinyl alcohol) gels were acetylated in
a mixture of pyridine-acetic acid-acetic anhydride at 90 °C.
After completion of the reaction the gels were washed and then
swollen to equilibrium in toluene at 25 °C (good solvent).

Methods

Swelling pressure of the gels was measured as a function
of the polymer concentration using a deswelling method.
Deswelling was achieved by equilibrating the swollen PDMS and
PVAc networks with PVAc-toluene solutions of known osmotic
pressure[30]. The gels were enclosed in dialysis bags to prevent
penetration of the dissolved polymer molecules[31].

Uniaxial compression measurements were performed on
cylindrical gel samples at constant volume. The stress-strain
data were evaluated using the Mooney-Rivlin equation. The C_1
term was identified with the shear modulus of the gel. In the
range of deformation ratio $0.7<\Lambda<1$, the C_2 Mooney-Rivlin term
was found to be equal to zero.

The small angle neutron scattering (SANS) measurements
were carried out on the D11 instrument at the Institut Laue-
Langevin, Grenoble. The neutron wavelength was 6Å. The Q range
explored was $0.003\text{Å}^{-1}\leq Q\leq 0.25\text{Å}^{-1}$, and counting times of between
twenty minutes and one hour were used. After radial averaging,
standard corrections for incoherent background, detector
response and cell window scattering were applied. Calibration
of the scattered neutron intensity was performed using the
signal from a 1 mm thick water sample in conjunction with the
absolute intensity measurements of Ragnetti et al.[32] The sample
cells consisted of 1 mm thick quartz windows with a 1 mm thick
annular Teflon spacer housing the swollen gel or solution.
All measurements were performed at T=25°C.

RESULTS AND DISCUSSION

Evaluation of the Macroscopic Osmotic and Elastic Measurements

Analysis of the swelling pressure, ω, vs polymer volume
fraction, φ, curves allows the separability of the two free
energy terms to be checked. In Fig.1 a typical set of swelling
pressure data are displayed. The experimental points are fitted
to the equation

$$\omega = A\varphi^n - G_O\varphi^m \qquad\qquad (12)$$

where A and G_O are constants, and the exponents n and m are
iteratively adjusted to minimize the variance for each set of
data points. The parameters obtained from the fit for chemically different network systems are summarized in Table I. The
values of n and m are in satisfactory agreement with the theoretical prediction (n=9/4 and n=3 in good solvent condition and
in the theta state, respectively, and m=1/3 according to the
theory of rubber elasticity).

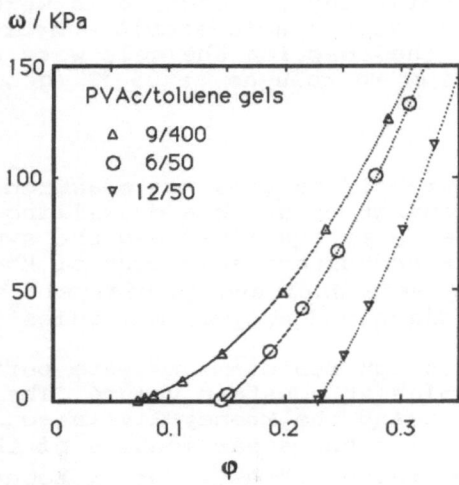

Fig.1. Swelling pressure, ω, vs polymer volume fraction, φ, for
       PVAc/toluene gels. The dotted lines show the least
       squares fit of eq.12 to the experimental data. Meaning
       of the codes: concentration of the polymer solution at
       cross-linking in percentage weight fraction/ratio of
       monomer units to cross-linker.

The validity of the Flory-Rehner hypothesis can be checked
by comparing the terms of eq.12 with directly measured quantities. From eqs.6 and 12 follows that $G_O\varphi^m$ can be identified
with the modulus of the network.

In Fig.2 the values of $G_O\varphi^m$ calculated for PVAc/toluene
gels swollen by toluene are shown as a function of the measured
shear modulus. The experimental points lie along the theoretical straight line of unit slope indicating the equivalence of
the two quantities.

Table 1. Fitting parameters of Eq.(12) to swelling pressure data for chemically different network systems

Sample	φ_e	G_0/kPa	n	m
PVAc/toluene (good solvent condition)				
6/50	0.146	59.7	2.29	0.331
6/200	0.078	17.1	2.22	0.340
9/50	0.208	123.6	2.27	0.355
9/200	0.112	33.9	2.27	0.326
12/50	0.229	168.3	2.35	0.383
12/200	0.133	50.1	2.26	0.335
PVAc/isopropyl alcohol (theta condition)				
6/50	0.253	66.0	2.78	0.326
6/200	0.149	26.6	2.61	0.413
9/50	0.330	95.2	3.70	0.252
9/100	0.237	53.5	3.00	0.331
9/200	0.201	37.5	2.85	0.321
PDMS/toluene (good solvent condition)				
unfilled 40	0.068	18.4	2.22	0.347
unfilled 60	0.095	35.2	2.21	0.343

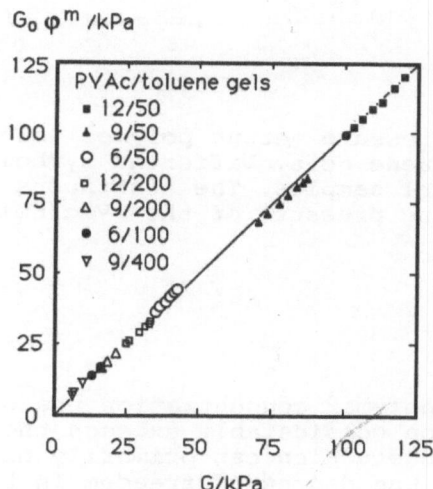

Fig.2. Plot of $G_0 \varphi^m$ versus G for PVAc/toluene gels.

The mixing pressure, Π_{mix}, of the cross-linked polymer is not directly measurable. The Flory-Rehner theory assumes that the interaction of the solvent with the cross-links does not differ considerably from that with the mid-chain segments. If this is the situation, the mixing pressure of the gel should be equal to the osmotic pressure of the solution of infinite molecular mass calculated by eq.4 using χ values obtained from measurements on the corresponding polymer solution. In Fig.3a and b Π_{mix} from swelling pressure and modulus measurements are presented together with the dependence calculated from the solution data.

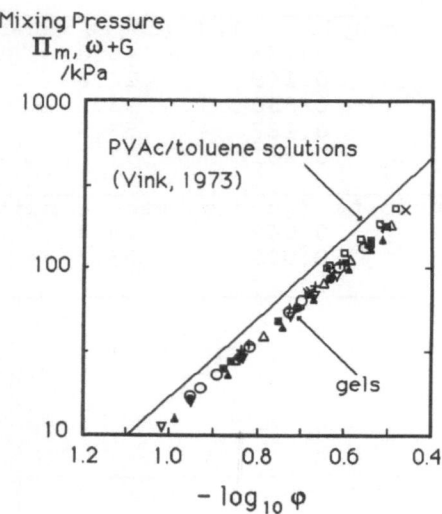

Fig.3.a. Mixing pressure versus polymer volume fraction for PVAc/toluene gels. Different symbols refer to different gel samples. The continuous curve represents the mixing pressure of the PVAc/toluene solution.

At identical polymer concentration the osmotic pressure of the polymer solution considerably exceeds the mixing pressure of the gel. This observation can primarily be attributed to the strong decrease in the degree of freedom in the swollen gel due to cross-linking[33].

It is also apparent that the mixing pressures in the cross-linked and the uncross-linked system exhibit similar dependence on the concentration. Agreement between the two terms is obtained if the polymer concentration in the gel is decreased by a factor 0.7-0.9 depending on the particular polymer/solvent system.

The relation between the measurable macroscopic quantities ω, G and φ is obtained from eq.12

$$\omega = G \ [(\varphi/\varphi_e)^{n-m} -1] \tag{13}$$

where φ_e is the volume fraction of the polymer in the freely swollen gel.

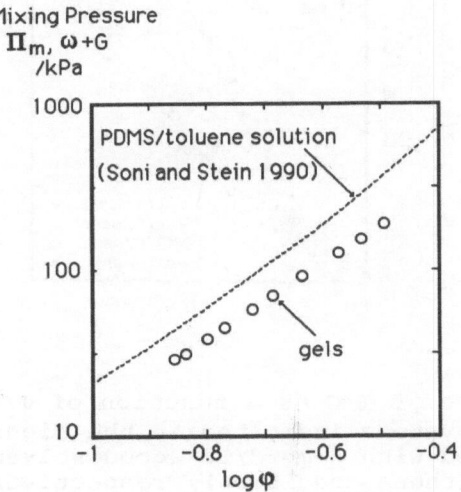

Fig.3.b. Mixing pressure versus polymer volume fraction for PDMS/toluene gels. The dotted curve represents the mixing pressure of the PDMS/toluene solution.

In Fig.4 experimental data are compared with theory. One can see that the points are scattered around the calculated curves. These results clearly demonstrate that the macroscopic thermodynamic properties of swollen polymer networks investigated here are satisfactorily described by eq.6. The total free energy of the gel can be separated into an elastic and a mixing free energy term. The separation yields an elastic contribution that is numerically identical to that deduced from direct modulus measurements.

The observed discrepancy in the mixing free energies between the solution and the swollen network is not

intrinsically surprising. In the cross-linked system the local
segment density is in general not completely homogeneous: in
the vicinity of the cross-link points the local polymer concentration is higher. Consequently the effective polymer concentration that controls the thermodynamic properties of the
gel is reduced.

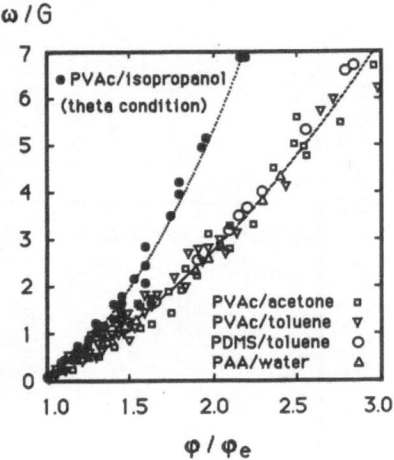

Fig.4. Dependence of ω/G as a function of φ/φ_e for different
    network systems indicated in the figure. Curves
    calculated with n-m=23/12 (good solvent condition) and
    n-m=8/3 (theta condition), respectively.

Scattering Properties of the Swollen Networks

In Fig.5a and b the neutron scattering spectrum of a
PDMS/toluene and a PVAc/toluene gel is shown together with the
spectra of the corresponding solutions. Inspection of the gel
and the solution spectra at low Q values in the gel reveals
excess scattered intensity. It is plausible to attribute the
excess intensity to the presence of nonuniformities in the network structure. Nonuniformities contribute to the scattering
spectra mainly in the low Q region because of their relatively
big size. In the high Q region the gel and the solution spectra
are practically indistinguishable. The continuous lines in the
figures show the fits of eqs.8 and 10 to the solution and the
gel spectra, respectively. The parameters obtained from the
least squares fits are displayed in Table II.

The data listed in the table demonstrate that the form of the static contribution strongly depends on the particular network system. In end-linked PDMS gels the spatial distribution of nonuniformities suggests[34] the value s=2, i.e. the Gaussian distribution seems a fair approximation. For this system, however, the range of Q over which this fit can be tested is rather limited. In PVAc gels best fits are obtained with s=0.7, in agreement with the much broader distribution of non-uniformities expected in randomly cross-linked networks[16]. The presence of static non-uniformities causes excess scattered intensity in the small Q region of the gel spectra but does not influence significantly the dynamic concentration fluctuations.

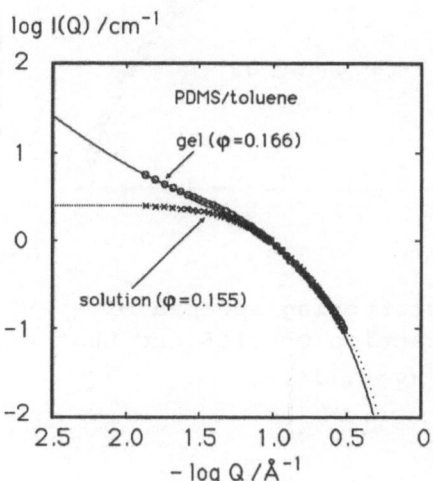

Fig.5.a. Neutron scattering spectra of a PDMS/toluene gel at volume fraction φ=0.166 and that of a PDMS/toluene solution (φ=0.155).

A critical test of the decomposition procedure is provided by comparing the swelling pressure calculated from the dynamic component of the gel scattering spectra with that obtained from swelling pressure measurements. In Fig.6 the swelling pressure of PVAc and PDMS gels determined from macroscopic osmotic measurements are presented as a function of the swelling pressure calculated from the dynamic component of the scattering spectra.

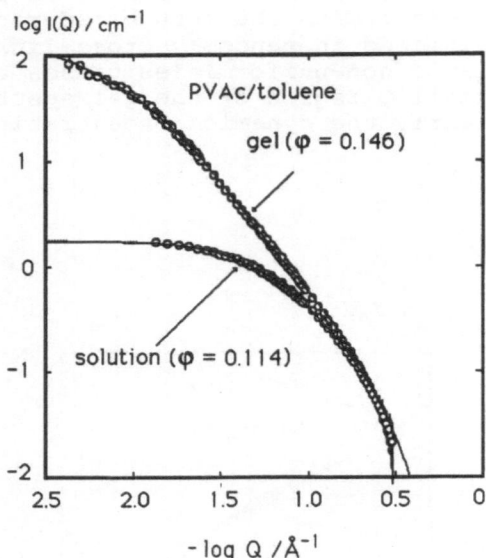

Fig.5.b. Neutron scattering spectra of a PVAc/toluene gel at
volume fraction φ=0.146 and that of a PVAc/toluene
solution (φ=0.114).

Table 2. Scattering parameters for PDMS/toluene and
PVAc/toluene networks and solutions

φ	s	$\Xi/\text{Å}$	$\xi/\text{Å}$	$\langle\delta\varphi^2\rangle$
PDMS/toluene				
Gel 0.166	2	59.5	16.7	$2.4\ 10^{-3}$
Solution 0.155	–	–	12.9	–
PVAc/toluene				
Gel 0.146	0.7	215	15	$7.5\ 10^{-3}$
Solution 0.114	–	–	18	–

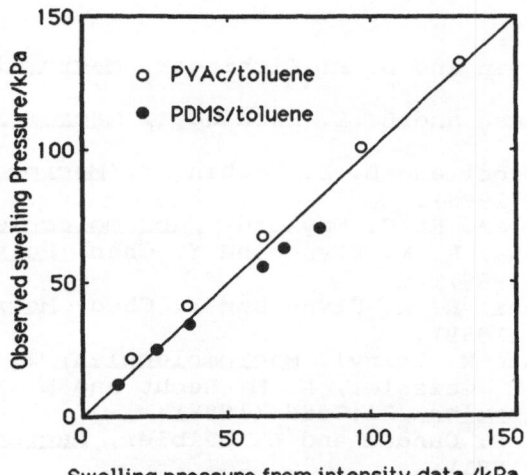

Fig.6. Swelling pressure, ω, of PVAc/toluene and PDMS/toluene gels obtained from macroscopic osmotic measurements as a function of the swelling pressure calculated from neutron scattering intensities.

It can be seen that the experimental points lie close to the theoretical straight line of slope unity.

CONCLUSIONS

The osmotic properties of chemically cross-linked polymer networks obtained from independent thermodynamic and scattering measurements are in reasonable agreement.

The thermodynamic behavior of the gels are satisfactorily described by a two-term equation containing a separable elastic and mixing contribution. In contrast to the classical Flory-Rehner theory the mixing term for the network polymer cannot be identified with that of the uncross-linked polymer of infinite molecular weight. The concentration dependence of the mixing pressure in the gel is close to that of the corresponding polymer solution.

The neutron scattering spectra of the gels at small wave vectors are dominated by the static nonuniformities of the gel structure. The form of the static contribution strongly depends on the cross-linking process. For end-linked polydimethyl-siloxane networks the spatial distribution of nonuniformities can be approximated by a Gaussian probability function. In randomly cross-linked poly(vinyl acetate) gels the distribution of the nonuniformities is certainly much broader. The thermodynamic information relevant to the gel systems are contained in the dynamic part of spectra.

ACKNOWLEDGEMENTS
This work is part of a joint CNRS-Hungarian Academy of Sciences project. We also acknowledge research Contract OTKA No.2158 from the Hungarian Academy of Sciences. We are grateful to the Institut Laue Langevin for beam time on the D11 instrument. F.H. acknowledges a research fellowship from the Alexander von Humboldt Stiftung.

REFERENCES

1. R. W. Brotzman and B. E. Eichinger, *Macromolecules*, 15:531 (1982).
2. R. W. Brotzman and B. E. Eichinger, *Macromolecules*, 16:1131 (1983).
3. N. A. Neuburger and B. E. Eichinger, *Macromolecules*, 21:3060 (1988).
4. M. Gottlieb and R. G. Gaylord, *Macromolecules*, 17:20 (1984).
5. G. B. McKenna, K. M. Flynn and Y. Chen, *Polym. Commun.* 29:272 (1988).
6. G. B. McKenna, K. M. Flynn and Y. Chen, *Macromolecules*, 22:4507 (1989).
7. F. Horkay and M. Zrinyi, *Macromolecules*, 21:3260 (1988).
8. F. Horkay, E. Geissler, A. M. Hecht and M. Zrinyi, *Macromolecules*, 21:2589 (1988).
9. J. Bastide, S. Candau and L. Leibler, *Macromolecules*, 14:719 (1981).
10. J. Bastide, F. Boué and M. Buzier, in "Molecular Basis of Polymer Networks" A. Baumgärtner and C. E. Picot, eds., Springer Proceedings in Physics 42; Springer, Berlin (1989).
11. A. M. Hecht, R. Duplessix , E. Geissler, *Macromolecules*, 18:2167 (1985).
12. N. S. Davidson, R. W. Richards and A. Maconnachie, *Macromolecules*, 19:434 (1986).
13. W. Wu, M. Shibayama, S. Roy, H. Kurokawa, L. D. Coyne, S. Nomura and R. S. Stein, *Macromolecules*, 23:2245 (1990).
14. W. Wu, S. Roy, H. Kurokawa and R. S. Stein, *Macromolecules*, 23:2245 (1990).
15. S. Mallam, A. M. Hecht, E. Geissler and P. Pruvost, *J.Chem.Phys.*, 91:6447 (1989).
16. F. Horkay, A. M. Hecht, S. Mallam, E. Geissler and A. R. Rennie, *Macromolecules*, 24:2896 (1991).
17. A. M. Hecht, F. Horkay, E. Geissler and J. P. Benoit, *Macromolecules*, 24:4183 (1991).
18. P. J. Flory and J. Rehner, *J.Chem. Phys.*, 11:521 (1943).
19. P. J. Flory, "Principles of Polymer Chemistry", Cornell University Press, Ithaca, NY (1953).
20. H. M. James and E. Guth, *J.Chem. Phys.*, 11:455 (1943).
21. L. R. G. Treloar, "The Physics of Rubber Elasticity", Clarendon, Oxford 3rd Ed. (1975).
22. P. G. de Gennes, "Scaling Concepts in Polymer Physics", Cornell, Ithaca, NY, (1979).
23. A. J. Pennings and W. Prins, *J.Polym.Sci.*, 49:507 (1961)
24. G. Gee, J. B. M. Herbert and R. C. Roberts, *Polymer*, 6:54 (1965).
25. R. F. Boyer, *J.Chem.Phys.*, 13:363 (1945).
26. K. Dusek, *Collection Czech.Chem.Commun.* 32:1554 (1967)
27. F. Horkay, E. Geissler, A. M. Hecht, P. Pruvost and M.Zrinyi, Polymer, 32:835 (1991).
28. F. Horkay, M. Nagy, M. Zrinyi, *Acta Chim.Acad.Sci.Hung.* 108:287 (1981).
29. F. Horkay and M. Zrinyi, *Macromolecules*, 15:1306 (1982).
30. H. Vink, *Europ. Polym. J*. 10:149 (1974)
31. M. Nagy and F.Horkay, *Acta Chim.Acad.Sci.Hung.* 104:49 (1980).

32. M. Ragnetti, D. Geiser, H. Höcker and R. C. Oberthür, _Makromol. Chem_., 186:1701 (1985)
33. F. Horkay, A. M. Hecht and E. Geissler, _J. Chem. Phys._, 91:2706 (1989)
34. S. Mallam, F. Horkay, A. M. Hecht, A. R. Rennie and E. Geissler, _Macromolecules_, 24:543 (1991)

CYCLIZATION DURING CROSSLINKING FREE-RADICAL POLYMERIZATIONS

Neil A. Dotson, Christopher W. Macosko, and Matthew Tirrell

Department of Chemical Engineering and Materials Science
University of Minnesota
Minneapolis, MN 55455

INTRODUCTION

Network formation by free-radical polymerization departs strongly from the behavior expected from the classical network theories.[1,2] The earliest manifestation of this discrepancy was the delay in the gel point from that predicted by one to two orders of magnitude.[3] One of the contributing factors to this delay has been shown in various studies to be the consumption of pendant double bonds by cyclization.

Cyclization is a reaction in which two functionalities, already connected, react so as to become further linked, forming a loop. Although possible in nominally linear stepwise systems, it is even more significant in nonlinear, network-forming systems of all types because of the increased possibilities for cyclization. Moreover, in these cases cyclization constitutes a severe departure from the classical Flory-Stockmayer theory[4-5] of gelation, severe because the cycles formed (at least in the pregel regime) are non-random. These cycles are thus difficult if not impossible to account for properly in a statistical branching theory.

Cyclization, and the attendant delay in gel point, is expected to be especially severe for nonlinear free-radical polymerizations for the following reason.[2,6] In a free-radical polymerization, high polymer of nearly infinitesimal concentration is formed at zero conversion, which because of connectivity allows neighbors of the active radical end to be from the same molecule, permitting cyclization if pendant double bonds are available. During a nonlinear stepwise (co)polymerization, on the other hand, at zero conversion potentially reactive neighbors must be from other molecules. Thus in a crosslinking free-radical polymerization, cyclization is expected to be significant at the beginning of the reaction, and crosslinking insignificant (because of the infinitesimal polymer concentration), while the reverse should hold for stepwise systems.

Cyclization during a crosslinking free-radical reaction can be categorized by how the radical and the pendant are attached, as done historically[7,8] and in the previous work of Landin[9,10] (see Figure 1). Cyclization within a primary chain is referred to as *primary cyclization* and occurs when a radical reacts with a pendant on its own primary chain. Thus, primary cyclization can occur during the formation of an isolated chain, or, in other words, at vanishing polymer concentration or conversion. This

Synthesis, Characterization, and Theory of Polymeric Networks and Gels
Edited by S.M. Aharoni, Plenum Press, New York, 1992

a   Primary Cyclization.

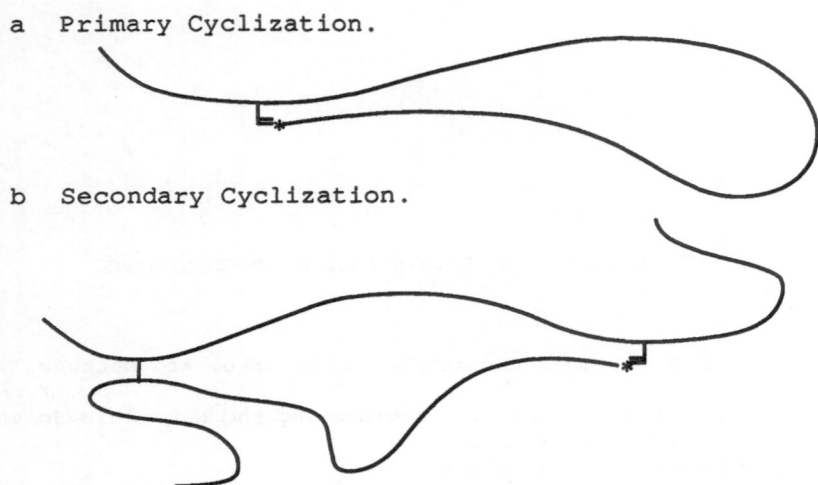

b   Secondary Cyclization.

Figure 1. Classification of Cycles Formed during Free-Radical
          Crosslinking Copolymerization.

distinguishes it from *secondary cyclization* which occurs between a radical
and a pendant on a chain to which that radical is attached by one or more
crosslinks, which is supposed only to occur at finite polymer concentration, and which should become increasingly more important as polymer concentration (or conversion) increases.

For cyclization to occur the two reactive groups, radical and pendant
double bond, attached to the same molecule need to come into a small reactive volume on the order of Ångstroms in dimension. If diffusion is not
controlling, then the chains equilibrate in configurational space between
reaction events, and the equilibrium concentration of the one group about
the other is controlling. This is true for free-radical polymerizations at
low conversion, since the time scale for propagation of an individual radical is generally on the order of 10^{-4} to 10^{-3} seconds while the time scale
for end-to-end cyclization is often on the order of 10^{-7} seconds (for short
chains) to 10^{-5} seconds (for moderately long chains, e.g. DP_n = 1000).[11]
(This latter time scale is that for the meeting of the two ends in space,
not necessarily for the reaction of functional groups at those ends.)

Thus, at low conversions static rather than dynamic configurational
properties of the polymer are controlling. If the polymer chain moreover
is unperturbed, then the distance r of any group about another on that
molecule is described by a Gaussian distribution, W(r).[4] The effective
concentration of that group about the other, located n bonds or monomers
away, is $C_n = \lim_{r \to 0} \{W(r)/(4\pi r^2)\}$, which scales as $n^{-\mu}$, where μ = 3/2. For
the formation of short cycles (less than 20 to 50 bonds, in general) for
which the short-range correlations which contribute to the characteristic
ratio are effective, Gaussian statistics are not, however, applicable and
the Rotational Isomeric State (RIS) model would need to be used.[12] Failure
to account for this would generally lead to an overestimation of the number
of these small cycles.[13] While performing the necessary RIS calculations
would be time-consuming, they present no theoretical difficulty.

Long-range excluded-volume effects which are generally present, however, do lead to theoretical difficulties. The segmental distribution of a
linear chain is altered, giving the familiar result that the radius of gyration scales like n^ν where ν = 3/5, rather than 1/2 as is the case for an

unperturbed chain.[4,14,15] More importantly for this problem, the end-to-end distribution of a linear chain is altered, especially about $r = 0$, so C_n depends more strongly on contour length ($\mu = 59/30$, rather than $\mu = 3/2$).[15-17] Furthermore, this end-to-end distribution does not adequately describe the relevant end-to-backbone distributions, as the expansion of the chain is not uniform, which has been shown theoretically[18-25] and experimentally.[26]

Of the previous theories addressing cyclization in free-radical systems,[6-9,27-30] only a few allow for prediction of the extent of cyclization.[6,9,27,29] That is the purpose of this paper: to make such predictions for primary cyclization, and to compare those predictions with the experimental results which are available. This will all be done assuming that intersegmental distances are described by the Gaussian distribution.

THEORETICAL PREDICTIONS

To determine the extent of primary cyclization, we look at a refinement of a model of Landin,[9] based on the much earlier ideas of Haward.[27] We consider a dead polymer chain of length ℓ at infinite dilution. The units in the chain may be bifunctional (mono-unsaturated) monomer A_2, tetrafunctional (di-unsaturated) crosslinker B_4, or the pendant B_2 of a previous crosslinker (i.e. a cycle). These are indexed down the chain from $i = 1$ (the initiated monomer) to $i = \ell$ (the terminated end). We will go as far as possible with a general chain length distribution $P(L=\ell)$, but will assume in the final results a geometric distribution:

$$P(L=\ell) = (1-q)\,q^{\ell-1} \tag{1}$$

where q is the probability of propagation[31,32] (i.e. the ratio of the rate of propagation to the combined rates of all reactions available to a radical). Because cycles are counted as reaction steps, L is no longer identical to the number of monomers (the degree of polymerization) but rather to the number of reaction steps, which we will call the chain length.

It will be assumed that, among sites which are not cycles, the crosslinker and monomer are distributed randomly, thus restricting ourselves to ideal copolymerization. Because this does allow unequal reactivity (with $r_A r_B = 1$) and thus the chain composition F_B to differ from the initial monomer composition f_B (the fraction of unsaturations in the initial mixture contributed by the crosslinker) both at the beginning and due to composition drift, we must denote the probability of a noncyclic unit being a crosslinker as F_B. (Even for the case of equal reactivity, the notation used must be F_B, since cyclization imposes a drift in composition.) Unfortunately, this means we use a notation which implies a broader applicability than intended, and this must be kept in mind.

Because we are interested in the cycles on the chain, we must observe the chain while it is growing to its final length ℓ. So, in addition to the details of the chain length distribution and the copolymer composition, we need to know the probability of the radical end (at position $i-1$) reacting with a pendant at position j, $j < i$, given that both the pendant and the radical still exist and that the radical does not terminate at this step. For chains of Gaussian configuration, we expect this to be given as P_{i-j}, i.e. a function only of the contour length between the two groups. (Note that this neglects the possible configurational influence of any intervening cycles, so that cycle formation is in this sense independent.) The probability P_{i-j} may be written as:

$$P_{i-j} = \frac{\text{rate of cycle formation between end at i-1 and pendant at j}}{\sum \text{rate of all addition reactions}} \qquad (2)$$

under the restrictions ('givens') above. We can rewrite eq 2 as follows:

$$P_{i-j} = rC_{i-j}/(M + rC_{i-j}) \qquad (3)$$

where r is the reactivity of the pendant relative to that of the monomer (assuming that f_B is small), C_{i-j} is the concentration of a pendant at j about the radical at i-1, and M is the overall monomer concentration. P_0 is obviously zero, but is included in certain summations to make the formalism plainer. (If we were dealing with chains with excluded volume for which the position of the pendant is important, then the single-indexed P_{i-j} would have to be replaced by the double-indexed $P_{i,j}$. A large share of the development that follows will be applicable for non-Gaussian chains if the replacement of P_{i-j} with $P_{i,j}$ is made; see reference 32 for details.)

What is the probability, $P(i=cycle)$, that the ith reaction formed a cycle? Or, equivalently, what is the probability that the ith unit is the pendant from a previous unit? Clearly this is related to P_{i-j}, $j=1,i-1$, defined above, but it is not identical to the sum of those, for a pendant need not be located at position j, and if one were it could have been consumed in some intervening cyclization reaction. To account for these complications we write:

$$P(i=cycle) = \sum_{j=1}^{i-1}(1-P(j=cycle))F_B\, P_{i-j} \prod_{k=0}^{i-j-1}(1-P_k) \qquad (4)$$

For the ith unit to be a cycle, it must react with one of the previous (i-1) units; hence, the summation over j. For the ith unit to react with the pendant of the jth, three conditions must be fulfilled: (1) the jth unit must not itself have been a prior pendant (as the closing of a cycle consumes the pendant, leaving no unsaturation on that unit) - hence the first term in the summation; (2) the jth unit must then be a crosslinker - with probability F_B; and (3) the reactions intervening between the jth and the ith must not have consumed the pendant of the jth unit - hence the product. Given these, cyclization will then occur with probability $P_{i,j}$. We thus obtain a recursive equation for $P(i=cycle)$ depending on all the previous $P(j=cycle)$, $j=1,i-1$, with $P(1=cycle) = 0$. Previous theories of cyclization in these systems have generally neglected the product term which prohibits multiple cyclization reactions of the same pendant,[6,29] although this is not the case with Landin's model[9] for which this development is an elaboration.

The expected or average number of cycles on a chain of length of length ℓ is given by:

$$E(N_{cycle}|\ell) = \sum_{i=2}^{\ell} P(i=cycle) \qquad (5)$$

and the expected number of monomers thus given by the difference:

$$E(N_{monomer}|\ell) = \ell - E(N_{cycle}|\ell) \qquad (6)$$

of which a fraction F_B are crosslinkers (bearing pendant units available for cyclization). The probability that a pendant on a chain of length ℓ is cyclized is thus:

$$P(\text{cycle}|\ell) = \frac{E(N_{\text{cycle}}|\ell)}{F_B(\ell - E(N_{\text{cycle}}|\ell))} \tag{7}$$

an equation about which we will say more later.

The fraction of pendants consumed in primary cyclization, or the probability of cyclization, P(cycle), can be written:

$$P(\text{cycle}) = \frac{\displaystyle\sum_{\ell=2}^{\infty} E(N_{\text{cycle}}|\ell)\, P(L=\ell)}{F_B \displaystyle\sum_{\ell=1}^{\infty} (\ell - E(N_{\text{cycle}}|\ell))\, P(L=\ell)} \tag{8}$$

i.e. an average over $P(L=\ell)$ which gives equal weighting to all pendants. Use of the geometric distribution simplifies this to the following:

$$P(\text{cycle}) = \frac{\displaystyle\sum_{i=2}^{\infty} P(i=\text{cycle})q^{i-1}}{\dfrac{F_B}{1-q} - F_B \displaystyle\sum_{i=2}^{\infty} P(i=\text{cycle})q^{i-1}} \tag{9}$$

This equation can be understood as follows. The numerator represents the number of cycles, which is the sum of the probabilities that an ith unit is a cycle, weighted by the probability that the chain survived to that unit. This is divided by the number of crosslinkers, equal to the number of monomers multiplied by F_B.

For the case of a position-dependent $P_{i,j}$, equation (9) is the simplest equation which can be written for P(cycle). For the case of Gaussian chains, however, equation (9) actually is not as straightforward as possible. To see this, consider an alternative argument for P(cycle), derived by taking the point of view of the pendant, rather than the radical end. Take a pendant on a growing chain, ignoring its position on the chain. The probability of forming a cycle of size s given that both the pendant and the radical still exist is simply qP_s, i.e. the probability that the radical adds another double bond multiplied by the probability that the double bond is the pendant. The probability that a cycle of size s forms, then, is given by:

$$P(\text{cycle of size } s) = qP_s \prod_{j=0}^{s-1} q(1-P_j) \tag{10}$$

where the product term is the probability that the pendant and the radical have both survived, which is equal to the product of the probabilities that at each previous step both have survived. The probability of a pendant being engaged in a primary cycle is simply the sum over all s, which yields (changing the index s to i):

$$P(\text{cycle}) = \sum_{i=1}^{\infty} q^i P_i \prod_{j=0}^{i-1} (1-P_j) \tag{11}$$

323

a much simpler equation. Now, nowhere in this argument have we assumed anything about F_B, and yet we find no dependence on F_B. Thus equation (9), despite its apparent complicated polynomial dependence on F_B, really has no F_B dependence, and the equivalence of equations (9) and (11) can be proven fairly simply.[32]

For position-independent cyclization, then, the overall probability of cyclization is independent of the fraction of crosslinkers on the chain. This is not true of the probability of cyclization on a chain of given length, somewhat paradoxically - the dependence of cyclization probability on chain length in the limit $F_B \to 0$ is given as:

$$P(\text{cycle}|\ell) = \frac{1}{\ell} \sum_{i=1}^{\ell-1} (\ell-i) P_i \prod_{j=0}^{i-1} (1-P_j) \tag{12}$$

which does not hold for finite F_B (the simple case $P_i = 0$, $i > 1$, is sufficient to show this). Since the distribution of cycles on chains of different lengths depends on F_B, so should the correction to the gel point prediction neglecting cyclization. This will be discussed further at the end of this paper. The fact that $P(\text{cycle})$ is independent of F_B, while a great mathematical simplification, is actually the result of one of the great weaknesses of this theory, which will be discussed when we compare with experimental values for the pendants wasted in primary cyclization. Finally, for the case of position-dependent cyclization, both the overall probability of cyclization and its distribution over chain length are dependent on the fraction of crosslinkers on a chain.

Before proceeding to the size distribution of these cycles, it is worth noting that for the case of infinitely long chains ($q = 1$), equation (11) reduces to

$$P(\text{cycle}) = 1 - \prod_{j=1}^{\infty} (1-P_j) \tag{13}$$

from which it can be proven[32] that $P(\text{cycle})$ is less than unity and that an upper bound can be written as

$$P(\text{cycle}) < 1 - e^{-a\zeta(\mu) r/M} \tag{14}$$

when $C_n \leq an^{-\mu}$, where $\zeta(\mu)$ is the Riemann zeta function[33] (for Gaussian chains, $\mu = 3/2$ and $\zeta(3/2) \approx 2.6124$).

COMPARISON WITH EXPERIMENT

Measurement of Primary Cyclization

Measuring the extent of cyclization poses a problem. Chemically, a crosslink and a cycle are equivalent; thus, with the possible exception of the smallest cycles, a doubly reacted crosslinker participating in a cycle is indistinguishable (by any spectroscopic technique) from one participating in a crosslink. The difference between the two is topological: it depends upon prior connectivity, possibly over large contour lengths. The only experiment which directly measures the amount of cyclization is that in which one hydrolyzes the crosslinker and compares the number of hydrolysis reactions to the decrease in DP_n.[34-38] Hydrolysis of a crosslinker participating in a crosslink will lead to a decrease in DP_n, while that of one participating in a cycle will not. Notice that this does not directly

324

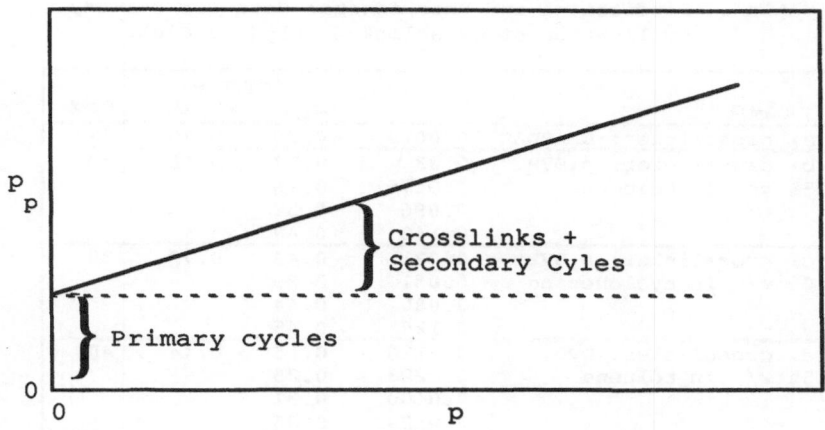

Figure 2. Typical Approach to Inference of Cyclization.

distinguish between primary and secondary cyclization.

Short of this method, then, one must *infer* the existence of cycles.
This is usually done by measuring the conversion of the pendants in the
polymer, p_p, and comparing with the mean-field value predicted (~p/2 at low
conversion, p). Distinguishing between primary and secondary cyclization
is then done by comparing the pendant conversion at different overall conversions (see Figure 2).[9,10] The intercept is attributed to primary cyclization and the slope to both crosslinking and secondary cyclization. Pendant conversion can be measured in any of a number of ways: infrared,[39] ul-
traviolet,[40] nuclear magnetic resonance,[6,9,10,41,42] or Raman spectroscopy,[43] or by bromination.[44,45]

Is it valid to equate the pendant conversion extrapolated to zero
conversion with the extent of primary cyclization, as Figure 2 implies?
The logic which has led to this equality in the past has been that at zero
conversion, the polymer concentration is zero, and thus a growing chain
cannot crosslink but can only propagate with monomer, or engage in primary
cyclization.[2,6] In fact, though, this situation never occurs. The zero-
conversion ensemble can essentially be equated with the ensemble of chains
made in the first radical lifetime, during which the radical concentration
increases from zero to nearly the quasi-steady state concentration. These
chains do not differ significantly from those formed immediately thereafter; the kinetic chain length is at most $\pi/2$ longer than that in the subsequent steady state.[32] This is necessary, because these chains terminate.
In the case for which transfer is negligible so that polymer chains end by
termination rather than transfer, each chain must, near the end of its
life, encounter another growing chain. It is entirely possible that before
the two chain ends meet to terminate, the two growing chains may crosslink
and form secondary cycles. Thus, even the zero-conversion ensemble may
contain contributions due to crosslinking and secondary cyclization between
growing chains. The extent of this contribution is uncertain, but none of
the conclusions of this paper is affected by this uncertainty. The measured extents of primary cyclization may be considered as upper bounds.

Crosslinked Poly(styrene)

The majority of studies of primary cyclization in nonlinear free-
radical systems have been from styrene copolymerized with small amounts of
crosslinker. The references analyzed are:

Table 1. Predictions and Experimental Data for Primary
Cyclization of Crosslinked Poly(styrene).

System	F_B	P(cycle) Exp.	P(cycle) Theo.	Ref.
(a) crosslinker: BMABDAE	0.0024	<0.20	0.25	37
(b) crosslinker: p-DVB 15% v/v in toluene	0.03	0.17	0.58	39
	0.0475	0.16		
	0.096	0.32		
	0.142	0.40		
(c) crosslinker: p-DVB 10% v/v in cyclohexane	0.021	0.48	0.72	39
	0.051	0.39		
	0.080	0.44		
	0.128	0.45		
(d) crosslinker: DVD 15% v/v in toluene	0.0116	0.15	0.44	40
	0.0223	0.25		
	0.0400	0.31		
	0.0528	0.36		
(e) crosslinker: DIPD 15% v/v in toluene	0.0076	0.079	0.29	40
	0.0132	0.061		
	0.0246	0.098		
	0.0339	0.11		
(f) crosslinker: EGDMA 15% v/v in toluene	0.0364	0.075	0.94	44
	0.0670	0.195		
	0.0900	0.22		
	0.1135	0.27		
(g) crosslinker: TEGDMA 15% v/v in toluene	0.032	0.13	0.74	44
	0.0565	0.24		
	0.079	0.255		
	0.107	0.30		
(h) crosslinker: PEGDMA 15% v/v in toluene	0.0135	0.10	0.53	44
	0.0285	0.16		
	0.0610	0.295		
	0.1010	0.36		

1. Wesslau.[37] Copolymers of styrene with small amounts of bis(4-methacryloxybenzilidene)-1,2-diaminoethane (BMABDAE) were analyzed after hydrolysis by photometry.

2. Soper, Haward, and White.[39] Copolymers of styrene and small amounts of p-divinylbenzene (p-DVB) were analyzed by IR spectroscopy to obtain pendant conversion.

3. Holdaway, Haward, and Parsons.[40] Copolymers of styrene with small amounts of 4,4'-divinylbiphenyl (DVD) and 4,4'-diisopropenylbiphenyl (DIPD) were analyzed by UV spectrometry to obtain pendant conversion.

4. Shah, Holdaway, Parsons, and Haward.[44] Copolymers of styrene with small amounts of three different ^{14}C-labelled dimethacrylates were analyzed by bromination titration to obtain the number of double bonds, and by radioactivity measurements to obtain the number of crosslinkers. Dimethacrylates used were ethylene glycol dimethacrylate (EGDMA), tetraethylene glycol dimethacrylate (TEGDMA) and poly(ethylene glycol) dimethacrylate (PEGDMA) with a poly(ethylene glycol) bridge of about 9 units.

No parameters were fit in the calculations based on equation (11). Characteristic ratios were taken as those of the chains in the absence of crosslinker. The relative pendant reactivity, r, was taken to be the inverse of the reactivity ratio r_A. (For more details about the sources of the values used for these parameters, see reference 32.)

Table 1 summarizes the results of this comparison. The agreement between theory and experiment for P(cycle) is quite poor; the theory overpredicts in the majority of cases (comparison is always made with experimental results at the lowest F_B). The most obvious fault with the theoretical predictions is that Gaussian statistics are assumed even for the smallest cycles, thus leading to an overprediction of the amount of cyclization.[13] While this is undoubtedly a source of error, it cannot be the only source of error, because in all cases above but two, at least the smallest 25 cycles need to be completely suppressed in order to match the predictions with experiment, and in three - (d), (e), and (h) - at least the smallest 50 or so cycles need to be discarded. This is far in excess of the number of bonds needed for the Gaussian distribution to be appropriate. This correction moreover assumes complete suppression of these smaller cycles, which is not necessarily the case, although it is certainly true for systems with p-DVB[39] and the biphenyl crosslinkers[40] for which the smallest cycle is definitely prohibited. The two exceptions to this disagreement are (a) and (c). Whether case (a) truly constitutes agreement is in question, since the extent of primary cyclization may be as low as 5%. Case (c), on the other hand, does present an exception, since only the three smallest cycles need to be excluded to match theoretical and experimental results. In this case, it is easily believed that non-Gaussian characteristics for the small cycles may be the sole cause of the disagreement.

Thus it appears that the non-Gaussian character of the small cycles is not sufficient to account for the discrepancy. That this is so except in case (c), for which the polymerization was performed in a poor (though probably better than θ) solvent, indicates that excluded volume plays a role. Although one data point is not sufficient to prove this point, there is no reason to assume that excluded volume is not effective at reducing the amount of cyclization. That excluded volume plays a role also is consistent with the fact that increasing the bridge length in dimethacrylates increases, rather than decreases, cyclization (see (f) through (h) in the table; Landin[9] has also noted this, see below). The longer the pendant bridge, the fewer intrachain contacts (which incur an energetic penalty) possibly entailed in forming a cycle. All of this suggests that experimental work in (mixed) solvents of different qualities, possibly with crosslinkers of different bridge length would be very enlightening.

Another interesting feature shown in Table 1 is that increasing the amount of crosslinker improves the comparison. This is surprising, since the theory should only be applicable for small amounts of crosslinker which ensures independence of cycle formation. Further support for this trend is given in Table 2, which shows data from the following:
    5. Dušek and Spěváček.[6] Styrene was copolymerized with modest amounts of EGDMA, and the resulting copolymer analyzed by ^{1}H and ^{13}C NMR, as well as IR spectroscopy.[41]
(The changes in the theoretical predictions with F_B are due solely to small changes in monomer concentration with monomer composition.) Here, we see a number of cases for which the theory *underpredicts*; in those cases in which overprediction is still the case, the difference is easily attributed to the small cycles. Clearly, with increasing amount of crosslinker, there is an additional effect which counteracts the discouragement of cycles by excluded volume.

With increasing amount of cyclization, do cycles still form independently? Figure 3 illustrates the effect of intervening cycles on a possible reaction between a radical end and a particular pendant. Intervening cycles constrain the radical end to regions of space nearer to the backbone and thus nearer to the pendant in question, so that the probability of cyclization is increased. Essentially, then, the intervening cycles act to reduce the apparent contour length between the radical and the pendant. Thus, contrary to theory, we expect P(cycle) to increase upon increasing F_B, rather than stay constant. This certainly accounts for the increase in P(cycle) with F_B in Table 1, and for the (fortuitous) agreement between theory and experiment at high F_B in Table 2; two effects not accounted for

Table 2. Predictions and Experimental Data for
Primary Cyclization of Highly Crosslinked
Poly(styrene). All from reference 6.

System	F_B	P(cycle) Exp.	P(cycle) Theo.
(a) crosslinker:EGDMA bulk	0.12	0.62	0.39
	0.18	0.60	0.38
	0.28	0.46	0.38
	0.43	0.31	0.37
	0.55	0.25	0.36
(b) crosslinker: EGDMA 80% v/v in toluene	0.11	0.65	0.46
	0.19	0.57	0.45
	0.26	0.52	0.45
	0.42	0.34	0.44
	0.55	0.25	0.43
(c) crosslinker: EGDMA 60% v/v in toluene	0.12	0.55	0.55
	0.18	0.53	0.55
	0.26	0.46	0.55
	0.40	0.34	0.54
	0.53	0.23	0.52
(d) crosslinker: EGDMA 40% v/v in toluene	0.11	0.70	0.70
	0.16	0.58	0.70
	0.25	0.51	0.69
	0.39	0.36	0.68
	0.53	0.21	0.67
(e)crosslinker: EGDMA 20% v/v in toluene	0.22	0.54	0.90
	0.46	0.30	0.89

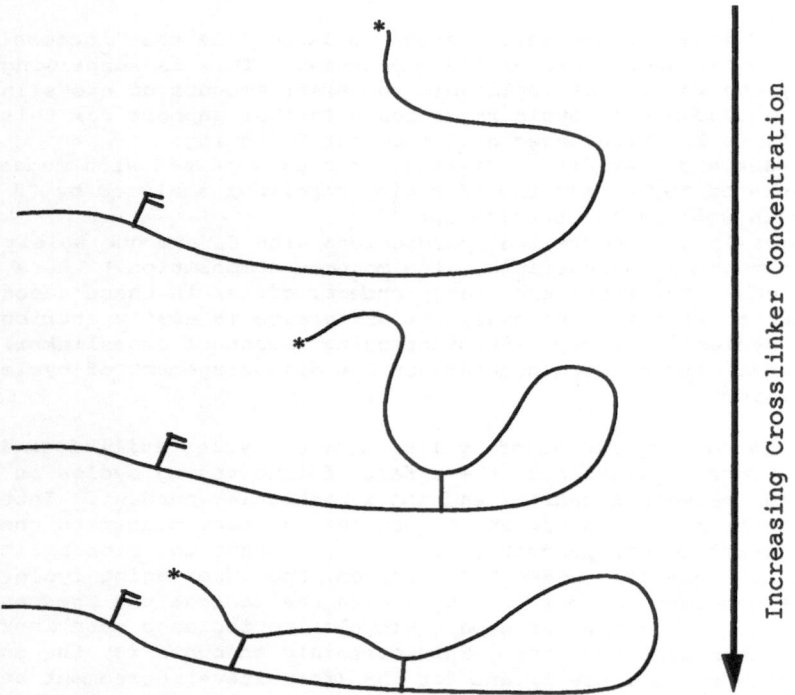

Increasing Crosslinker Concentration

Figure 3. Cyclization in the Presence of Intervening Cycles.

in the theory - excluded volume and correlation between cycles - have
cancelled each other out. But Table 2 shows that at yet higher amounts of
crosslinker, P(cycle) decreases, so that yet some third effect is
influencing events. Such an effect may be the formation of heterogeneity,
and a resulting shielding of pendants.

Another complaint that could be made against the theory is that it
assumes random placement of crosslinkers among non-cycled sites, which is
only true for ideal copolymerization, and none of the systems studied are
ideal. Nonetheless, in a copolymerization, as the amount of one comonomer
vanishes, its placement will be essentially random because most radicals
will be those on the other monomer. Thus the effect of the strict inappli-
cability of this theory should not matter for small amounts of crosslinker.

It is worthwhile noting that the expected size of a cycle is always
quite large, often on the order of one hundred monomers.[32] This, along
with the fact that the square root of the variance of the cycle size distribution is often twice this, shows that the distribution of cycle sizes
is quite broad, and that a treatment like the one in this paper (accounting
for cycles of all sizes) is necessary unless strict cyclopolymerization is
occurring. Results for the expected cycle size are in disagreement as was
the case for P(cycle). This may be due to the approximate nature of the
experimental measurement of the average cycle size,[39] as well as the fact
that theoretically it is much more sensitive to q and the shape of P_i than
is P(cycle). Thus these predictions are more sensitive to the parameters
than are the predictions for P(cycle) discussed above.

Crosslinked Poly(methyl methacrylate)

To a large extent work subsequent to that with poly(styrene) backbones reviewed above has been done with crosslinked poly(methyl methacrylate). These studies are:
1. Minnema and Staverman.[36] Methyl methacrylate (MMA) was copolymerized with small amounts of N,N'-bis(β-methacryloxyethyl)-p-xylylenedi-
urethane (MXU), and the resulting copolymers analyzed by hydrolysis.
2. Whitney and Burchard.[42] MMA was copolymerized with ~15% EGDMA,
and the resulting copolymer was analyzed by ^{1}H NMR.
3. Mrkvičková and Kratochvíl.[38] MMA was copolymerized with BMABDAE
(as used by Wesslau), and the resulting copolymer was analyzed, after
hydrolysis, by elemental analysis, NMR, and UV spectroscopy.
4. Landin and Macosko.[9,10] MMA was copolymerized with small amounts
of EGDMA, and the resulting copolymer analyzed by ^{1}H NMR.
The results of these predictions are shown in Table 3. (Again, see
reference 32 for details as to the values of the parameters used.)

The results here for the most part substantiate the conclusions
above, although there are more exceptions to the trend of disagreement. Of
the exceptions - (a), (b), (g) and (h) - only one, (b), can be explained by
the relatively large amount of crosslinker. No satisfactory explanation
can be found for the agreement for the others. For the remaining cases at
least ~20 of the smallest cycles must be eliminated in order to establish
agreement, probably more than can be justified theoretically.

Although there is no large body of data like that of Dušek and
Spěváček[6] for higher amounts of crosslinker in MMA polymerizations, it
should be pointed out that in the work of Galina et al.[43] on polymerization
of EGDMA, pendant conversion was measured to be ~30%, independent of conversion and fairly independent of dilution, in agreement with the earlier
results of Aso[28] on EGDMA and with the results of Dušek and Spěváček[6] already discussed. A value of 30% is probably close to that which would be
predicted by theory, in agreement with the supposed opposing influences of
excluded volume and correlation between cycles.

Table 3. Predictions and Experimental Data for Primary
Cyclization of Crosslinked Poly(methyl methacrylate). All systems bulk unless otherwise noted.

System	F_B	P(cycle) Exp.	P(cycle) Theo.	Ref.
(a) crosslinker: MXU	0.013	0.25	0.19	36
	0.026	0.25	0.19	
(b) crosslinker: EGDMA	0.15	<0.10	0.17	42
(c) crosslinker: BMABDAE	0.0085	0.07	0.11	38
	0.0151	0.06		
	0.0243	0.14		
	0.0218	0.31		
(d) crosslinker: EGDMA w/o LM	0.01	0.028	0.19	9,10
	0.02	0.035		
	0.03	0.030		
(e) crosslinker: EGDMA w/ LM	0.01	0.046	0.18	9,10
	0.02	0.039		
	0.10	0.23		
(f) crosslinker: EGDMA w/o LM; 50% v/v in toluene	0.02	0.072	0.35	9,10
(g) crosslinker: EGDMA w/o LM; 25% v/v in toluene	0.02	0.35	0.57	9,10
(h) crosslinker: TEGDMA w/o LM	0.02	0.81	0.13	9

Table 4. Predictions and Experimental Data for Primary
Cyclization of Crosslinked Poly(acrylamide).
All systems 6% v/v in H_2O. All from refer-
ence 45.

System	F_B	P(cycle) Exp.	P(cycle) Theo.
(a) crosslinker: BAM; w/o isopropyl alcohol(IPA)	0.07	0.81	0.96
	0.15	0.83	
(b) crosslinker: BAM; w/ IPA	0.07	0.84	0.95

It should be pointed out that the data of Landin is questionable on
two grounds. First, with such minute quantities of crosslinker, the
measurement of pendant conversions of 10% is very difficult by ^{1}H NMR.
Second, the calculation of pendant conversion used is based on the possibly
flawed assumption of equal reactivity (justified on the basis of NMR data
which is elsewhere claimed to be unreliable). Until data is published that
contradicts that of Landin, though, it is reasonable to assume it valid.

Crosslinked Poly(acrylamide)

The extent of cyclization has been found for other systems, particularly polyallyl systems[34,35,46] for which unfortunately not enough is known
(e.g. characteristic ratios) to make theoretical predictions. This is not
the case for crosslinked poly(acrylamide) systems, for which Tobita and
Hamielec[45] have recently published data (see Table 4). Here, acrylamide
has been cross-linked by N,N'-methylenebisacrylamide (BAM) and the

resulting copolymer analyzed by bromination to obtain the surviving pendant double bonds. In this case the theoretical predictions are essentially in agreement with the experimental results, but this is probably due to the high level of cyclization promoted by extreme dilution. Because of this, the effect of correlations between cycles may be expected to influence the amount of cyclization at lower levels of crosslinker.

CONCLUSIONS

Predictions for the amount of primary cyclization in crosslinking free-radical polymerizations are, for the most part, consistently higher than the experimental results under conditions at which the theory should be most applicable (i.e. low fraction of crosslinker, and bulk monomer). Overestimation due to the neglect of special treatment of the small, non-Gaussian cycles is partly to blame, but does not seem to be sufficient to explain the entire difference. The role of excluded volume in suppressing the larger cycles is suggested by data in a poor solvent[39] for which the predictions more nearly match experiment, and by the fact that increasing the size of a crosslinker often seems to increase cyclization, contrary to predictions.[9,44] Thus it is clear that much work needs to be done in order to improve the predictions, in particular: (1) use of RIS theory to predict the small cycle formation; and (2) incorporation of excluded volume effects, after the influence of solvent quality has been demonstrated more adequately by experiment.

Correlation between cycles, which contributes increasingly as the amount of crosslinker increases (see Figure 3), appears to improve the comparison between theory and experiment, but this is merely the fortuitous cancellation of two errors. Accounting for these correlations between cycles is difficult, primarily because the topology or 'cycle structure'[47] of these molecules can be very complex. The structure of these molecules is thus far more complicated than the cases treated in the past of chains with pendant cycles,[48] instead being more closely related to more densely intramolecularly crosslinked molecules,[49] although even here there is a distinction since in the case at hand the chains are being crosslinked while being formed. In any case, it is difficult to know what the equilibrium configuration of these molecules would be, although the first approximation would be to assume a Gaussian end-to-end distribution along the shortest path between the pendant double bond and the radical. Corrections along these lines should not be attempted before the predictions which should be valid in the limit of no crosslinkers are brought into agreement.

Further complicating matters is the fact that as the amount of crosslinker further increases, a third effect seems to come into play which decreases the amount of cyclization, contrary to the trend established by correlations between cycles.[6] It is very likely that this effect may be linked with heterogeneity and hence will be very difficult to describe.

The discussion thus far has focused entirely on the extent of cyclization and (to a much lesser degree) on the size of the cycles thus formed. These are clearly important, and provide predictions which can be directly compared with experiment results. These predictions alone, however, are not sufficient to tell how the branched structure is affected by these cycles, at least not beyond obvious observations such as, "cyclization will delay the gel point." Incorporation of cycles into the statistical (recursive) description is the subject of this section, which will be limited to a qualitative discussion rather than rigorous math.

Incorporation of primary cyclization into the recursive description is relatively straightforward, since all of the cyclization occurs within a primary chain, which can be considered as a superspecies. Unfortunately, despite the complicated appearance of the development for primary cyclization in this paper, it is not sufficiently detailed to allow for predicting the appropriate chain statistics for deriving the weight-average molecular weight, M_w. What is lacking is the joint distribution of chain length, number of crosslinkers, and number of cycles.

Despite the lack of detailed development, some qualitative comments can be made. Restricting attention to the gel point, for a general copolymerization it can be shown[32] that the gel point is given as:

$$P_{B,c} = 1/E(N_{B4|B4}{}^{out})^0 \qquad (15)$$

where $P_{B,c}$ is the critical conversion of B_2-type (crosslinker) unsaturations, and $E(N_{B4|B4}{}^{out})^0$ is the expected number of crosslinkers (B_4) on a primary chain (the polymer considered with crosslinks and secondary cycles cut) looking out of a crosslinker. This is the proper extension of the Stockmayer criterion[5] for gelation when crosslinkable sites are not necessarily distributed randomly. In the presence of cycles[56] it is clear that the gel point will be given by the following equation:

$$P_{B,un,c} = 1/E(N_{B4,un|B4,un}{}^{out})^0 \qquad (16)$$

where $P_{B,un}$ is the conversion of B_2 groups not involved in cycles (uncyclized), and $E(N_{B4,un|B4,un}{}^{out})^0$ the expected number of uncyclized B_4 groups on a primary chain looking out of an uncyclized B_4 group. The first approximation that can be made is to write the following:

$$E(N_{B4,un|B4,un}{}^{out})^0 = (1-P(cycle))F_B(DP_w{}^0-1) \qquad (17)$$

where $DP_w{}^0$ is the weight-average degree of polymerization of the primary chains. This approximation is based on the assumption that cycles are randomly distributed on chains, and as such represents a *spanning tree approximation*.[51-55] The extent to which this approximation is valid can only be known by knowing the joint distribution mentioned above, from which the true $E(N_{B4,un|B4,un}{}^{out})^0$ can be calculated. Even in the absence of this, though, we may observe the following. In Figure 4 is shown $P(cycle|\ell)$ for the theoretical modelling of the bulk MMA/EGDMA polymerizations of Landin and Macosko.[9,10] The probability of cyclization as a function of chain length approaches its asymptotic value fairly slowly (three decade of chain length is apparently not sufficient to have reached this). If the chains are infinitely long, the spanning tree approximation will be exact; but if the chains are of finite length, the shorter chains in the distribution will have fewer cycles per monomer unit than the longer chains, and the following inequality will hold (since $P(cycle|\ell)$ grows slower than ℓ):

$$E(N_{B4,un|B4,un}{}^{out})^0 < (1-P(cycle))F_B(DP_w{}^0-1) \qquad (18)$$

Thus, the gel point predicted by the spanning tree approximation will be lower than the actual gel point; i.e. primary cyclization delays the gel point more severely than the spanning tree approximation predicts.

Prediction of secondary cyclization can also be made by arguments similar to those for primary cyclization, although in a much more approximate manner.[32] Comparison with experiment in the manner illustrated in Figure 2 is problematic, though, because of the difficulty in discriminating between secondary cycles and crosslinks. If one assumes, however, the mean-field value for the number of crosslinks, the comparison is quite poor. While it is clear that the number of crosslinks is probably less than that predicted by the mean-field theory,[1] this only makes the comparison worse.[56] Incorporation of secondary cycles into the statistical

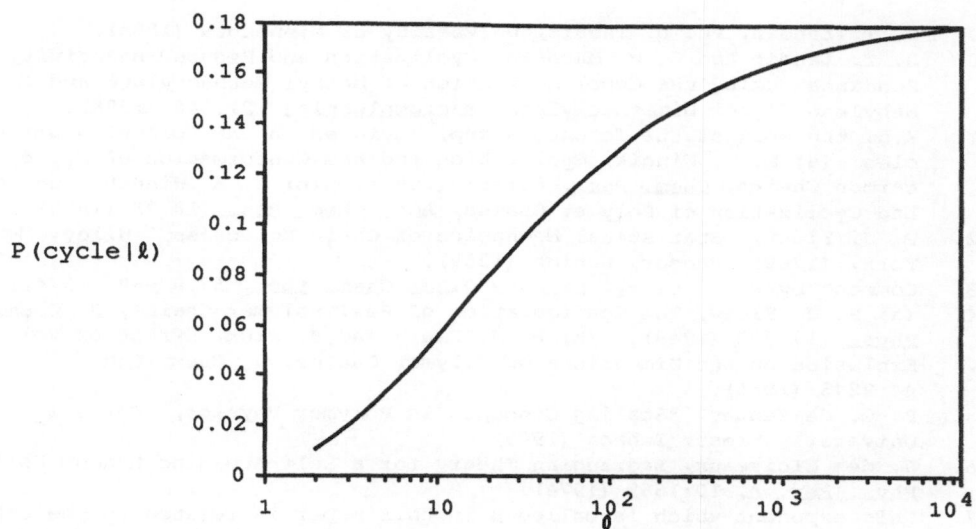

Figure 4. Predicted P(cycle|ℓ) versus ℓ for Bulk MMA/EGDMA Polymerization
at F_B = 0.01.

description of the polymerization, in an exact way, is moreover impossible
due to the long-range structural correlations those cycles impose.[57]

ACKNOWLEDGEMENTS

N.A.D. acknowledges the following sources of support: the Army
Research Office (grant #DAAG29-83-K-0149), Chevron, the University of
Minnesota Graduate School, and the Office of Naval Research.

REFERENCES

1.   (a) N. A. Dotson, T. Diekmann, C. W. Macosko and M. Tirrell, Non-
      idealities Exhibited by Cross-linking Free-Radical Copolymerization,
      talk presented at ACS National Meeting, San Fransisco, April 6, 1992.
      (b) N. A. Dotson, T. Diekmann, C. W. Macosko and M. Tirrell, Non-
      idealities Exhibited by the Crosslinking Copolymerization of Methyl
      Methacrylate and Ethylene Glycol Dimethacrylate, submitted to
      Macromolecules, October, 1991.
2.   K. Dušek, Network Formation by Chain Cross-linking (Co)Polymerisa-
      tion, in: "Developments in Polymerisation - 3"; R. N. Haward, ed.,
      Applied Science, London (1982).
3.   C. Walling, Gel Formation in Addition Polymerization, J. Am. Chem.
      Soc., 67:441 (1945).
4.   P. J. Flory, "Principles of Polymer Chemistry," Cornell University
      Press, Ithaca (1953).
5.   W. H. Stockmayer, Theory of Molecular Size Distribution and Gel
      Formation in Branched Polymers. II. General Crosslinking, J. Phys.
      Chem., 12:125 (1944).
6.   K. Dušek and J. Spěváček, Cyclization in Vinyl-Divinyl Copolymeriza-
      tion, Polymer, 21:750 (1980).
7.   M. Gordon, Network Theory of the Gel Point and the "Incestuous"
      Polymerization of Diallyl Phthalate, J. Chem. Phys., 22:610 (1954).
8.   M. Gordon and R.-J. Roe, Diffusion and Gelation in Polyadditions. IV.
      Statistical Theory of Ring Formation and the Absolute Gel Point, J.
      Polym. Sci., 21:75 (1956).

9.   D. T. Landin, Ph. D. Thesis, University of Minnesota (1985).

10. D. T. Landin and C. W. Macosko, Cyclization and Reduced Reactivity of Pendants during the Copolymerization of Methyl Methacrylate and Ethylene Glycol Dimethacrylate, Macromolecules, 21:846 (1988).

11. E.g. the work of the Toronto group, reviewed in the following articles: (a) M. A. Winnik, Cyclization and the Conformation of Hydrocarbon Chains, Chem. Rev., 81:491 (1981). (b) M. A. Winnik, End-to-End Cyclization of Polymer Chains, Acc. Chem. Res., 18:73 (1985).

12. P. J. Flory, "Statistical Mechanics of Chain Molecules," Wiley, New York, (1969); Hanser, Munich (1989).

13. Comment by P. J. Flory, Faraday Disc. Chem. Soc., 57:87-88 (1974).

14. (a) P. J. Flory, The Configuration of Real Polymer Chains, J. Chem. Phys., 17:303 (1949). (b) P. J. Flory and S. Fisk, Effect of Volume Exclusion on the Dimensions of Polymer Chains, J. Chem. Phys., 44:2243 (1966).

15. P.-G. de Gennes, "Scaling Concepts in Polymer Physics," Cornell University Press, Ithaca (1979).

16. J. des Cloizeaux, Lagrangian Theory for a Self-Avoiding Random Chain, Phys. Rev. A, 10:1665 (1974).

17. This exponent which is called μ in this paper is related to the other exponents of self-avoiding walks by the following: $\mu = -\nu d - \gamma + 1$. Here, ν is the scaling exponent on the end-to-end distance, d is the dimensionality of space, and γ is an entropic exponent (the total number of walks is proportional to $z^N N^{\gamma-1}$). From simple scaling arguments one obtains the following: for d = 2: $\nu = 3/4$, $\gamma = 4/3$, and so $\mu = 11/6$; for d = 3, $\nu = 3/5$, $\gamma = 7/6$, and so $\mu = 59/30$; for d $\geq$ 4, $\nu = 1/2$ and $\gamma = 1$, and so $\mu = d/2$ (since linear chains are ideal at d $\geq$ 4). It should be noted that these results for μ assume the simple scaling results for ν and γ; slightly different results are obtained from different arguments; e.g. $\mu = 1.9175$ instead of 59/30 (= 1.96) if values of ν and γ from ε-expansion are used (P.-G. de Gennes, Exponents for the Excluded Volume Problem as Derived by the Wilson Method, Phys. Lett., 38A:339 (1972)). None of the conclusions of this paper depends on such small differences.

18. M. Kurata, H. Yamakawa, and E. Teramoto, Theory of Dilute Polymer Solutions. I. Excluded Volume Effect, J. Chem. Phys., 28:785 (1958).

19. F. T. Wall and J. J. Erpenbeck, New Method for the Statistical Computation of Polymer Dimensions, J. Chem. Phys., 30:634 (1959).

20. Z. Alexandrowicsz and Y. Accad, Monte Carlo of Chains with Excluded Volume: Distribution of Intersegmental Distances, J. Chem. Phys., 54:5338 (1971).

21. A. J. Barrett, Investigation of the Moments of Intrachain Distances in Linear Polymers, Macromolecules, 17:1561 (1984).

22. J. des Cloizeaux, Short-Range Correlation between Elements of a Long Polymer in a Good Solvent, J. Phys., 41:223 (1980).

23. Y. Oono and T. Ohta, The Distribution Function for Internal Distances in a Self-Avoiding Polymer Chain, Phys. Lett., 85A:480 (1981).

24. W. L. Mattice, Subchain Expansion in Generator Matrix and Monte Carlo Treatments of Simple Chains with Excluded Volume, Macromolecules, 14:1491 (1981).

25. F. Ganazzoli, Excluded Volume Effect on Polymer Chains: Internal Parts vs. Free Chain Expansion, Makromol. Chem., 187:697 (1986).

26. Y. Matsushita, I. Noda, M. Nagasawa, T. P. Lodge, E. J. Amis and C. C. Han, Expansion Factor of a Part of a Polymer Chain in a Good Solvent Measured by Small-Angle Neutron Scattering, Macromolecules, 17:1785 (1984).

27. R. N. Haward, Polymerization of Diallyl Phthalate, J. Polym. Sci., 14:535 (1954).

28. C. Aso, Polymerization of Ethylene Glycol Dimethacrylate, J. Polym. Sci., 39:475 (1959).

29. (a) K. Dušek and M. Ilavsky, Cyclization In Crosslinking Copolymerization. I. Chain Polymerization of a Bis Unsaturated Monomer (Monodisperse Case), J. Polym. Sci. Symp., 53:57 (1975). (b) K. Dušek and M. Ilavsky, Cyclization In Crosslinking Copolymerization. II. Chain Polymerization of a Bis Unsaturated Monomer (Polydisperse Case), J. Polym. Sci. Symp., 53:75 (1975).

30. A. B. Scranton and N. A. Peppas, A Statistical Model of Free-Radical Copolymerization/Crosslinking Reactions, J. Polym. Sci. Polym. Chem., 28:39 (1990).

31. N. A. Dotson, R. Galván and C. W. Macosko, Structural Development during Nonlinear Free-Radical Polymerizations, Macromolecules, 21:2560 (1988).

32. N. A. Dotson, Ph.D. Thesis, University of Minnesota, 1991.

33. M. Abramowitz and I. A. Stegun, "Handbook of Mathematical Functions," Dover: New York (1972).

34. W. Simpson, T. Holt and R. J. Zetie, The Structure of Branched Polymers of Diallyl Phthalate, J. Polym. Sci., 10:489 (1953).

35. W. Simpson and T. Holt, Gelation in Addition Polymerization, J. Polym. Sci., 18:335 (1955).

36. L. Minnema and A. J. Staverman, The Validity of the Theory of Gelation in Vinyl-Divinyl Copolymerization, J. Polym. Sci., 29:281 (1958).

37. H. Wesslau, Zur Kenntnis der vernetzenden Copolymerisation, Angew. Makromol. Chem., 1:56 (1967).

38. L. Mrkvičková and P. Kratochvíl, Analysis of the Products of Copolymerization with Chain Branching in the Pregelation Stage, J. Polym. Sci. Polym. Phys., 19:1675 (1981).

39. B. Soper, R. N. Haward and E. F. T. White, Intramolecular Cyclization of Styrene-p-Divinylbenzene Copolymers, J. Polym. Sci. A-1, 10:2545 (1972).

40. I. Holdaway, R. N. Haward and I. W. Parsons, Copolymerisation of 4,4'-Divinylbiphenyl and 4,4'-Diisopropenylbiphenyl with Styrene in Solution, Makromol. Chem., 179:1939 (1978).

41. J. Spěváček and K. Dušek, Manifestation of Microgel-like Particles of Styrene-Ethylene Dimethacrylate Copolymers in Solution in ^{1}H and ^{13}C NMR Spectra, J. Polym. Sci. Polym. Phys., 18:2027 (1980).

42. R. S. Whitney and W. Burchard, Molecular Size and Gel Formation in Branched Poly(methyl methacrylate) Copolymers, Makromol. Chem., 181:869 (1980).

43. H. Galina, K. Dušek, Z. Tuzar, M. Bohdanecký and J. Štokr, The Structure of Low Conversion Polymers of Ethylene Dimethacrylate, Eur. Polym. J., 16:1043 (1980).

44. A. C. Shah, I. Holdaway, I. W. Parsons and R. N. Haward, Studies of Dimethacrylates as Crosslinkers for Styrene, Polymer, 19:1067 (1978).

45. H. Tobita and A. E. Hamielec, Crosslinking Kinetics in Polyacrylamide Networks, Polymer, 31:1546 (1990).

46. O. V. Pavlova, S. M. Kireyeva, I. I. Romantsova and Yu. M. Sivergin, Intramolecular Cyclization during the Radical Polymerization of Diallyl Monomers, Polym. Sci. USSR, 32:1187 (1990) (translated from Vysokomol. Soyed. A, 32:1256 (1990)).

47. H. M. J. Boots and N. A. Dotson, The Simulation of Free-Radical Cross-linking Polymerization: the Effect of Diffusion, Polym. Commun., 29:346 (1988).

48. (a) G. Allen, J. Burgess, S. F. Edwards and D. J. Walsh, On the Dimensions of Intramolecularly Crosslinked Polymer Molecules. II. The Theoretical Prediction of the Dimensions in Solution of Intramolecularly Crosslinked Polystyrene Molecules, Proc. R. Soc. London, A334:453 (1973). (b) J. E. Martin and B. E. Eichinger, Dimensions of Intramolecularly Cross-Linked Polymers. 1. Theory, Macromolecules, 16:1345 (1983).

49. E.g.: M. Antonietti, D. Ehlich, K. J. Fölsch, H. Sillescu, M. Schmidt and P. Lindner, Micronetworks by End-linking of Polystyrene. 1. Synthesis and Characterization by Light and Neutron Scattering, Macromolecules, 22:2802 (1989).

50. The equivalent equation in reference 32 (equation (4.7.1)) incorrectly includes a factor of (-1) in the denominator.

51. M. Gordon and G. R. Scantlebury, Theory of Ring-Chain Equilibria in Branched Non-Random Polycondensation Systems, with Applications to POCl$_3$/P$_2$O$_5$, Proc. Royal Soc. (London), A292:380 (1966).

52. M. Gordon and G. R. Scantlebury, The Theory of Branching Processes and Kinetically Controlled Ring-Chain Competition Processes, J. Polymer Sci. C, 16:3933 (1968).

53. K. Dušek, M. Gordon and S. B. Ross-Murphy, Graphlike State of Matter. 10. Cyclization and Concentration of Elastically Active Network Chains in Polymer Networks, Macromolecules, 11:236 (1978).

54. C. Sarmoria, E. Vallés and D. R. Miller, Ring-Chain Competition Kinetic Models for Linear and Nonlinear Step-Reaction Copolymerizations, Makromol. Chem., Macromol. Symp., 2:69 (1986).

55. C. Sarmoria and D. R. Miller, Validity of Some Approximations Used to Model Intramolecular Reaction in Irreversible Polymerization, Macromolecules, 23:580 (1990).

56. If consulting reference 32 with regards to secondary cyclization, it should be noted that the logic of the central paragraph of page 166 is in error. Decreased number of crosslinks makes the comparison worse.

57. While certain long-range correlations arising from local effects (such as a first-shell substitution effect on the crosslinker) can be exactly described statistically (N. A. Dotson, Correlations in Nonlinear Free-Radical Polymerizations: Substitution Effect, Macromolecules, 25:308 (1992)), in the case of secondary cyclization the source of the correlation itself is long-ranged, and so statistical description fails.

NOVEL POLYFUNCTIONAL ISOCYANATES FOR THE SYNTHESIS OF MODEL NETWORKS

Werner Mormann* and Barbara Brahm

Universität Siegen, Laboratorium für Makromolekulare Chemie
Adolf-Reichwein-Straße 2
D-5900 Siegen, Germany

INTRODUCTION

Polymer networks are of great importance for technical applications as well as in the scientific field. Thermoset polymers like epoxy resins, macroporous polymeric gels as used for polymeric reagents or catalysts, vulcanized rubber, and thermoplastic elastomers are examples for the wide spread use of crosslinked polymeric structures. Theories have been derived to predict gelation behavior, the built up of molecular weight, and mechanical properties as a function of the extent of reaction (1-3). Structure-property-relationships, a deeper understanding of elasticity, of the process of network formation, and of network theories, however, can be derived only from networks with well defined structure. The manufacture of such well defined or model networks, on the other hand, depends on the availability of suitable monomeric or oligomeric precursors and a clean, complete reaction by which chain extension and crosslinking is achieved. This will allow to control network parameters like crosslink density or the molecular weight between crosslinks.

High purity monomers of equal functionality and uniform reactivity are required in the case of thermoset networks, and, in addition, uniform length of the oligomers is a prerequisit for the synthesis of elastic model networks.

Poly(urethane)s are probably the most important class of crosslinked polymers which are obtained from a non-linear step polyaddition reaction (4). Quantitative understanding of these networks and of their formation is limited (5) because the available building blocks have serious drawbacks as far as their suitability for model networks is concerned. No trifunctional and higher functional polyisocyanates have been reported, to our knowledge, that fulfill the requirements of both equal functionality and reactivity. Only a few higher functional isocyanates are known to date. The biuret-modified 1,6-diisocyanatohexane 1 (Desmodur N™) (6) has an average functionality of nearly 6. If one assumes that structures higher than biuret are not formed, the structure is similar to that shown in scheme 1. Another trifunctional isocyanate often used is tris(4-isocyanatophenyl)thiophosphate 2

Synthesis, Characterization, and Theory of Polymeric Networks and Gels
Edited by S.M. Aharoni, Plenum Press, New York, 1992

(Desmodur RF™), which is a technical grade isocyanate with a rather high melting point of 90 °C. The tetraisocyanate 3 has been described recently (7). Its purity and functionality depends on the quantitative nature of each of the reactions involved in the synthesis since it cannot be purified by distillation.

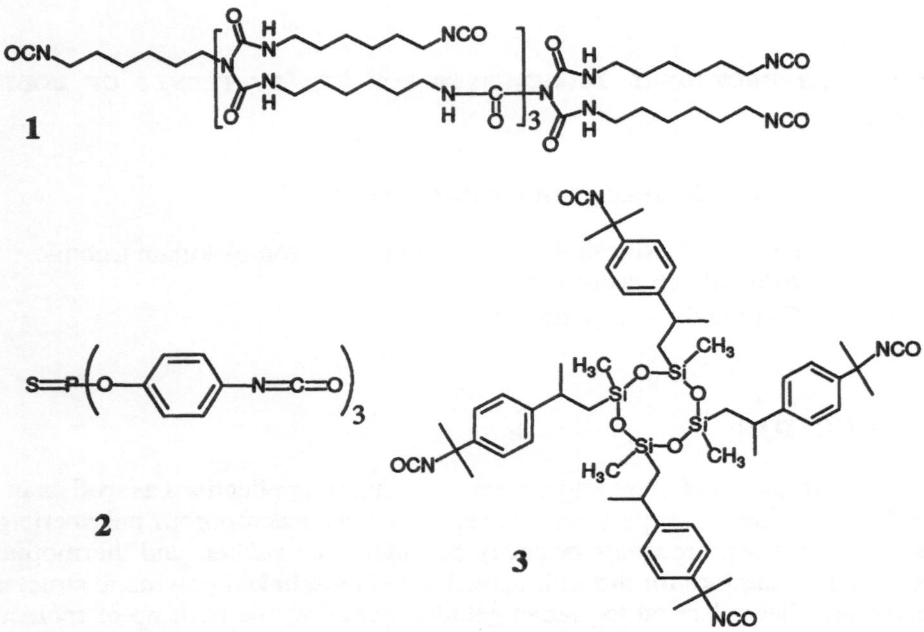

Scheme 1.   Plurifunctional isocyanates

This paper reports on the synthesis and properties of two types of isocyanates suitable for the synthesis and investigation of model networks:

1. Novel plurifunctional ester group containing isocyanates obtained by reaction of trialkylsilyl protected tri- and tetrahydroxy compounds with isocyanatoacyl chlorides.

2. α,ω-Diisocyanatotelechelics from commercially available oligomeric diols or triols and isocyanatoacyl chlorides.

Finally, the ability of the macrodiisocyanates to form networks by cyclo-trimerization is proven.

EXPERIMENTAL

α-Hydro-ω-hydroxy-poly(oxytetramethylene), mw 1000 was a gift from Quaker Oats, α,ω-dihydroxypoly(oxypropylene), mw 2000, poly(oxypropylene) triol mw

3000, pentaerythritol, 2,2-bis(hydroxymethyl)butanol (trimethylolpropane), and hexamethyldisilazane were gifts from the Bayer AG, Leverkusen.

Trimethylsilyl protected hydroxy compounds were obtained by reaction of the hydroxy compounds with hexamethyldisilazane as described in the literature (8) and purified by distillation or by removal of the volatiles with a thin film evaporator. Isocyanatoacyl chlorides were formed either by reaction of the corresponding amino acids with carbonyl chloride (9) or from the reaction of a cyclic anhydride with trimethylsilylazide and conversion of the resulting ω-isocyanatocarboxylic acid trimethylsilylester with thionyl chloride (10). The isocyanato acyl chlorides were purified by vacuum distillation.

Plurifunctional isocyanates were made from the trimethylsiloxy compounds by reaction with a 5% excess of the appropriate ω-isocyanatoacyl chloride in the presence of 0.1 mol % of 4-dimethylaminopyridine without a solvent. The mixture was heated to 130 °C in a flask equipped with a micro distillation apparatus until the calculated amount of chlorotrimethylsilane had distilled off. Low molecular weight compounds were purified by molecular distillation, isocyanatoprepolymers were purified by removal of the volatiles.

Characterization was made by ir-spectroscopy, nmr-spectroscopy, elemental analysis, determination of isocyanate content, and viscometry.

RESULTS AND DISCUSSION

Synthesis of ester isocyanates

The general synthetic route applied in industry for the synthesis of isocyanates uses di- or higher functional amines as substrates in the final step which are converted into isocyanates by reaction with a carbonic acid derivative (carbonylchloride, dimethylcarbonate). The amines in turn are obtained by reduction of nitro or nitrile groups as in TDI (diisocyanatotoluene) and HDI (1,6-hexamethylenediisocyanate) or by linking monoamine precursors as in MDI (methylene-bis(phenyl isocyanate)). The high reactivity of the isocyanate group towards a variety of functional groups prevents the synthesis of plurifunctional isocyanates by reaction of monoisocyanate precursors. One of the rare examples of this approach is compound 3 in scheme 1, which is obtained by a tetrafold hydrosilylation of 4-(2-isocyanato-2-propyl)-α-methylstyrene and tetramethylcyclotetrasiloxane (7).

The selective reaction of the chloroformyl group with silyl protected nucleophilic compounds (amines, alcoholes phenols) is another effective and versatile route to isocyanates, which works by linking together monoisocyanate precursor molecules (cf. scheme 2). Trimethylchlorosilane, which is inert towards isocyanates as well as ester groups is formed in this reaction and can easily be removed from the reaction mixture by distillation.

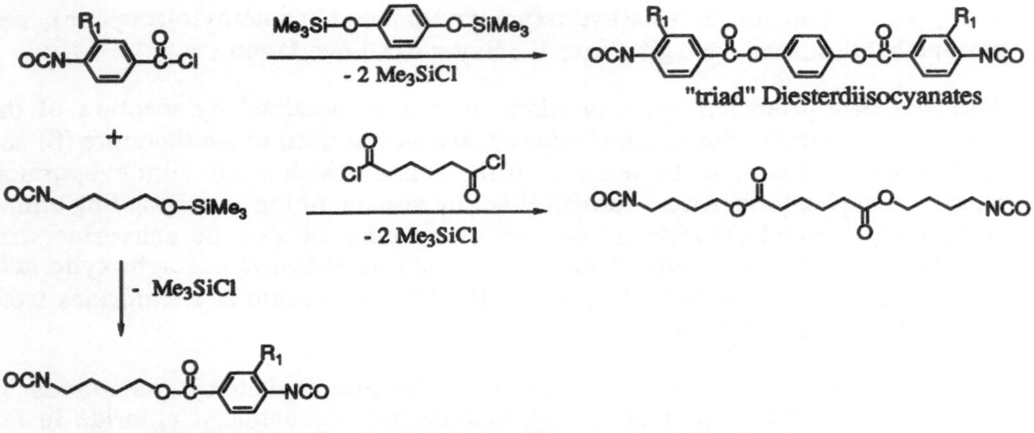

Scheme 2. Synthesis of ester group containing isocyanates

In a previous paper we have shown that amide group containing isocyanates can be obtained in high yields from monosilylated aliphatic or aromatic amines and acyl chlorides (11). Ester group containing isocyanates are formed in an analogous reaction of silyl protected alcoholic and phenolic hydroxy compounds with aliphatic isocyanatoacyl chlorides. The esterification reaction has to be catalyzed by acidic or basic catalyst in order to proceed at reasonable rate and temperature (100 - 150 °C). Under the conditions applied, no reaction occurs between the isocyanate group and the trimethylsiloxyalkane or -arene. Scheme 2 shows some of the synthetic possibilities opened by this synthesis. We have published the synthesis of aliphatic esterisocyanates (12) of fully aromatic esterdiisocyanates and the corresponding polyurethanes which form liquid crystalline phases (13, 14).

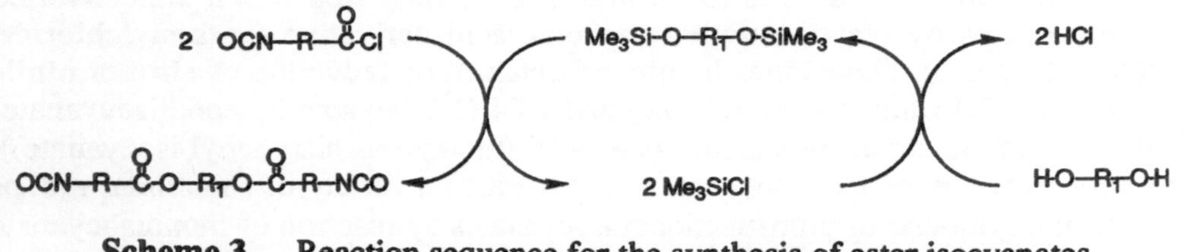

Scheme 3. Reaction sequence for the synthesis of ester isocyanates

Silyl protected alcohols and phenols are synthesized most conveniently by reaction of the hydroxy groups with hexamethyldisilazane (8). Trimethylchlorosilane also is an effective silylating agent for hydroxy groups and affords silylethers in high yield. Unlike in the silylation of amines, addition of a base as a hydrogen chloride acceptor is not necessary. Hence trimethylchlorosilane can be recycled since it is consumed in the formation of silylethers and set free in the esterification step. Thus the overall reaction is that of an isocyanatoacyl chloride with an alcohol to the isocyanatoester and hydrogen chloride as indicated for a diesterdiisocyanate by the reaction steps in scheme 3. This makes the synthesis of plurifunctional isocyanates from monoisocyanate precursors an interesting approach also on a technical scale and under commercial aspects.

Tri- and tetrafunctional esterisocyanates

In the study presented here we have extended the synthesis of esterisocyanates to examples of tri- and tetrafunctional isocyanates. Two different isocyanatoalkanoyl chlorides were used to vary the length of the acid part from three carbon atoms (3-isocyanatopropanoyl chloride) to six (6-isocyanatohexanoyl chloride). Isocyanato-butanoyl chloride was not used since it has the tendency to undergo an isomerization accompanied by cyclization to form an N-chloroformylbutyrolactam under the reaction conditions.

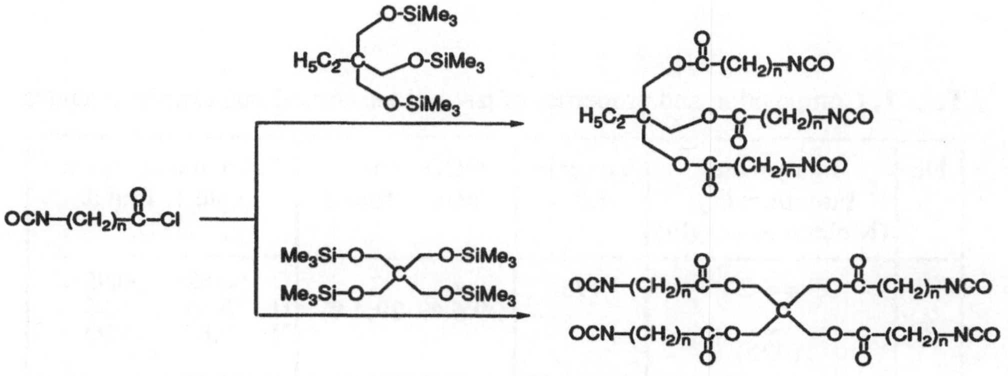

Scheme 4. Tri- and tetrafunctional esterisocyanates

2,2-Bis(hydroxymethyl)-butanol (trimethylolpropane) and pentaerythritol were chosen as hydroxy compounds. Glycerol, which has an even lower molecular weight, was not used. Elimination of trimethylsilanol most likely would occur as a side reaction during the esterification which was observed with trimethylsiloxy-cyclohexane (14). Thus the molecular weights of the tri- and tetrafunctional isocyanates were sufficiently low to allow purification by short path distillation in a Kugelrohr-apparatus or by molecular distillation without decomposition in order to ensure the theoretical functionality.

It is crucial, however, to work under anhydrous conditions if really pure compounds are desired. Traces of moisture react with either the chloroformyl group or with the trimethylsiloxy group and the selectivity of the esterification reaction is lost. The chloroformyl group is more reactive than the isocyanate group and the hydroxy group is more reactive than the trimethylsiloxy group. Hydrogen chloride, which is formed by the reaction of the acid chloride with water, can add to the isocyanate moiety and give a carbamoyl chloride which now is able to react with a siloxy group. Hydrogen chloride may also cleave the silicon oxygen bond by formation of a new hydroxy group and chlorotrimethylsilane.

Some representative properties of the esterisocyanates are given in table 1. A remarkable feature of the novel esterisocyanates is their extremely low viscosity compared to plurifunctional isocyanates containing hydrogen bonds like 1. Linked with their high boiling points are the low vapor pressures. The esterisocyanates do not have the typical pungent odor of isocyanates and in contrast to 1 there is no tendency of a back reaction towards isocyanate fragments since the former are made by a condensation reaction whereas 1 is obtained by an addition reaction.

The infrared spectra of the ester isocyanates show the characteristic absorptions of the isocyanate group at 2270 cm^{-1} and of the estercarbonyl at 1720 cm^{-1}. A representative example is the infrared spectrum of the triisocyanate 6 shown in fig. 1. Further evidence of the proposed structure and purity is obtained from the amount of chlorine and silicon respectively present in the distillates. In good runs these are in the range of less than 100 ppm for both.

Tab. 1. Composition and properties of tri- and tetrafunctional esterisocyanates

No	Isocyanate, Functionality, (Molecular weight)	Viscosity mPas	NCO content calc. found	Elemental analyses calc. found	
4	n = 2 3 (425)	170	29,6 % 29,4 %	C 50,82 H 5,45 N 9,88	50,95 5,55 9,95
5	n = 2 4 (524)	210	32,1 % 31,8 %	C 58,79 H 7,49 N 7,62	58,95 7,70 7,75
6	n = 5 3 (551)	160	18,2 % 17,9 %	C 50,82 H 5,45 N 9,88	50,95 5,55 9,95
7	n = 5 4 (694)	190	24,2 % 23,8 %	C 50,82 H 5,45 N 9,88	50,95 5,55 9,95

The proton nmr-spectrum of the tetrafunctional esterisocyanate 5 in fig. 2. shows the expected signals for the methylene groups of the pentaerythritol part at 4.1 ppm, the absorptions of the methylene groups attached to the ester group at 3.5 ppm and to the isocyanate group at 2.55 ppm respectively both as triplets. The molecular weights of the polyfunctional esterisocyanates (cf. tab. 1) is too high to allow distillation under normal high vacuum conditions. Molecular distillation is required to obtain the desired compounds in good purity.

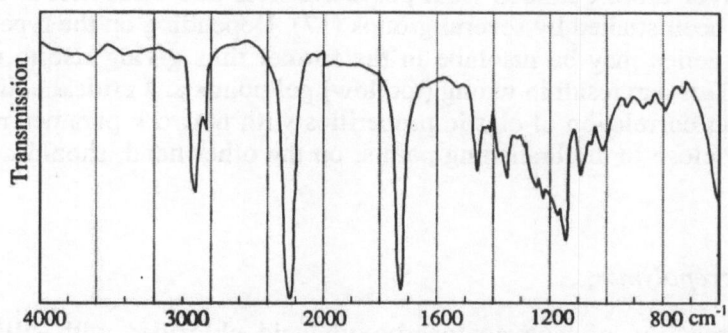

Fig. 1. Infrared-spectrum of 6.

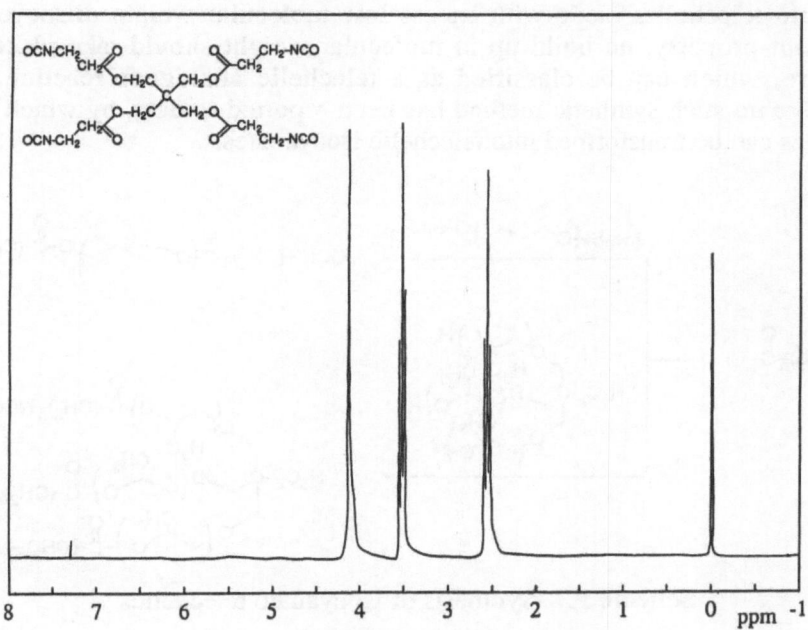

Fig. 2. Proton nmr-spectrum of 5.

The low molecular weight plurifunctional isocyanates are well suited to form networks by step polyaddition to α,ω-dihydroxy compounds. Thus they should allow to study the effect of three and four armed branches on the properties of networks with identical molecular weight between crosslinks and almost identical chemical constitution and size of the crosslinks. This type of network has further advantage over those obtained from polyethertriols and difunctional isocyanates, which have been studied by several groups (17). Depending on the type of triol the urethane moieties may be insolube in the former thus giving rise to microphase separation. This can result in wrong (too low) gel points and critical conversion and also give bad correlation of elastic properties with network parameters. Urethane groups very close to the branching points, on the other hand, should not have this effect.

Isocyanate prepolymers

The clean reaction of isocyanatocarboxylicacid chlorides with silyl protected hydroxy compounds encouraged us to apply it also to the synthesis of α, ω-isocyanato functional telechelics from the corresponding hydroxy compounds. In analogy to low molecular tri- and tetrahydroxy compounds one should be able to obtain higher functional isocyanate prepolymers without the drawbacks of isocyanato telechelics made with excess low molecular weight diisocyanates. If carried out properly, no build up in molecular weight should take place by this procedure, which can be classified as a telechelic analogous reaction. To our knowledge no such synthetic method has been reported to date, by which hydroxy telechelics can be transformed into telechelic isocyanates.

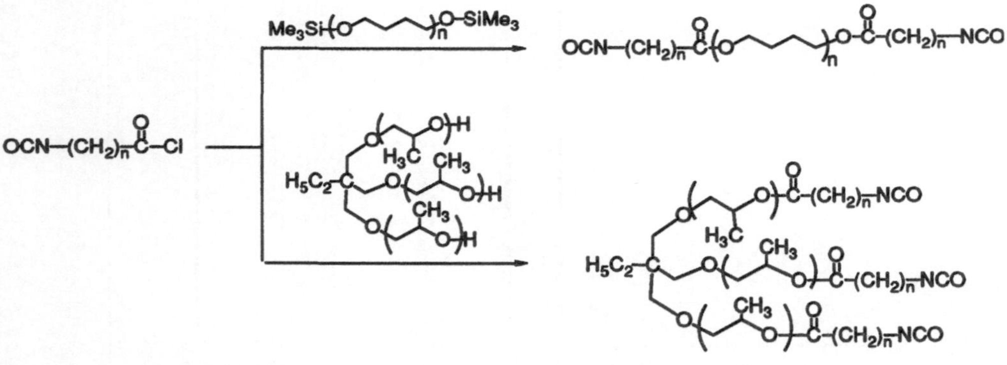

Scheme 5. Synthesis of isocyanato telechelics

In the normally applied method (15) telechelic hydroxy compounds (polyols) are reacted with a large excess of a diisocyanate, which has to be removed in a second step. Due to side reactions of the isocyanates - e.g. formation of allophanates or reaction of a residual isocyanate group with another hydroxy group - no proper control of molecular weight and functionality is possible by this method. Free radical polymerization with isocyanate group containing azoinitiators has only limited applicability (16).

As hydroxy compounds we used α-hydro-ω-hydroxy-poly(oxytetramethylene) with a molecular weight of 1000, α,ω-dihydroxypoly(oxypropylene) with a molecular weight of 2000, and poly(oxypropylene) triol with a molecular weight of 3000. Silylation of the hydroxy compounds was performed by reaction with hexamethyldisilazane, which was used in a 15 % stoichiometric excess thus allowing to use the polyols as received, since traces of water were removed by formation of hexamethyldisiloxane.

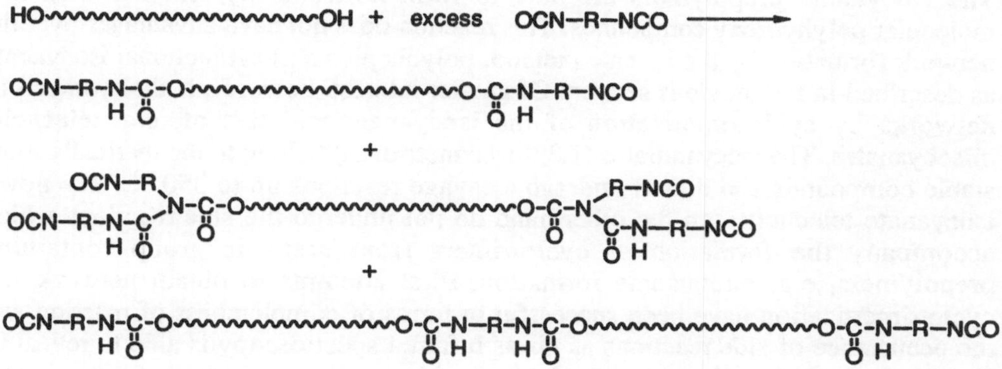

Scheme 6: Synthesis of isocyanato functional prepolymers from polyols with excess diisocyanate

The purity of the trimethylsiloxy functional oligomers was checked by infrared and nmr-spectroscopy (absence of hydroxy groups). Care has to be taken in the purification step of α-trimethylsilyl-ω-trimethylsiloxy poly(oxytetramethylene). The silylation renders the compounds more volatile and in the thin film distillation step the lower homologoues can be removed. This can be seen in the hplc-chromatograms where the four lowest homologues have been removed by molecular distillation at 180 °C.

Tab. 2. Diisocyanate telechelics (prepolymers) based on 6-isocyanatohexanoyl chloride

Basic polyol (Functionality)	Viscosity (23 °C) mPas	Viscosity of polyols mPas	Molecular weight	NCO content calc.	found
Poly(oxypropylene) (2)	450	300	2280	3,7 %	3,6 %
Poly(oxypropylene) (3)	650	460	3420	3,7 %	3,8 %
Poly(oxytetramethylene) (2)	850	solid	1400	5,9 %	5,7 %

Isocyanate prepolymers were formed by bulk reaction of the trimethylsiloxy compounds with 6-isocyanatohexanoyl chloride in the presence of small amounts of sulfuric acid as catalyst. The isocyanatoacyl chloride was used in 10 % stoichiometric excess to ensure complete conversion of the trimetylsiloxy groups. The reaction was monitored by ir-spectroscopy (disappearence of the carbonyl absorption of the acid chloride and increase of the ester carbonyl band) and the amount of chlorotrimethylsilane formed. The prepolymers were purified by thin film distillation to remove excess isocyanatoacyl chloride.

The isocyanate prepolymers are able to form networks by reaction with low molecular polyhydroxy compounds. This reaction does not have advantages over the network formation by the inverse method, polyols plus a plurifunctional isocyanate as described in the previous section. Of greater interest, however, is the synthesis of networks by cyclotrimerization of the isocyanate moieties of, e.g. telechelic diisocyanates. The isocyanurates (1,3,5-triazinetriones) belong to the thermally most stable compounds and do not undergo cleavage reactions up to 350 °C. The novel isocyanate telechelics on the other hand do not undergo the side reactions which accompany the formation of cyclotrimers from urethane group containing prepolymers, e.g. allophanate formation. First attempts to obtain networks by cyclotrimertization have been successful in terms of completeness of reaction and the occurrance of side reactions as far as infrared spectroscopy is able to reveal it. Fig. 3 schematically shows a network obtained by cyclotrimerization of diisocyanate telechelics in which no functional groups except the isocyanate moiety itself and the isocyanurate rings are present.

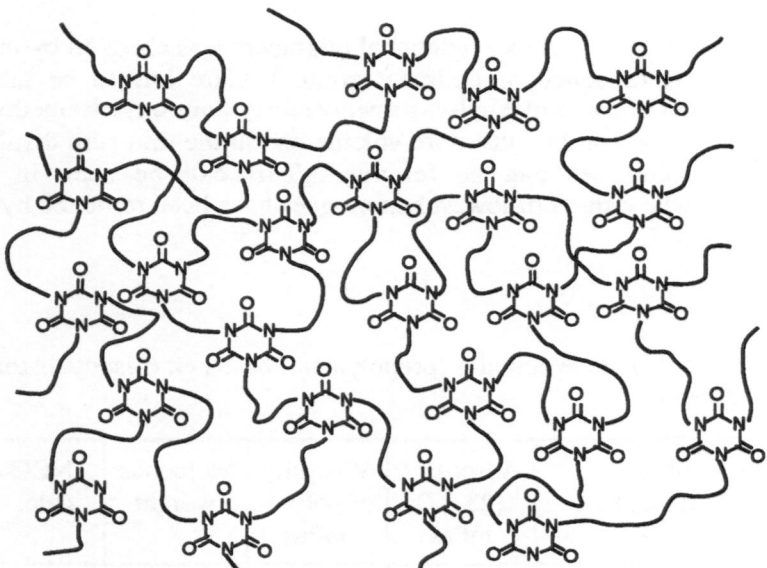

Fig. 3. Network from cyclotrimerization of telechelic isocyanates

The combination indicated by the vertical branch in scheme 2, namely the reaction of an aromatic isocyanato acid chloride with an aliphatic trimethylsiloxy compound, was the only one to cause major difficulties. When performed with 4-isoyanato-benzoyl chloride and 1,4-bis(trimethylsiloxy)-butane, this reaction results in an infusible mixture of obviously crosslinked material.

CONCLUSION

Novel building blocks for networks were obtained by reaction of isocyanatoalcanoyl chlorides with silyl protected polyhydroxy compounds. Trifunctional and tetrafunctional isocyanatoesters from low molecular weight polyhydroxy compounds can be purified by distillation and thus represent a new class of building blocks with uniform functionality and equal reactivity for the synthesis of model networks by step addition reactions. Conversion of silylated hydroxy telechelics into isocyanate telechelics can be achieved without undesired coupling and increase of molecular weight. The α,ω-isocyanate oligomers are starting materials for network formation by step addition and cyclotrimerization reactions.

REFERENCES

1) Stauffer, D.; Coniglio, A.; Adam, M.; *Adv. Polym. Sci.* **1982**, *44*, 103.

2) Flory, P. J.; *Principles of Polymer Chemistry*, Cornell University Press, Ithaca, New York, **1953**.

3) D. R. Miller, C. W. Macosko, Macromolecules **1980**, *13*, 1063,

4) Oertel, G., Ed.; *Polyurethanes Handbook,* Hanser, Munich, **1985**.

5) Dusek, K.; in *Telechelic Polymers: Synthesis and application* Ed. E. J. Goethals,CRC Press, Boca Raton **1989**.

6) Gnanou, Y.; Hild, G.; Rempp, P.; *Macromolecules* **1984**, *17*, 945.

7) Zhou, G; *Polym. Bull. (Berlin)* **1989**, *22*, 85.

8) Bruynes, C. A. Jurriens, T. K.; *J. Org. Chem.* **1982**, *47*, 3966.

9) Iwakura, Y.; Uno, K.; Kang, S.; *J. Org. Chem.* **1966**, *33*, 142.

10) Kricheldorf, H. R.; *Justus Liebigs Ann. Chem..* **1975**, 1387.

11) Mormann, W.; Hohn, E.; *Makromol. Chem.* **1989**, *190*, 1981.

12) Mormann, W.; Hißmann, E.; *Tetrahedron Lett.* **1987**, *28*, 3087.

13) Mormann, W.; Brahm, M.; *Makromol. Chem.* **1989**, *190*, 631.

14) Mormann, W.; Brahm, M.; *Macromolecules* **1991**, *17*, 945.

15) Zalipsky, S.; Gilon, C.; and Zilkha, A.; *Eur. Polym. J.* **1983**, *19*, 1177.

16) Ghatge, N. D.; Vernekar, S. Pl; and Wadgaonkar, P. P.; *Makromol. Chem. Rapid Commun.* **1983**, *4,* 307.

17) a) Stepto, R. F. T.; *Developments in Polymerisation*, Vol. 3, Haward, R. N., Ed., Applied Science Publishers, London, **1982**, 81.
    b) Matejka, L.; and Dusek, K.; *Polym. Bull.,* **1980**, *3*, 489.

COMPARISON BETWEEN LIGHTLY CROSSLINKED IONOMERIC MATERIALS AND

HIGHLY CROSSLINKED MATERIALS DERIVED FROM POLY(ACRYLIC ACID) WITH

ORGANOSTANNANES AS THE CROSSLINKING AGENT

Charles E. Carraher, Jr., Fengchen He and
Dorothy Sterling

Florida Atlantic University
Department of Chemistry
Boca Raton, FL 33431

INTRODUCTION

Ionomers are ion-containing polymers in which the ion-containing segment is present in only a small amount (1). Here the ionomer is a carboxylic ester or acid-containing polymer derived from the copolymerization of the acid-containing monomer and ethylene. Ionomers are usually neutralized by addition of a mono or divalent cation. Here the carboxylic acid is neutralized through reaction with sodium hydroxide with subsequent reaction with organostannane halides(2). While they can be considered to be crosslinked, they are thermoplastics, melting under application of heat and pressure.

Ionomeric materials are employed in a wide variety of applications including golf ball coverings, packaging, automotive parts and in footwear.

Carraher and Piersma(3) reported the synthesis of organostannane esters through reaction of organostannane halides employing the interfacial condensation technique about two decades ago. In this report little other than simple structural analysis employing infrared spectroscopy and elemental analysis was reported. Esters of poly(acrylic acid), known collectively as polyacrylates, are synthesized in quantities near a half million tons per year in the USA. They are used in coatings, as adhesives, and in fiber modification.

Reaction of polymers containing acid groups and organostannane halides produce crosslinked products if the organostannane contains two or three halides. The use of monohalo-containing organohalides will give products without an increased amount of crosslinking. Thus crosslinking can be varied from zero to high through varying the number of halo-functional groups and through the use of polymers containing varying amounts of acid-containing functional groups. The affect of varying crosslinking is investigated through the use of organostannanes with varying number of functional groups and through the use of polymers containing varying proportions of acid groups is reported in this study.

Organostannane halides were employed in this study for a number of reasons including the following:

Ready availability of numerous
organostannane halides

Synthesis, Characterization, and Theory of Polymeric Networks and Gels
Edited by S.M. Aharoni, Plenum Press, New York, 1992

349

- Tin-containing products are generally
  biologically active and may provide
  materials with useful biological
  properties.

- Organostannane halides are readily
  available that provide functionalities
  of one, two and three.

EXPERIMENTAL

The ethylene-acrylic acid copolymer (Aldrich Chemical Company,
Milwaukee, WI.; 9010-77-9; 20 weight-% acrylic acid content) was used as
received.

Poly(acrylic acid) was synthesized from freshly distilled acrylic acid.
The acrylic acid ((150 ml) was added to 200 ml distilled water. To this
solution was added 0.1% (by weight) a,a-azodiisobutyronitrile. The resulting
mixture was flushed with nitrogen to remove oxygen and heated to 65°C. It
was maintained at 65°C for 30 min. After an initiation period, the
polymerization occurred rapidly and was very exothermic. The resulting
solution was dried. The product was purified by washing it several times
with water. Further purification was accomplished by dissolving the polymer
in water and drying it, thus allowing unreacted acrylic acid to be removed.

The poly(acrylic acid) exhibited a limiting viscosity of 7.2 (dl/g) in
DMSO, which corresponds to a weight-average molecular weight of 1.9×10^5 by
light-scattering photometry.

The modifications were carried out using the interfacial technique.
Briefly, poly(acrylic acid) (or the ionomer) was dissolved in water to which
an equivalent amount of sodium hydroxide had been added. The aqueous
solution was added to a 1-pint Kimax emulsifying jar fitted on top of a
Waring blendor. (A detailed description of the polymerization apparatus is
given elsewhere(3)). The cap is screwed on and the blender is turned on.
The organic phase containing the organo- metallic reactant was added through
a hole in the jar lid. Stir-time began after all the organic phased had been
added. Rapid precipitation of the product occurred. After an appropriate
time, the stirring was stopped. The resulting mixture was filtered, washed
and dried.

A wide variety of solvents were investigated with respect to their
ability to dissolve the modified products. Infrared spectra were obtained
using a Nicolet 5DX FTIR, Beckman IR-10 and Perkin-Elmer 237-B IR
spectrophotometers using films and KBr pellets. Elemental analysis was
carried out at Gailbraith Laboratories, Knoxville, TN.

Thermal gravimetric analysis (TGA) was carried out using a du Pont 950
TGA. Differential scanning calorimetry (DSC) was conducted employing a du
Pont 900 Thermal unit with a duPont DSC cell. A linear baseline compensator
was used with the DSC cell to insure a constant energy baseline. A Mettler
H20T semimicro balance was used for the DSC sample weighings. DSC
measurements were obtained on samples contained in an open aluminum cup to
allow the free flow (away from the solid) of volatilized gases, thus more
closely stimulating the conditions under which TGA studies were conducted. A
gas flow rate (air and nitrogen) of about 0.3 l./min was maintained for both
the DSC and the TGA studies. The samples were ground to aid in obtaining
reproducible results.

High resolution electron impact (HREI) positive ion mass spectral
analyses were carried out at the Midwest Center for Mass Spectrometry,
Lincoln, Neb. using a Krates MS-50 mass spectrometer and at FAU employing a
DuPont 21-496-B double focusing mass spectrometer with a resolution of 1/1500
at 0% valley. The samples were inserted as solids in a glass ampule using a
direct insertion probe (DIP-EI-MS).

RESULTS AND DISCUSSION

Substitution for products derived from PAA is in the mid-range with
about 30-60% of the possible sites containing organostannane moieties(3).
For the ionomer products sub- stitution is somewhat higher with substitution
ranging from mid-range to almost 100% (Table 1). This is consistent with

steric, space factors being a substitution limiting factor. Thus, there is more "available" space between substitution sites in the ionomer allowing greater degrees of substitution.

Table 1. Proportion of Substitution for Ionomeric Materials

	%-Sn Found	%-Sn Max.	%Substitution
$BuSnCl_3$	17.63	24.0	63
Bu_2SnCl_2	13.88	21.5	51
Bu_3SnCl	7.85	19.5	26
Ph_3SnCl	18.43	17.3	100
$PhSnCl_3$	16.43	23.0	61
$MeSnCl_3$	9.14	26.2	25
Bz_2SnCl_2	15.37	19.2	68
La_2SnCl_2	11.68	15.4	57

Yields are based on assumed structures. Even so general comparisons can be made if the yield data is treated similarly. Thus, general yields for PAA-modified products are in the mid-range varying from about 40 to 80%. By comparison, yields for the ionomer-based products are generally higher typically being in the 70% and greater range.(Table 2) It appears that the critical aspect for substitution is the initial substitution on a chain. Once substitution begins on a chain, it appears to continue with a high degree of substitution. Reaction is believed to occur within the organic phase but near the interface since neither the organostannane or the acid salt are appreciably soluble in the "other phase". By comparison the ionomer, with a higher proportion of "organic units", is more soluble in the organic phase than the PAA allowing a larger portion of ionomer chains to come in contact with the organostannane.

Table 2. Yields for the Synthesis of Ionomer Materials

Organostannanes(mmole)	Yield(g)	Yield(%)
$Bu_2SnCl_2(1.00)$	0.683	79
$Bz_2SnCl_2(1.00)$	0.888	96
$La_2SnCl_2(1.00)$	0.884	82
$Bu_3SnCl(2.00)$	0.110	90
$Ph_3SnCl(2.00)$	trace	–
$BuSnCl_3(0.67)$	0.797	90
$PhSnCl_3(0.67)$	0.864	96
$MeSnCl_3(0.67)$	0.738	105

Reaction conditions: Organostannane solution (50ml $CHCl_3$)
added to a stirred (18,000 rpm, no load) aqueous solution (50 ml) containing ionomer (2.00 mmole of acrylic acid units) and
sodium hydroxide (2.00 mmole) at 25_oC. with 30 secs. stirring.

For a given organostannane with a functionality greater than one, extent of crosslinking should increase as the functionality increases. Also, the relative extent of crosslinking should be greater for PAA-derived products in comparison to the ionomer-materials for a given functionality greater than one because of the much greater density of acrylic acid groups in PAA.

The PAA-associated products are powdery and hard in appearance and do not melt to 300°C though they begin to discolor by about 250°C. The discoloration corresponds to an initial weight loss (by TGA) of only several percent. While a small endotherm occurs in both air and nitrogen atmosphere (by DSC) about 100° they do not melt sufficiently to allow film formation under applied pressure.

By comparison, the ionomer-associated products melt giving tough and flexible films below 150°C. For the dibutyl and tributyl stannane derived products, film formation can occur in the 30-50° range. For products derived from the monobutyltin trichloride melting under pressure occurs about 150°C. The ionomer itself exhibits endothermic peaks just above 50°C and about 90°C. The sodium salt gives a small endotherm just above 50°C., a large endotherm about 100°C and a sharp endotherm about 170°C. The dibutyltin product exhibits two endotherms, at 70 and 80° C while the dilauryltin product shows endothermic peaks at about 75, 85 and 90°C. The dibenzyltin product shows endothermic peaks between 80 and 100°C. As noted before, all of these ionomeric products, except for the dibenzytin product, melt under applied pressure to give clear, strong, tough and flexible films similar to other ionomeric products.

The textures of the products vary. In general, flexibility decreases as the amount of crosslinking increases. For instance, the butyltin-containing products from PAA are powdery whereas the corresponding products from the ionomer vary from powdery for the butyltin trichloride derived product to being spongy and flexible for the dibutyltin dichloride and tributyltin chloride-derived products.

Analysis by electron impact mass spectrophotometry, ei-MS, is consistent with the proposed structures including ion fragments containing both tin and polymer backbone-associated fragments. All of the spectra contain the characteristic ion fragments from the breakdown of the main chain with series containing C_1H_x, C_2H_x, C_3H_x, C_3H_x, C_4H_x, etc. For products derived from the ionomer and dihaloorganostannanes, ion fragments containing unreacted chloride are present whereas those derived from the poly(acrylic acid) do not contain such chloride-containing fragments. This may be due to the greater accessibility of acetate groups in the PAA in comparison to the ionomer. In point of fact, the ionomer contains, on the average, only one acid group for each ten repeat units compared with PAA where each unit contains an acid group.

For comparison the mass spectral results for the two monomers are given in Tables 3 and 4. The significant ion fragments for the products derived from the condensation of dibutyltin dichloride with the ionomer and with PAA are given in Tables 5 and 6. Both products degrade similarly yielding groupings, as noted before, containing four, five, six, etc., carbons and the appropriate numbers of hydrogens. Also as noted before, ion fragments containing Sn-Cl groups are present in the ionomeric product but absent in the PAA-derived material. Both show less intense higher mass ion fragments that contain both tin and portions of the polymer backbone. For both the ionomeric and poly(acrylic acid) products derived from reaction with dibutyltin dichloride butyltin-containing ion fragments are found at (all ion fragments are described in terms of m/e or m/z=1) 499 assigned to $BuSnCO_2CHCH_2CHBuSnCO_2CH_2$; at 468-475 assigned to $BuSnCO_2CHCH_2(CHCO_2CH_2)_2$ $(CHCH_2)_3$; 443 $Bu_2SnCO_2CHCH_2CHCO_2CH_2CHCH_2CHCH_2CH$; 403-399 assigned to $Bu_2SnCO_2CHCH_2CHCO_2CH_2(CHCH_2)_3$; etc.

There are seven naturally occurring isotopes of tin. Some tin-containing fragments are present in sufficient abundance to allow a comparison of the ion abundance as a function of the isotope of tin. Table 7 contains one such comparison for the product derived from dibutyltin dichloride and the ionomer.

It may be possible to evaluate, in a secondary manner, the steric environment of the tin-containing moiety using the location of the carbonyl stretching vibration. Tin-containing moieties connected to a carboxyl group can be either bridging or non-bridging. In general, for group IV A organometallics, the tendency towards bridging, and consequently towards octahedral formation, is Si>Ge>Sn>Pb for analogous products (for instance 4). It can be argued that there is a greater space requirement for octahedral formation compared to simple tetrahedral formation, ie, that the geometrical arrangements of the octahedral form are more restrictive. Thus the majority of the linear tin-based polyesters reside as bridged structures since there are not specific steric restrictions that can be present in crosslinked materials.

Table 3. Yields for the Synthesis of Ionomer Materials
Ion fragments derived from the Ionomer Salt (400°C.)

M/e	Relative Intensity	Assignment
30	20	
35	9	
37	32	
40	16	
41	92	C_3
42	34	
43	100	
44	90	CO_2
45	8	
51	8	
53	14	
54	14	C_4
55	90	
56	62	
57	89	
58	6	
60	9	
67	28	
68	17	
69	54	C_5
70	43	
71	51	
73	9	
81	23	
82	21	
83	46	C_6
84	19	
85	22	
95	12	C_7
96	13	
97	38	
109	18	
110	14	C_8
111	25	
112	12	
123	9	
124	9	C_9
125	12	
126	8	

Table 4. Significant (>10%) ion fragments derived from poly(acrylic acid).

m/e	Relative Intensity	Assignment	
41	85		
42	15		
43	22	C_3	CO_2
44	21		
45	66		
55	70		
56	23	C_4	
57	68		
61	12		
67	18		
68	18		
69	100	C_5	
70	22		
71	34		
73	58	M	
21	39		
22	14	C_6	
23	28		
25	20		
95	15	C_7	
97	18		
129	19	C_9	
137	14	C_{10}	
149	22	C_{11}	

M=acrylic acid monomer

Table 5. Significant Ions (≥ 10%) Derived from the Condensation Product of Dibutyltin Dichloride and the Ionomer (from m/e >50).

m/e	Rel. Int. (%)	Assignment
53	12	
55	71	
56	73	C_4H_x
57	100	
58	42	
67	10	
69	29	C_5H_x
70	24	
71	25	
83	21)	C_6H_x
85	15)	
97	17	C_7H_x
154	11	SnCl
175	12)	
177	17)	SnBu
211	10)	
213	12)	SnBuCl
265	17	
266	11	
268	16	SnBu$_2$Cl
271	18	

Table 6. Significant Ion Fragments (≥0.9%) Derived from the Condensation
Product of Dibutyltin Dichloride and Poly(acrylic Acid)
(from m/e > 50)

m/e	Rel. Int. (%)	Assignment
55	20	
56	8	C_4H_x
57	17	
60	10	
67	7	
69	13	C_5H_x
70	6	
71	8	
73	10	
81	6)	
83	9)	C_6H_x
97	5	C_7H_x
101	0.9	$C_4H_9CO_2$
105	1	
107	0.9	
109	2	
110	1	C_8H_x
111	2	
112	0.9	
115	0.9	
129	3	
137	0.9	$C_7H_9CO_2$
185	0.9	$C_7H_{13}(CO_2)_2$
241	1	$C_8H_{13}(CO_2)_3$
248	0.9	$BuSnCO_2CHCH_2$
256	1	$BuSnCO_2CHCHC$

Table 7. Comparative ion fragments derived from $HCCO_2Sn(C_4H_9)_2$ as a function
of tin isotope.

Tin Isotope	116	117	118	119	120	122	124
$HCCO_2Sn(C_4H_9)_2$	287	288	289	290	291	293	295
Natural %							
Abundance	14	8	24	9	33	5	6
Found	14	9	24	11	32	4	6

Bridging Group IV A-based polyesters exhibit two carbonyl-associated
stretching vibrations, the symmetrical occurring about 1430 to 1410 and the
asymmetrical about 1570 (all assignments given in cm^{-1}). Analogous
nonbridging structures show bands around 1370 to 1350 for the symmetrical
stretch and about 1650-1610 for the asymmetrical stretch.

Products derived from poly(acrylic acid) exhibit bands characteristic of
nonbridging consistent with a tight fit for the products. By comparison the
ionomeric-based materials exhibit bands characteristic of both bridging and
nonbridging. For instance, the product from the ionomer and dibutyltin
dichloride exhibits bands at about 1580 and 1420 indicative of bridging and
consistent with the availability of more space allowing formation of the
octahedral structure.

One common view of ionomer bonding is that the ionized carboxyl groups form "ionic" crosslinks with the metal-containing moiety (for instance 1). Further, that there exists two general types of bonding groups - one that is small and involves one or only a few metal atoms while the second type of grouping involves larger clusters of the reactive groups (for instance 1). These ideas associated with ionomer bonding may well be true for ionomers containing Group IA metal atoms such as sodium especially for industrially produced ionomers. It is important to remember that industrially produced ionomers are generally synthesized through simple mixing, and heating of the reactants. In this case, aggragates of metal-containing clustes probably occur. By comparison, many in academics employ synthetic procedures, as done in the present study, that require the reactants to be dissolved in a solvent. In these cases, aggravation of large clusters of metal containing reactants is less likely to occur and consequently the second type of grouping less likely to occur, or at least the proportion of the second type of clustering should less.

In both synthetic cases, bonding with metal ions containing so called "low lying vacant orbitals" such as transition elements, probably involves, to some extent, polar, directional bonding. Such polar bonding is probably the predominant, if not the exclusive type of bonding found for the reaction products of organostannane halides and salts of carboxylic acid. Since the presently described
ionomeric materials formed through reaction with organostannane dichlorides do melt upon addition of heat and pressure similar to that of conventional ionomers, the discussion of bonding by at least some ionomeric materials should be enlarged to include polar bonding components. It is likely that the melting and film forming properties of the present materials is due, at least in part, to chain slippage between methylene segments. It is also possible that chain slippage may be assisted by the somewhat spherical bonding that can be envisioned as occurring with the organostannane and two carboxyl groups forming an octahedral geometry as described before. Such bonding might allow the tin-containing moiety to act as a slippery sphere that can "roll" between aggregates of carboxyl groups. It is not surprising that the high concentration of crosslinks does not allow ready slippage or "rolling" to occur for the analogous poly(acrylic acid) products.

ACKNOWLEDGEMENTS

We are pleased to acknowledge partial support of this project from ACS-PRF#19222-87-C.

REFERENCES

1.   C. Carraher and C. Pittman, Polymer News, 15(1), 7(1990).

2.   C. Carraher, D. Sterling, F. He, F. Nounou, R. Pennisi, J. Louda and L. H. Sperling, Polymer P., 81(2), 430 (1990).

3.   C. Carraher and J. D. Piersma, J. Applied Polym. Sci., 16, 1851(1972).

4.   C. Carraher and C. D. Reese, Angew. Makromol. Chemie, 65, 95(1977).

CONTRIBUTORS

INDEX